WAR ON THE MIND

The Military Uses and Abuses of Psychology

PETER WATSON

Basic Books, Inc., Publishers

NEW YORK

Library of Congress Number: 77-75237
ISBN: 0-465-09065-6

Printed in the United States of America
10 9 8 7 6 5 4 3 2 1

WAR ON THE MIND

TO CHRISTINE

Contents

Preface

To most of us 'the military uses of psychology' means first and foremost propaganda. At some time or another we have all heard of the leafletting operations carried out in recent conflicts – such things as safe-conduct passes dropped by the million from aircraft. Similarly, both the overt and covert activities of propaganda radio stations during the Second World War have received their share of publicity. Even the efforts of armies (principally of the western powers) to influence the 'hearts and minds' of their potential adversaries by such propaganda exercises as building bridges, installing plumbing and so on, have become a familiar sight in contemporary news-clippings.

Outside the field of propaganda the work of psychologists in helping to ensure that the design of tanks, helicopters, gun-sights and radar screens takes into account the limitations and variations in human abilities has received some notice, as have attempts to improve selection procedures. But in the military context the work of psychologists has generally taken a back seat in comparison with that of other scientists. Their results, set beside the achievements of physicists, mathematicians and biologists, have often seemed to be small, uninteresting and their effectiveness, in the nature of things, difficult to assess.

This, at any rate, was the rough picture I had about military psychology when, in 1973, I began an inquiry for the Insight team of the *Sunday Times* into the uses of psychological warfare by the British in Northern Ireland. As someone who had read psychology at university and who, through journalism, had the opportunity to keep in touch with developments in research, I was also aware of a wider interest among some psychologists and anthropologists in the nature of aggression for what light this might throw on the way wars occur, the way soldiers fight, and the ways peace might either be achieved or maintained. But in general I shared the view of Lord Chalfont, writing in the London *Times*, that the whole field of conflict research, the psychological side included, though it promised a lot has actually delivered very little. Intellectually, it is a barren field. So far as I then knew, psychological research into attitude formation and the mechanisms of persuasion was the only aspect of psychology to be seriously adapted for military purposes.

The first books I read as background did nothing to change these ideas: Daniel Lerner's two-volume study of propaganda and 'sykewar' in the Second World War, Terence Qualter's *Propaganda and Psychological Warfare*, and *Strategic Psychological Operations and American Foreign Policy* by Robert Holt and Robert van de Velde all concentrated basically on propaganda and theories of persuasion, albeit through fascinating details of such aspects of the subject as war information policy, the disintegration of morale in the *Wehrmacht*, how propaganda themes were thought up and the delicate art of rumour-mongering.

So when, in the course of my Insight research, I found myself for a week at Fort Bragg in North Carolina, the home of the US Army's special warfare school and which has an entire section devoted to 'psychological operations' (the peacetime equivalent of psychological warfare), I was totally unprepared for what I found.

Fort Bragg is an enormous base, being also the home of the 82nd Airborne, one of the USA's tactical nuclear strike forces. The John F. Kennedy Center for Military Assistance, of which the psychological operations school forms a part, is a gleaming white complex of modern offices lost amid a delightful pine forest north of Fayetteville, a typical southern states town. The centre is fully equipped with all modern teaching devices – language laboratories, darkrooms, visual aids and so on – but its main attraction for me was its unique library. This was where, with the invaluable help of Mrs Jacqueline Baldwin, the head librarian, I discovered an enormous and extraordinary collection of papers and documents. Here was row upon row of (largely unpublished) reports of military experiments which together showed the full extent of the way psychology had in recent years been adapted for the purposes of war-making.

In London, Hutchinson, and in New York, Basic Books, had both expressed an interest in a new book on psychological warfare provided enough new material could be found on recent campaigns such as the Vietnam and Yom Kippur wars and the continuing troubles in Ulster.

But within a day of sifting through the Fort Bragg library I knew that I wanted to change my plans. Psychological warfare would still be a large part of what I wanted to look at; but in a much wider context and, on top of that, many of the documents in Fort Bragg described military adaptations of psychological research that went far beyond anything I had till then conceived.

What became very clear was that, during the past twenty years and with hardly anyone in the outside world noticing, the military uses of psychology had come of age. Delving through literally yards of documents at Fort Bragg, I could now see that these military uses were no longer confined, as I and many others had thought, to the general use of broad theories of communication or attitude formation, or to the use of social psychological concepts to help understand how soldiers adjust, for example, to being away from home during battle or on manoeuvres. Everything you could think of – from the psychology of the cell structure of underground insurgencies to the psychological effects of weapons, from the selection of men to work behind enemy lines to the ways to induce defection, from the way to stop men chickening out of battle to how to avoid being brainwashed, from tests to select code-breakers to the use of ghosts to harry tribal peasants – had been investigated in remorseless detail and the relevant psychological research drained of any military application it might have.

Not always with success – far from it. But the very fact that it had been done at all seemed to me of far greater general interest than my initial aim of providing what in comparison would have merely been up-dated examples of traditional concepts.

After my initial visit to Fort Bragg, therefore, I proposed to Hutchinson and to Basic that the outline of my book be changed to cover these developments. It needed, for example, an entire section on the psychology of counter-insurgency and guerrilla wars. It meant that the book would be much more factual, less a discursive foray into policy; it meant that there would be many more institutions for me to visit, that the book would take longer to write and be far longer than originally agreed, in view of the amount of material that I could now see was waiting to be collated. Fortunately for me, both publishers agreed and the book took on the form presented here.

A good idea of the size of what must rate now as a burgeoning division of psychology can be had from the references listed in this book. But the number of institutes specializing in this work also shows the character of the field.

I visited eight countries in North America, Europe and the Middle East in order to research this book, but even so did not get to more than a small proportion of the places where military psychology is carried on. I counted 146 *separate* institutes of one sort or another, the overwhelming majority (130) being in the United States. Only eighty are actually located within the armed forces, however, the remainder being divided between universities, a couple of specialist hospitals and private research institutes and 'think-tanks'. (The main organizations are described in Appendix II.)

The fragmented nature of the field, as exemplified by the large number of institutes involved, together with their small average size, has, I believe, played an important part in determining the poor quality of much of the research which in turn has had an impact on the overall character of military psychology. I discuss this at length in the first chapter and again in the conclusion. But however critical I am there of these institutes, I want *here* to thank the many scientists who gave me their time to explain their work. I was able to visit no more than about thirty of the most important ones but have corresponded with psychologists in many others. Without their kind consideration and prompt assistance much of this book certainly could not have been written. If, in view of the fragmented nature of the field, the book helps to put some scientists in touch with other workers whom they might otherwise not have known about, then I shall be pleased to have had an opportunity to repay some of the help they so generously gave.

One of the problems of trying to write a book which attempts to reshape people's ideas of a particular subject is the extent to which traditional, more well-known material should be incorporated. In my case, this meant to what extent should I try to cover the three areas of psychological warfare, leadership and selection, and combat psychiatry. These topics, as I said at the beginning, have each received quite a bit of attention especially with regard to their operation in the Second World War. I chose to do something different with all three.

Psychological warfare: I have concentrated on new *techniques* developed in the last twenty years and have tried to use documents that have not been available, or used, before. This has meant that I tend not to discuss general policy (since in my view there have not been many developments since the Second World War) and I have instead included far more descriptions of actual psywar techniques as used in Malaya, Vietnam, the Yom Kippur War and so on. I do not, for instance, dwell on information *policy* in Vietnam because the principles used were not substantially different from earlier conflicts, which have been well described in earlier works. What were new were many of the techniques to get information and propaganda across – for example, the armed propaganda teams. My aim has been to describe these in detail that is interesting to the layman and full enough for the professional soldier and to try to show why certain techniques were successful when others were not.

Combat psychiatry: here I have concentrated on the *scientific* advances made since the Second World War. In order to understand these, however, a certain amount of basic information is necessary. Even so I have tried to dwell mainly on new material, such as the new techniques for treatment in the battle zone and, particularly, on the problems which face psychiatrists when armies are either just about to go into battle or are engaged in withdrawal. Since many conflicts build up slowly and conclude very messily this is coming to be recognized as a special problem and a lot of recent scientific effort has looked into these aspects. I have tried to reflect this interest.

Leadership: this, too, has been well covered in the past. I set myself one major aim: to point up what I see as an important conceptual difference between 'leadership' and 'command'. This has been a dominant trend in military research in leadership since the height of the Vietnam War, but appears to have been little heard of outside a small group of specialists.

The book is designed to be read either from the first chapter on or, for specialists, the sections may be regarded as self-contained entities, to be read in any order. A real problem in writing a book of this sort, however, is the tone in which the material is presented, the attitude of the author which the choice of material and its organization reflect. Books about police, military and security affairs occupy a special position, for the time being anyway, following the deception which appears to have dogged public policy-makers in the last decade: the bombing of Cambodia and other Vietnam-based crimes; the Watergate scandal; the proven excesses of the British Army in Ulster; and the weird exploits of the CIA in a number of places.

It is difficult not to feel critical of many of these activities and I can quite see that a book of this sort, detailing as it does a variety of equally weird activities either by psychologists or on their behalf, could easily have been written with a strong 'anti-military' strand running through it. Since, however, a primary aim of the book is to introduce to the general reader as well as to the specialist a great number of studies in a field that he or she may not have been aware of before, and because I could not hope to explore the proper military and scientific background of all or even most of them, I opted instead for a different approach.

In a first chapter which is deliberately long I discuss what to me are the most important issues raised by many of the most controversial studies reported here for the first time. The implications of some of them are explored in what I think is justifiable detail in order to show fully how psychology can be a worrying science in the hands of the military and how it can actually change military life in important respects. I also try to define what I mean by studies which are 'provocative', and make wars or further conflict more likely, and those which are not. This seemed to me the best way to explain the attitude which I think I share with many others on military affairs, including the application of science in this field. This attitude may be broadly summed up by saying that extreme attitudes or forms of behaviour are rarely useful. I do not think, for instance, that unilateral pacificism works. No one really knows – or wants to take the risk – and social science studies of the 'pacifist strategy' do not help either. Conflict studies are also pretty inconclusive on whether disarming disarms potential adversaries or whether threatening behaviour actually provokes conflict.

We are left then with a more moderate attitude which, as I see it, views the military use of science as justifiable but only when it is used to conserve life or if it is in response, direct or anticipated, to some new threat. The deliberate development of weapons of unnecessary suffering, on the other hand, is out. I make no pretence that this is a comfortable distinction. Indeed in several places throughout the book I am sure I break out of it. But some attitude of this kind does seem to be needed, some code among scientists, if their imaginations are to be kept in check. New weapons, new developments, can always be stolen, 'leaked' or captured, the consequences of which may be incalculable. Also, without such a code new scientific concepts arise which owe more to the scientist's fertile brain than to the needs of the military situation and, as I think is shown on several of the pages which follow, the scientists end up leading the politicians in directions they cannot know and with implications that cannot be foreseen. I believe the opposite should obtain: scientists should serve the politicians. That at least should be a minimum code.

But in any case the book is intended, as much as anything, to provide the first airing of a great number of subjects and studies. Others will have different interpretations to put on the events I describe and greater opportunity to explore the political background to many of the episodes. In time, therefore, judgements I have formed in the course of writing the book will no doubt come to be modified, or changed completely.

For all these reasons the main chapters of the book are descriptive only. With the important issues raised at the beginning, I try then to take the reader as clearly and as concisely as possible through a vast amount of material, stopping only at a few highly controversial studies to discuss the rights and wrongs, the implications and the possible consequences, of these activities. In the conclusion I make a set of different, more general points about the direction in which military psychology appears to be going. It is a kind of futuristic scenario, an attempt to encourage the reader to extrapolate from what he has read and to decide for himself whether he thinks that is worrying. In this way I hope to provide the reader with all the information and many of the ideas he needs to make up his own mind on the issues raised in the subsequent pages.

I am grateful to Charles Clark, Managing Director of Hutchinson, and Erwin Glikes, President of Basic Books and Publisher, Harper and Row Trade Group, for agreeing to change horses in mid-stream; and for publishing a much longer work than they – or I – at first expected. To the many scientists, colleagues on the *Sunday Times* and friends, who discussed the issues and studies with me and who provided journalistic guidance and encouragement, I owe a lasting debt.

And without Miss Joan Thomas, who so cheerfully deciphered my handwriting and speedily typed up large chunks of the manuscript, I doubt whether any deadlines, however belated they were, would have been met.

Highgate Village, London
1977

1 Introduction

COLOURFUL CAPERS

As 1965 drew to a close a secret research project for the United States Department of Defense was concluded by a 'think-tank' of seven psychologists at the American Institutes for Research, in the Maryland woods outside Washington, DC. Their final report, *Psychological Phenomena Applicable to the Development of Psychological Weapons*, was a confidential document, with the following security classification stamped across it: 'Distribution of this report is limited to DoD activities because the information is not considered of general interest to the public, scientific and technical community.' However, the document was then, and still is, of the greatest possible interest not only to the scientific community but to the general public as well (1).

Its purpose was to examine how weapons designers could 'enhance the psychological effects' of new weapons. Its seventy pages detail, among many other things, how fears can be manipulated, how perceptions of military significance can be deliberately distorted, how the stresses of war can be systematically increased and even how the psychological characteristics of different racial groups can be exploited.

At much the same time as this document appeared four psychologists were working on an equally unusual enterprise on the other side of Washington at the Human Resources Research Office in Alexandria, Virginia.[1] They found that the shooting effectiveness of combat platoons could be improved by more than 50 per cent by a simple psychological ploy (2). All that had to be done was to redistribute the available ammunition so that the men at the end of the firing line – in 'trigger happy' positions – had less than those in the centre. This, it turned out, made the men at the end of the line far more careful about using up

1. For a description of HumRRO and the other main military psychology think-tanks, see Appendix II.

their ammunition and upped the 'kill ratio' of the platoon as a whole. The only drawback was that although the platoon as a group was safer, the end men were now more likely to get killed.

Fifteen thousand miles away two other military psychologists were at this time beginning to take a fresh interest in the pigeon. This bird has a long military history – not just as a messenger. For example, towards the end of the Second World War the famous behaviourist, Professor B. F. Skinner, was given a defence grant to develop a scheme whereby pigeons could be trained to guide missiles onto targets. Skinner had found as early as 1944 that pigeons in flight could distinguish man-made objects from naturally occurring ones. This finding began to interest the Israelis towards the end of the sixties. An American psychologist who spent some time in Israel developed an adapted system in which the pigeons could effectively be used as spies against the Arabs. They were trained to fly over Arab-held territory, kitted out with small radio direction-finders, and land on or near man-made objects. They could even follow moving columns, looking for food. The radio direction-finders of course told the Israelis exactly where the birds, and whoever they were following, were located (3).

But perhaps the most unusual plan was under development at the US Air Force Aeronautical Systems Division, at the Wright–Patterson Air Force Base in Ohio. There, a detailed cross-cultural study was made by social psychologists and anthropologists of the olfactory senses of various races, particular emphasis being placed on which smells particular races found offensive. The research was carried out by R. Huey Wright and Kenneth Michels. The idea, it seems, was to develop 'smell bombs' for use in the jungle. These would 'flush out' guerrillas who would otherwise be difficult to get at with more conventional weapons. Against this, even Russian plans to harness ESP as a weapon lose their imaginative sparkle (4).

Exotic-sounding stuff, this military psychology. Not all the tactics and techniques described in this book are – unfortunately perhaps – as flamboyant as these. Nevertheless, these four capers (there is no other word) do illustrate the four main points I wish to bring out in the pages which follow.

In the first place, each of the experiments was carried out in the last fifteen years. The military uses of psychology were first taken seriously during and immediately following the Second World War. Only in the early 1960s, however, did the subject really take off; relative to other military sciences, therefore, it is a new field.

Second, these four examples, unusual as they appear, between them do give a good inkling of the surprisingly wide range of studies which have been completed in such a short time and the extent to which the military potential of psychology has been ruthlessly exploited in recent years – even as much, perhaps, as the military potential of biology or chemistry, which have received far more publicity.

Which brings us to point three. Many psychological studies, like the four examples, are – or have been – either secret or, if not actually classified, circulated only among a relatively small handful of specialists. The closed-world nature of this branch of science has both determined its unique character and prevented a wider discussion of the various issues – scientific, military, political and ethical – which are raised by many of the studies.

My fourth point – and the main aim in this book – is to bring to a wider public what I see as the more important and controversial of these studies and to stimulate an open discussion of the issues raised by them: for example, the exploitation of racial differences for military advantages (5); the study of public attitudes towards war episodes and acts of terrorism so that armies know what counter-measures are acceptable and when (6); or the use of behaviour-modification techniques to make soldiers less worried about killing (7).

Nowadays, perhaps, we are familiar with developments in science which provoke ethical and political problems for scientists and governments. But psychology is in some ways unique among the military applications of science in two important respects. First, it produces information which helps to change the relationships *between soldiers* (as in the ammunition experiment) and in the process changes certain traditional aspects of military affairs, such as rank, discipline and the concept of bravery. This, therefore, makes military psychology controversial *for soldiers* as well as civilians in a way that ballistics – say – or engineering could never be. And second, one aim of much military psychology is to change the way people *feel* towards certain aspects of war and warfare. Fear, weapons, killing all begin to mean different things when seen through the eyes of the military psychologist.

These wider issues are intriguing and controversial. Before discussing them further, however, it is worth going back over the first three points in more detail. They help in understanding particularly how some of the more extraordinary efforts of the military psychologists came to be undertaken – experiments which outrank most others when it comes to cruelty, deception, ingenuity and sheer absurdity.

GODS AND GUERRILLAS: THE GROWTH OF MILITARY PSYCHOLOGY

A world-wide change took place in the collective mind of the military in the early 1960s. It suddenly seemed to click with the generals, the air-commodores and the admirals that all the important developments in warfare since 1945 – the advent of guerrilla-type insurgencies, the greater use of more intricate hardware, the more complex tactical situations arising from greater mobility and easier communications, and not least the enormous stresses stemming from the prospect of nuclear war – pointed to the need for a much greater understanding of the human factors involved in armed conflict. Several aspects of military psychology had already been explored by this time but in the sixties a more coordinated effort got under way. Government spending doubled during the period, as did the number of studies completed.

In 1963 the US Department of Defense held its first Worldwide Psyops Conference which outlined twenty-eight specific tasks. The new programme came into effect late in 1964. In 1963 also US Army Contract AR 70–8 authorized a project 'Human factors and non-material special operations research' – 'the foundation', according to Lynn Baker, the Army's Chief Psychologist, 'on which all [military psychology] rests'. The period also saw many symposia on the links between academic psychology and military needs, and the war colleges also turned their interests this way. The first NATO symposium on defence psychology was held in Paris in 1960 and a year or two later F. H. Lakin, from the Army Operational Research Establishment in Britain, went to Fort Bragg to address a conference on human factors in military affairs on British psychological warfare techniques in Malaya. In 1964 the US Director of Defense Research and Engineering commissioned a secret study by the Advanced Research Projects Agency (ARPA) entitled: 'Research in the behavioral and social sciences relevant to counterinsurgency and special warfare' (8). The US Army even established a Human Factors and Operations Research Unit in the Far East – at Seoul in Korea.

A clear discussion of the reasons for this change appeared in 1964 in the *American Psychologist*, a monthly journal for professional psychologists. An article entitled 'The future of military psychology: paramilitary psychology' (9) forecast that military psychology was about to shift gear decisively and would no longer simply be concerned with its three traditional areas – matching soldiers to their military machines, training military specialists, and selection. Instead, military psychology would increasingly concern itself with 'paramilitary' issues: for example, studies relevant to counter-insurgency operations, understanding the political motivations of guerrillas, the social and psychological effects of economic underdevelopment and their relation to conflict, the human factors involved in clandestine organizations, and so on. The growing recognition of the cultural influences on man would mean a switch from an emphasis on human engineering to cultural engineering. Communication would become even more important as a weapon.

The authors of the article were two psychologists, Charles Windle and Theodore Vallance. Until this time both had been on the staff of the Human Resources Research Office, HumRRO for short, an organization which features prominently in many sections of this book. HumRRO works mainly on the more traditional, human-engineering problems in a military context. The psychological difficulties presented by night operations, cramped flightdecks or a well-camouflaged target are the kinds of thing which HumRRO exists to overcome. Windle and Vallance, however, at the time they wrote their article for the *American Psychologist*, had just moved over to SORO – the Special Operations Research Office. This organization, as its name implies, was concerned with studies relevant to special operations. And that, in the early sixties, meant mainly guerrilla-type insurgencies.

Windle and Vallance were among the first psychologists to move into this growing field and they were quick to see the extent to which psychologists (and social scientists in general) would become involved in the study of war in the years which followed. Indeed, they were themselves largely responsible for much of this involvement.

The full story of that involvement has never been told but, since that change of emphasis in the early sixties, the military uses of psychology have been pursued with ever more energy and increasing imagination. For example, the last few years have seen such innovations as a direct computer link between the brain of a pilot and the missiles located under the wings of his aircraft (10). This means that pilots will eventually be able to fire their weapons and then guide them onto target merely by thinking 'Fire! . . . Left a bit, down, right . . . ' and so on. Equipment has been developed to use low-lying cloud as a screen off which to bounce huge propaganda shows. Tape-recordings of primitive gods have been prepared, to be played from helicopters, thus frightening tribes and keeping them in their villages – a sophisticated form of population control. Details of such cultural 'weaknesses' as prostitutes, pickpockets and fraudsters in several countries are now on file at centres in the United States, Germany and the Far East – possible tools for creating dissension in a potential enemy country should the need arise. There is even a world-wide list of propitious and non-propitious dates available at the Pentagon to enable generals to time their bombing campaigns to coincide with unpropitious days, thus 'confirming' the forecasts of local gods (11).

As ambitious and controversial as many of these psychological weapons are, none of them has ever had any publicity. You will not find any details of them in the professional psychological journals. Look through these and indeed, in any one year, you will be lucky to spot as many as a dozen articles which fit readily into the bracket of 'military psychology'. On the official surface, as reflected in the journals, military psychology appears to be a small and not very enterprising aspect of the science.

The extent to which this belies the true picture, however, can be seen from the fact that the research for this book has unearthed many organizations around the world sponsoring no less than 7500 studies. Many are small-scale; but taking them together, it means that, since Japan surrendered and the Second World War came to an end on 15 August 1945, one investigation has been completed *every day*. Expenditure by the US Department of Defense on psychological research increased by $4 million from 1973–4, consuming 12 per cent (as opposed to 9 per cent for 1973) of the federal research budget in this area.

Clearly, military psychology is now an enormous field. But – equally clearly – it is closed to most of us. No one can know, of course, just how much secrecy there is. Many studies are not secret in the technical sense – it is just that the reports arising from them are duplicated and circulated only among the handful of specialists working in that particular area of inquiry. Undoubtedly, however, a large proportion of the studies is officially secret. A little-known (and privately circulated) United States Department of Defense bibliography, *Military Psychology, Propaganda and Psychological Warfare*, prepared a few years ago, contains 761 items and is far from complete (12). But no fewer than 322 of the studies listed are classified; some of the unclassified ones have other restrictions placed upon them.

SOME SINS OF SECRECY

Secrecy, although it has its uses (surprise and deception, especially where psychological weapons are concerned, can be all-important), has been at least partly responsible for some of the most intriguing yet highly questionable studies carried out in this field in recent years. This, I hope, will help explain why I discuss openly documents the secrecy of which is supposed to preserve the national security of either Britain, the United States, Israel, or some other country. I am sure that many readers will agree with me that some of the reasons for these secret tactics and techniques are very lame – in some cases absurd. But that, in itself, would not be enough reason for publishing their details. What matters is that secrecy has spawned its own difficulties in this field which have led to major errors in fighting strategy. Repetition of such errors can only be avoided, surely, by more open discussion.

Detailed accounts of the way secrecy affects science and the institutions where science is carried out are somewhat thin on the ground. One suspects that such accounts would not be very flattering and so tend not to get written. One man who has attempted such an account, however, is Ritchie Lowry, a social scientist now at Boston University but who was formerly attached to SORO (13):

> My experience in doing research for the federal government on military problems suggests that three of the most important latent functions of secrecy in that context are the production and protection of relatively useless and unreliable knowledge, the consequent guarantee of individual and organizational job security and a resulting extension of security into areas involving sensitivity.

For Lowry the most important *direct* effect of secrecy is that it prevents the normal interchange between scientists, the legitimate criticism of methods and deductions which normally lead to the refinement, modification and even the abandonment of accepted ideas. Poor work is the result.

Lowry quotes a number of examples, the funniest of which was a study of cross-cultural persuasion, involving considerable cost, which concluded: 'Influence attempts [propaganda] are most successful when the objectives of both the initiator and the recipient are most similar.' As Lowry somewhat drily observes: 'In other words, after much time and expense the study concluded that persuasion attempts requiring the least amount of persuading are the most likely to succeed.'

Shoddiness is protected from the cold light of day by elaborate rationalizations. One of these, Lowry quotes, is the pressure – quite unlike that in universities – *not* to publish. 'If publication is bad the project may be cancelled; if it is good, the project may be terminated as successfully completed.'

The other rationalization is simply to label everything as 'secret'. This process becomes so abused, says Lowry, that security inevitably comes simply to mean 'sensitivity'. Things are kept secret not to preserve national security but so as not to embarrass someone, either because the methods or conclusions of the study are controversial or simply because the shoddiness would be exposed for all to see. Lowry quotes one example of an article written by a 'Soron' (as the employees of SORO were nicknamed; it rhymes with 'moron') for publication in a magazine. The magazine agreed not to publish any organizational identification (Army, SORO, etc.), but the Army Research Office classified the paper, giving as the reason the fact that it was critical of foreign policy.

Of course, these various aspects of secrecy feed upon one another and in many instances prevent the quality of work ever improving. It is, however, not only the quality of science that is affected. Much more important is that, almost inevitably, bad policies must follow.

Probably the most notorious example of this arose out of the study carried out between 1964 and 1969 by the famous US Air Force think-tank, the Rand Corporation, into the morale and motivation of the Viet Cong (14). Through what can only have been poor methods, the study – especially two documents prepared by the project head, Leon Goure – concluded that the Viet Cong were near to breaking-point. All that was needed, ran the thrust of the documents, was a few weeks' heavy bombardment and the VC's will would break, their resistance collapse and the war would be over. With hindsight it is clear how far out these conclusions were. But at the time, according to a presidential aide, this study lay 'at the heart' of President Johnson's strategy for the continued heavy bombing of large areas of Vietnam.

This action, mistakenly based on erroneous science, cost many lives and millions of dollars and may actually have lengthened the war.[2]

Other cases of bad policies arising out of bad science may be cited. During the 1973 Yom Kippur war, for example, the Israelis introduced a device, produced in collaboration with psychologists, which enabled tank crews to know when an enemy Arab tank had 'locked on' to them with its gun sights (15). It was, in other words, supposed to be

2. Some of these documents were produced as part of a contract between the Advanced Research Projects Agency and Rand, and as such, were private property and therefore unavailable to the United States Congress (16).

an early warning system to enable the tank crews to take evasive action. Now there is quite a bit of psychological research on the way warnings do or do not exert their effects and it is clear that the Israeli psychologists had not properly done their homework. For this would have told them that the production of the device could provide as many headaches as it solved in a rapidly changing and highly stressful situation like a tank war (British psychologists, for example, have been busy in recent years making the instrumentation of their tanks *less* complicated). And so it proved in the Yom Kippur war. The main effect of the device appears to have been to transfix the crews the second they realized they had become targets. The gadget was just another piece of apparatus they had to contend with and for many it was too much – they simply froze. It was hurriedly removed.

Returning to Vietnam, in another SORO study the researchers spent several months and many dollars trying to find out why AWOL rates were so much lower there than they had been before in other wars. They wanted to know what policies the army was adopting to keep its men so highly motivated. The secrecy with which the study was carried out could not have helped for no one stopped to ask the simpler – and cheaper to answer – question: where could a white American go AWOL in a steaming Asian jungle (17)?

Bad policies arising from inadequate research will be seen to crop up regularly in the later chapters of this book and in many the secrecy surrounding the studies must take at least some share of the blame for the low quality of thought that lies behind these enterprises. This, as I have said, is my chief excuse for openly discussing classified documents. Before we move on from secrecy, however, there are three other side-effects which it has and are worth mentioning.

Inevitably, with many secret studies there will always be some sort of leakage – a few, often garbled, details dribble out and may feed a rumour. Even in official circles this happens, according to Ritchie Lowry (18). He quotes a formal briefing which took place inside the Pentagon at which it was said that a SORO study had shown that Latin American officers who received training in France were more 'fascistic' than those trained in the United States. 'In reality,' says Lowry, 'the study had dealt with only 70 Venezulean officers, had not mentioned France, and carried one table comparing those officers receiving some overseas training with those receiving training only in Venezuela.' The results, apparently, showed that those with overseas training were more likely to be engaged in coups or to become politically active later in their careers. But nothing in the study indicated whether such activity was rightist or leftist so far as political content was concerned. Thus secrecy allows rumour and prejudice to pass themselves off as science.

Much the same sort of thing happened in 1965 when SORO planned Project Camelot, a major inter-disciplinary study of Chile; this was to investigate the factors which gave the country its particular social complexion and assess which factors made for greater stability and which made for greater change. According to Theodore Vallance, writing a couple of years later, there were no sinister intentions, but while it was still in the confidential planning stage, news of the proposals leaked out in Chile itself. An uproar was created in that country and the embarrassed US Ambassador in Santiago, who knew nothing of these plans, protested to the State Department which, being the administrative head of SORO, had the project cancelled (19).

A second side-effect of secrecy is that the methods by which many studies are carried out do not always conform to the accepted scientific standards. In 1962, for example, Mitchell Berkun, working at the Human Resources Research Office, conducted a series of experiments aimed at exploring whether troops could be battleproofed by making their training so stressful that they would enter battle inured to any further fears (20). In one experiment men flew in an aircraft which 'developed' an 'engine fault'. In another, men were 'accidentally' led into a 'shelling zone'. (Berkun's experiments are considered in detail in chapter 9.)

In the course of these experiments some men experienced real terror and were exposed to actual physical danger. The HumRRO scientists appeared to think that the ends – training the men to cope with fear – justified the means – scaring the wits out of them when it was not strictly necessary. The public took a different view: when news of the experiments broke, reaction was swift and they were condemned as inhumane. They were soon stopped, but had secrecy not prevailed at an earlier stage, it is probable that they would never have taken place at all.

A final unwelcome side-effect of secrecy applies not just to psychological studies but to all military operations. Military secrecy is usually based on the argument that if the enemy is deprived of important information, it may be possible to keep him guessing to the point where one's own chances of success are maximized. On the other hand, several (admittedly artificial) studies have shown that a precise knowledge of the comparative strengths and weaknesses of the opposing parties in a dispute is the most effective deterrent to conflict. In these cases, it could well follow that secrecy may actually increase the likelihood of war (21).

There is a good case for bringing the activities of the military psychologists right out into the open. It is, however, easier said than done. Having convinced myself that a true picture of work in the field could only be given by discussing at least a sizeable proportion of the classified studies that have been undertaken, the next problem was actually to get at them. In the main, four 'detective techniques' have been used in an attempt to ensure that the picture given is a reasonably comprehensive one.

THE DOCUMENTARY EVIDENCE

I have been extremely fortunate in having had a limited amount of classified information and documents made available to me. I would like to acknowledge my gratitude to those individuals who did this but who, for obvious reasons, must remain anonymous.

A sizeable proportion of the classified documents quoted later in the book I have seen in their entirety – but my access to many more has been via official limited-distribution bibliographies, usually containing short summaries of the work done. They provide an indication of the chief areas of classified study. For example, the following studies are (at the time of writing) all classified as 'secret', 'confidential', or 'for official use only':

The spread of information following an atomic manoeuvre (22);
Psywar in a tactical nuclear war (23);
Chinese and Korean methods of motivating riflemen for combat (24);
Use of hypnosis in intelligence and related military subjects (25);
Aspects of air power in morale-oriented tactics (26);
Communist vulnerabilities to the use of music in psywar (27);
Exploratory efforts concerned with a study of the interrogation process (28);
Psychological strategy and tactics in countering insurgency (29).

By comparing complete documents with summaries of others by the same scientists or the same institutions, it has often proved possible to read between the lines of many abstracts and to arrive at some understanding of the overall picture in a specific area of interest. For example, by putting the classified work by the US Army on hypnosis together with unclassified work carried out for the Office of Naval Research and with advances made by the Israelis, who are known to have collaborated with some of the US Army scientists, it is possible to build a picture of the role of hypnosis in combat. This has two main elements: training people to resist interrogation; and use with witnesses – someone returning from a reconnaissance mission can produce more information if he is hypnotized and then asked questions about what he has seen than if he simply writes down, consciously, what he saw.

In several cases, particularly with United States research carried out for the military, a rather strange procedure is followed with the security classification of certain studies. For example, some years ago a socio-psychological analysis of China was carried out by an army social science research outfit and on the basis of this several 'targets' for psychological warfare were identified. Now although the document containing the 'targets' themselves is classified, its sister-document, containing the analysis leading to the choice of targets, is not. Anyone with a training in this sort of social science analysis can extrapolate from the unclassified research to the secret and get a pretty good idea of the official 'targets' (30).

Many allies exchange scientific information which has military significance and this makes full secrecy harder to maintain. The military authorities of one country do not always seem to attach the same security label to a piece of research as the original country did. For example, the details about British psychological operations in Ulster are top secret in Britain (the Ministry of Defence does not even acknowledge that the army has any psychological operations); yet they are not impossible to come by in Washington, including the names of the London advertising agency and the officials of the British Psychological Society who help train the troops (see page 379) (31). Another example is the training of pilots in Australia; certain aspects of this are restricted in that country but are more easily available in Britain.

Access to some classified documents also enables one to compare the classified version of a study with the version that is released to the public (when it is released, that is). In this way one learns what kind of things get left out and to recognize when a published version is in fact rather more than it appears. One series of experiments which exists in two versions, for example, relates to the performance of men in remote Arctic sites for extended periods (32). Mainly carried out by a particular navy, these studies – publicly – deal with the types of individual who get on well together with the aim of assessing just how they can be chosen in advance so as to minimize friction. The classified side of the studies, however, also looks at how good men with different kinds of personality are at keeping secrets from one another. In other words, the experiment is also a training ground for budding spies, a study of the reactions to being (literally) left out in the cold and how well they keep quiet, without divulging crucial information, at times of stress.

With experience this approach can be cautiously applied to other studies where the published details do not quite seem to add up, where there is reason to doubt that all the aims of the study are being made public. Philip Zimbardo's prison study (described in detail in chapter 13) falls into this category. Zimbardo's work was published in the *Naval Research Review*, the official journal of the US Office of Naval Research, under the title, 'A study of prisoners and guards in a simulated prison'. The stated purpose of the experiment was to investigate those factors 'which spawn disruptive interpersonal conflict in Naval prisons', but certain features of the experimental design suggest that it was more concerned with reactions of prisoners of war or with the selection of guards for POW camps (see pp. 259–64). The case for reading more into the study than was actually published was strengthened when a spokesman at the Pentagon, in response to another inquiry of mine, said that in 1973 – the year in which Zimbardo's study was published – the Navy had actually closed down all its prisons.

PR AND PAY: CONTROVERSIAL FINDINGS

Although secrecy brings with it some dubious methods and questionable conclusions, military psychology can be quite controversial enough without secrecy. In a new field, the mere acquisition of new knowledge or techniques presents both the soldiers and the scientists with fresh responsibilities. I hope that most of these are identified at appropriate places in the text – certainly it has been my aim to do so where it seems necessary. But some of the main points are worth examining now to give some flavour of the difficulties raised by the study of military psychology, secret or otherwise.

Take first the army's relationship with the public – especially during a conflict or in the period leading up to it. It is arguable that one guide to the public's view as to the rightness or wrongness of an army's involvement in a particular war is the number of conscientious objectors (draft-dodgers) who show themselves. The American draft-dodgers of the late sixties and early seventies are probably viewed in a very different light now since the United States has been so ignominiously defeated in Vietnam.

The tools of advertising and public relations are, however, now available to the military no less than to anyone else; and from these – and several studies of conscientious objectors (see pages 177 ff.) – armies especially have learned a great deal about how to present themselves at a time of crisis so that recruitment does not pose as many problems as it might and about how to retain the sort of image necessary to enable them to take what their commanders see as appropriate action. Specific studies in this area have included the differences between the public's view on defence and disarmament and that of political leaders, the particular role of women's attitudes towards the fact that their men may have to leave them, which information about the victims in a war (numbers killed or numbers made homeless) is most effective in eliciting support for more military involvement, the effects of anti-war demonstrations on public opinion, and many more (33).

A related matter is the public impact of all sorts of violent events. The Israelis are probably unique in this respect in that they have a special institute (run by Louis Guttman, a well-known attitude researcher) which samples Israeli opinion every week and is geared to do this every day if ever war breaks out again in that part of the world (34). This may seem a luxury. On the other hand, Guttman's institute is able to sample the public's mood one day and, in times of war, have its report ready for the next day's war cabinet – a useful service if the government and its chiefs of staff are thinking of introducing harsh measures at a time when the political climate is likely to be controversial. The British Home Office was contemplating a research project into the public impact of terrorism at the beginning of 1976, so the idea is not confined solely to the Israelis (35).

There are strong arguments for such a systematic approach to difficult problems at a time of crisis as in a war or when armed terrorism is growing. On the other hand, the production of such knowledge does provide the opportunity for those in authority to push through what may be unnecessarily illiberal measures (say, detention without trial) just because the opinion polls show that these measures would be popular and elicit political support even though they might not be effective. The only safeguard against misuse of this sort of information seems to be for its collection and collation to be public at all times. The Israelis, at the time of writing at least, seem more inclined to do this than either the Americans or the British, for example.

A different matter, one briefly alluded to earlier, concerns the effect which the study of military psychology has upon the relationships *between* soldiers. The more war and soldiers are studied the more their view of themselves is liable to change. With the trend towards low-intensity operations, as Brigadier Kitson calls guerrilla wars, at much the same time as communications have grown more efficient, it is now highly likely that the average soldier – and certainly the average officer – is more aware of the detailed political issues surrounding any particular conflict. This no doubt creates specific difficulties for some soldiers fighting for causes they cannot believe in (witness the young Portuguese officers whose growing hate for their colonial wars in Africa led directly to the coup which toppled Caetano's government in April 1974). But a more general spin-off is that psychological research carried out among the military is now much more likely to be interpreted by the men in a political context.

Take, for example, the research by psychiatrist Peter Bourne among US counter-insurgency troops in Vietnam in the late 1960s (see chapter 9) (36). By analysing the chemicals in their urine, Bourne found that, as the opportunity for action, or the threat of attack, approached, there was a distinct difference in the behaviour and the feelings of the men, depending primarily on their position in the squad. Action was clearly much more stressful for officers and for radio operators than anyone had ever thought. What should the man-managers of the army therefore do? Is there a value to be placed on stress? If so, is it perhaps to be reflected in higher pay or a higher rank, for example, in the case of the radio operator?

Then there is Catch-23. Catch-22 is well-known. A military man, theoretically, can get a medical discharge from combat if he is mad. But to want to escape war is judged a mark of sanity. So everyone who goes mad is, by definition, sane and stays put: at the front line.

Catch-23 is the way some soldiers get back at this system. Recent studies by psychologists have shown that many soldiers manage to avoid being sent anywhere near the front line long before war breaks out (37). The better soldiers – the more intelligent, the more able, the better fighters – cluster in the safer jobs away from the front line. Should the military accept this, and simply look upon the less intelligent men as unfortunate cannon fodder? Or should it ensure that more intelligent people take their turn nearer the front line? Either way it is going to court unpopularity from a large number of men.

Other precise statistics produced by psychologists pose similar difficulties. The rates of mental breakdown are related to rank or military speciality in wartime. Some ranks or positions, such as rifleman, are three times as likely to breakdown as a gunner, mortar crewman or ammunition handler (38). Men in armoured vehicles are more than twice as likely as infantry to break down and three times as likely as airmen. Should these differentials be accepted as facts of military life? Or should they be reflected in pay differentials or different job-rotation policies – say, shorter tours of duty for the more dangerous specialities?

It is in these sorts of ways that military psychology is beginning to change the relationships between soldiers, relationships which in many cases have been fixed for centuries. Danger and stress are being quantified now and though the military is not yet sure how to cope with this, there are bound to be changes. And changes to the structure of armies or navies (as happened in Britain when various regiments were disbanded or merged) are resisted much more than technical changes – in new vehicles, say, or new weapons. It will be painful for many of the traditionalists in the military to experience: but fascinating for the rest of us to watch.

THE SCIENCE OF SUFFERING: DEHUMANIZATION AND FEAR

A great deal of military psychology is concerned with the feelings and thinking of people in war; the manipulation of these feelings and the attempts to alter these thoughts hold perhaps the greatest fascination.

A lot has been said about the dehumanizing effects of modern weapons. Usually this means that these weapons have such powers of devastation that the victims can only be thought of in terms of huge numbers – with 'megadeath' as the unit of counting. On the other hand, the sheer size of the numbers involved is impressive and, arguably, has played an important part in deterrence. On the psychological side of warfare, however, dehumanization means something rather different – and is possibly without any positive aspects to it.

There are three ways in which psychologists help to dehumanize the enemy. Perhaps the most important arises, paradoxically, from the fact that psychologists at war, unlike other military scientists, are less concerned with the cold finality of death and instead have to concentrate on what is often the lingering business of suffering. Fear, hate, deceit, pain, humiliation, loneliness, homesickness, envy, jealousy – this black side of human nature is the currency in which the psychological warfare specialist (or 'psywarrior' as he is sometimes known) trades.

One might think that such an approach would lead to a greater appreciation on the part of psychologists for the enemy as people. Unlike military chemists or biologists, who need only the crudest data about the enemy population (such as its density, how it gets its water and so on), psychologists need to see the opposing side as living souls – with emotions, with families, with neighbourhoods, and maybe even with a case for fighting the war they all happen to be involved in. In fact, the opposite seems to occur. Precisely because he has to explore the habits and customs, the likes and dislikes, and the personal characteristics confined to the enemy, *and then take advantage of them,* the psywarrior simply seems to give himself and others more ammunition to fuel dislike of their opponents and more reasons to consider them lesser human beings.

A startling example of this was revealed in the House of Representatives by Congressman Cornelius Gallagher in 1971. He questioned the intentions underlying a naval research project being carried out by psychologist Sigmund Streufert (39). The research seemed designed to measure how different individuals value human life; in other words to screen for those who, attaching little value to life, might make good killers. The work also appeared to measure the value various groups of peoples place on human life – an attempt to assess whether some nations object less to being killed than others. Such work, where it is not absurd, is a disturbing example of dehumanization. (A detailed investigation of Streufert's work and its relation to acts of atrocity appears in chapter 12). The many studies of primitive gods, witchcraft and sorcery, and how these may be used to control people by the military, are another example (40). Inevitably soldiers come to see themselves as superior to those who are susceptible to such ploys.

A second way in which psychology helps to change feelings about conflict and to dehumanize the enemy is by making brutality less visible. Psychological techniques of interrogation and torture, for example, are more popular these days for precisely this reason.

A mental scar is preferable to a physical scar because it takes time to show itself. Whether this has an effect on the interrogator or torturer (as some evidence now seems to show) it certainly seems to ensure that worse punishment is handed out than anyone realizes at the time. The power the guards had over the prisoners in Zimbardo's prison study (p. 259) appeared to be addictive (for the guards) so long as there were no physical signs of maltreatment to be laid at the door of the custodians. There have been many real-life cases where this has happened, among them the treatment handed out by the Portuguese secret police before the coup in April 1974 and the even more notorious example of the treatment given by the British Army to the fourteen IRA suspects picked up in Ulster in August 1971 (41) (see chapter 14). There were never many physical scars, but some of the men have suffered lasting mental damage. Professor R. J. Daly, head of the Department of Psychiatry at University College, Cork, was *still* seeing some of the men in early 1976 – nearly five years after the episode (42).

The third method of dehumanization goes even further. This works by actually taking – or at least attempting to take – the feeling out of killing. The work of Lieutenant Commander Thomas Narut is described in detail in chapter 12. He employs 'Clockwork Orange' techniques, in which men have their head clamped in a vice, their eyelids propped open, and they are then shown horrific films. With such techniques the aim is to completely desensitize the men to pain or suffering, to remove any emotion associated with it that might interfere with killing. According to Commander Narut, men have been trained in the US Forces in this way – for use as special combat units for rapid assassinations. Can it be that they are also trained not to mind too much if they have to kill themselves should they get caught?

But there are other feelings in war which the psychologists are interested in changing and which do not involve dehumanizing the enemy or making brutality less loathsome. For example, the most natural emotion of all in warfare is fear. The changes in this area are subtle. For a start, psychologists have stopped calling it 'fear' and now refer to it instead as 'stress'. Stress is a psychological jargon word: most important, it is 'value-free' – neither good nor bad but something which, like any coldly clinical entity, can be manipulated, experimented with and, given luck, smoothed out of the picture entirely.

One of the effects of this is that the legitimate moral and indeed philosophical role of fear in conflict is robbed of much relevance. Fear is hardly ever studied by the psychologists as a legitimate response to a frightening situation, or as something which plays a part in the very sensible process of deterrence (without fear there could be no such thing). Instead, stress is looked upon as a physiological response which produces psychological effects. Once the physiology is understood, this approach implies, we shall eventually be able to regulate it so as to eliminate the psychological end-product most of us call fear. No attempt at all, by the military psychologists, to try to change the situations which are frightening.

The kinds of study I am referring to here include Peter Bourne's work on the stresses encountered among counter-insurgency forces and Mitchell Berkun's series of experiments where he tried to battleproof soldiers (see chapter 9). Of course, in one way they do make sense. If you *have* to fight a war, then it is better to have your men prepared as well as possible. On the other hand, if it actually proves possible to battleproof troops beforehand (and this has not really happened yet), would it make for more aggressive soldiers? Would it mean that on occasions they would attack when prudence told them to hold back? Or to fight when retreat would be the better part of valour? These are questions which have not yet arisen in a big way; however, the thrust of the research by such people as Drs Bourne, Berkun and Commander Narut does seem to be leading towards a situation where hazardous missions will be undertaken more readily. Does this mean, therefore, that, tactically, we can expect more 'aggressive' manoeuvres in limited wars in the future or that escalation is more likely?

A lot of unanswered and, for the moment perhaps, unanswerable questions. Each of these developments, however – the dehumanization of the enemy, the debrutalizing of killing and torture, and the tampering with fear – reflect an attitude that everything, even in war, can be solved by the appropriate piece of psycho-technology. It is a seductive approach but not without its drawbacks, as I have been trying to suggest. The most important of these drawbacks, however, has still to be tackled. It is the confusion which many military psychologists appear to have over the links between psychology, war and politics.

PEACE AND PERSONALITY: POLITICAL IMPLICATIONS

In chapter 17 I describe a study by Andrew Molnar of the Special Operations Research Office (43). Molnar looked at more than twenty insurgencies that have taken place since the Second World War and concluded that people become guerrilla fighters mainly for personal and psychological reasons rather than for political ones; and that even where motives are political then they are so because of some psychological characteristics in the individual which leads him to identify with a particular cause.

Molnar's aim, in his study, was to justify the argument that psychological offensives against insurgencies are what is needed to win these limited wars. But in doing so, he entirely ignored the question of the legitimacy of a particular political viewpoint, and whether action should therefore be taken against people supporting that viewpoint. A society, in Molnar's eyes, becomes simply a 'social system' that is neither right nor wrong, neither just nor unjust, but only something which takes a particular form, the most relevant characteristic of which is its stability or otherwise. The moral question relating to intervention is in this way side-stepped. War becomes simply a mechanical approach to human affairs on a grand scale: the social system (a country) is not ticking over in a regular (i.e. stable) manner, as a result of which it does unpredictable things. The military (social) mechanic simply gets out his psychological tool-kit and tightens up a few nuts and bolts here and there. Stability returns. The case for a new engine is never debated.

Much the same attitude lay behind the Rand Corporation's motivation and morale study, which as we have already seen is now reckoned to have had tragic effects on US policy in Vietnam. In that study the clear indication is that motivation and morale were seen by the Rand scientists to depend on personal matters: physical discomfort, separation from loved ones, the attitude of superior officers towards subordinates were three frequently quoted. There are, however, good grounds for thinking that it was the other way round – that political fervour, at least in some, kept them going in adverse conditions when others might well have caved in without any sustaining ideology. And clearly one of the fears of the Chileans, in taking so much objection to Project Camelot, was that the study of the country as a 'social system' would provide America with a great advantage in imposing its own will on the country under the guise of 'stability operations'.

This is not to say, of course, that politics and personality are not linked in many intimate ways. Indeed, several studies have shown that they are (though not necessarily in an insurgency or war any more than in peacetime). Nor is it to deny that an understanding of the 'Human factors in underground insurgencies' (the title of Molnar's study) helps in controlling them. If you know what type of people insurgent organizations select as couriers, or as *agents provocateurs*, and what type of people become its leaders, then you have at least made a beginning in understanding how the underground movements operate and this helps the security forces to know who to keep an eye on and when.

My aim in discussing the confusion that all too easily arises over the psychology-versus-politics question is to point out simply that an intimate link between someone's personal make-up and his ideology in no way makes his set of beliefs any less valuable or legitimate. It is a dangerous confusion for it can lead to premature involvement, by a major power, say, in the affairs of a smaller one, if the political beliefs of insurgents are regarded merely as an expression of some sort of personality deviation. I return to this important question in more detail in the concluding chapter.

When all is said and done, however, it may be argued that the psychological side of warfare, be it the study of insurgencies, computer links between soldiers' brains and their weapons, or the re-design of maps so that they are easier to be read in a hurry or at night, is at least somewhat less lethal than chemical, biological or nuclear war. And obviously true, as far as it goes.

Several studies have shown, however, that when one side in a dispute has available a *number* of sanctions of *varying* severity which can be brought to bear on the other side, then the more sanctions there are, the less steep is the gradient from the least severe to the most and it is *more* likely that the whole gamut of sanctions will be used. It is as if, having used one sanction, the others – still to come – appear less severe. This makes use of the next more likely – after which the remaining ones appear even more devalued. An interesting concept, only tested so far in the laboratory. But it is relevant here. In coming after the refinement of the terrible weapons devised by nuclear physicists, biologists or chemists, psychological warfare techniques may well have lessened the gradient leading to the use of these more destructive devices. In that sense, the activities of the military psychologists may be more lethal than we – or they – think.

DANGER, DECODING AND THE DARK: POSITIVE ASPECTS OF MILITARY PSYCHOLOGY

Let us not get too apocalyptic, however. Despite all the difficulties we have been discussing so far, it is also true that military psychology also provides some fascinating insights into many specific aspects of behaviour in war that are not especially controversial in the way that 'smell bombs', kamikaze pigeons or tape recordings of terrifying gods are.

For instance, which type of person is good at keeping secrets and can therefore be regarded as an acceptable security risk (44)? How do you train combat platoons in 'stealth behaviour' so that the men are able to spot and avoid bombs and boobytraps in the trail ahead (45)? What personalities should be selected for the most dangerous job of all – the bomb disposal officer (46)? And how do you train these men to recognize the odd screw or wire out of place which signals that a bomb may be boobytrapped (47)? Why are some people better at breaking codes than others and how can you spot this ability (48)? Why are some people particularly adept at judging imminent danger and what use can be made of them in a military context (49)? How does night affect the senses and the performance of various military operations (50)?

These questions appear to me more legitimate ones for the operational side of the military to ask of its scientific advisers. The answers provided by the scientists have not always been what the generals, the admirals or the air-commodores expected. They have not always been that useful. And they have not always been taken notice of, even when they have been useful.

But by and large these studies are defensive in character, aimed at conserving life rather than the reverse, at making conventional war safer rather than exploring new weapons concepts with unpredictable side-effects. I have thus sought a balance in this book. Believing that conflict is endemic in human affairs, it follows that the more defensive uses of military psychology have a crucial part to play in helping to ensure that the effects of conflict are no worse than they have to be. Therefore, the rest of the book, if not this introduction, devotes as much space to these more positive studies as it does to the more provocative, aggressive aspects. Which are offensive in all senses of that word.

PART ONE *Combat*

2 The Individual Soldier: Selection and Training

FIGHTER VERSUS NON-FIGHTER

Systematic psychological studies of soldiers in combat first began during the Second World War. In *The American Soldier,* Samuel Stouffer studied the attitudes of soldiers to many aspects of their military lives – to their units, their uniforms, to their weapons. He also studied racial differences and – briefly – surveyed the men's fears (1). In *Men Against Fire,* S. L. A. Marshall explored combat behaviour in the Second World War rather more thoroughly. For instance, he reported that, in general, only about 15 per cent of the men available in a company actually pulled the trigger against the enemy. Even in exceptional companies, he said, the number of men who fired only went as high as 25–30 per cent. Marshall did not impute cowardice to those who did not fire, since he found that for the most part they exposed themselves to just as much danger as those who did fire. This simply meant, he said, that there were psychological differences between fighters and non-fighters. For instance, neither the type nor the length of the action seemed to make much difference to the proportion of men firing. If a man was going to fight at all, concluded Marshall, he began firing early on in the fight and went on more or less continuously throughout. Furthermore, in repeated actions the same men did the firing. Only in exceptional cases did men who had previously been inactive later start fighting (2).[1]

Marshall's work proved controversial and was heatedly contended

1. British studies in the Second World War were not so much psychological as psychiatric. R. A. C. Ahrenfeldt's official history of this side of the war, for instance, concentrated on the rates of mental breakdown rather than the difference between an effective and ineffective soldier (3).

by other military historians. Nevertheless, his argument – or part of it – had made its mark for when the Korean War came along Marshall was called on again to repeat his study. This time he found that the proportion of non-firers was much smaller. In most actions, he said, approximately 50 per cent of the men fired and in some perimeter defence situations almost everyone pulled the trigger and aimed in the right direction (4). Even so, it was clear by now that there was much greater scope for improving the killing efficiency of army units than anyone had ever dreamed of. What was needed was a much better understanding of the psychological differences between 'fighters' and 'non-fighters', as the two types came to be called, so that changes either in selection or training might improve the ratio of fighters to non-fighters. Thus, since the Korean War there have been continuing attempts to define ever more deeply the personality of the good combat soldier.

A few questions had, in fact, been asked of Second World War veterans on their return from duty in the late forties. But the two most ambitious studies were carried out in Korea, one by the Personnel Research Branch (PRB) of the US Army's Adjutant General actually on the front line (5), and the other by the Human Resources Research Office (HumRRO, an 'independent', Army-funded organization) which carried out its project just behind the battle lines. The PRB research was hampered by front-line conditions and only fifty-seven men were interviewed. The HumRRO research has therefore been taken as the most reliable (6).

The HumRRO team interviewed men from three infantry divisions who had recently been involved in one or more of the following activities: assault, combat patrols, repelling an enemy attack, enduring artillery and mortar barrages. The research team wanted to distinguish good fighters from poor ones and this was based on both the number and nature of first-hand observations reported by other men. A man was selected for the fighter groups if: (a) two or more other men reported specific examples of good performance; or (b) if one other man reported a specific instance of good performance and if it was known that the individual in question had received or been recommended for a Bronze Star for valour or some higher decoration. A man was rated a non-fighter: (a) if two or more men reported specific instances of poor behaviour; (b) if the man himself admitted to inadequate performance; or (c) if one person only rated another poorly and if, in the opinion of the HumRRO interviewer, that man was an impartial and competent observer, and no one else could have witnessed the incident anyway.

Good combat performance included remaining calm under fire and performing specific actions of daring or bravery. The 'always cool' individual was not selected for testing unless specific actions were cited that could be seen to indicate that his performance was exceptional *and* appropriate to the situation. For example, the following incident, was rated as good fighting behaviour:

On the night of 6 July 1953, Able company of the 17th Infantry Regiment was attacked by the Chinese and some of its positions were overrun. A Private First Class from the company, accompanied by another man, went up to the trench toward the Chinese and set up a barbed wire block. The next morning, he and two other men knocked out three enemy-held bunkers with a 3·5 rocket launcher, which none of them had fired since basic training.

From the reports the experimenters were able to classify the good fighting behaviour as follows:

The good fighter exposes himself to enemy fire more than others in order to:

provide leadership, either as normal function or in lieu of the designated leader,
take aggressive action, exclusive of leadership role,
perform supporting tasks.

Under the same conditions of exposure to fire as others the good fighter:

leads men effectively, either as normal function or in lieu of designated leader,
takes aggressive action, exclusive of leadership role,
exhibits high degree of personal responsibility,
remains calm and cool.

The soldiers interviewed also described a variety of incidents to illustrate behaviour which they considered inappropriate for the combat soldier. These ranged from seeing or hearing imaginary things to 'bugging out' (actively withdrawing) under enemy attack. Here is an example of the kinds of actions which would qualify a man as a non-fighter:

Soldier B was said to be very nervous. When a flare was fired, he jumped and fired into the air. On a patrol he had the Browning Automatic rifle and the leader could hardly keep him in position. He was jumpy and fired at imaginary objects.

On the basis of episodes like these the following classification of poor combat performance was derived:

The non-fighter is the man who, under the same conditions of exposure to fire as others in his unit:

actively withdraws, usually under fire,
withdraws psychologically (freezes, stays in bunker, obeys orders only at gun-point),
malingers,
defensively over-reacts – sees and hears things,
becomes hysterically incapacitated – trembles, cries, etc.

In all no less than eighty-six psychological tests were given to 310 soldiers, surely making them the most tested individuals in history. The results were surprisingly consistent.

Fighters had a significantly higher military rank than non-fighters; they had also been in the army and in the combat zone longer. Regular soldiers made better fighters than draftees (see Table 1).

Table 1 Percentage of volunteers and conscripts among fighters and non-fighters

Component of soldiers	*Fighters*	*Non-fighters*
Regular army	30	23
Draftees	69	75
Other	1	2

Race was found to be important in several ways. First, it looked very much as though Negroes were much worse fighters than whites (see Table 2).

Table 2 Percentage of fighters and non-fighters by racial background

	White	*Negro*	*Other*	*Total*
Fighters	67	21	59	55
Non-fighters	33	79	41	45

This finding could be the result of bad methodology. No note was taken of the race of people making the evaluations so it proved impossible to say how valid their judgements were; they could have been expressing racial prejudice in the close quarters of combat. Nevertheless, the *impression* created by the research team was that Negroes made poorer fighters than whites in Korea.

Secondly, for whites the amount of education someone had received was found to be related to his effectiveness as a fighter – the more education, the better fighter someone made. But for Negroes this relationship did not hold up. There was also evidence that Negroes from the Deep South were less likely to be rated as fighters than Negroes from the North. It may be, concluded the HumRRO investigators, that

> the Southern Negro's early training in the 'virtues' that make existence easier for this minority group, such as suppressing overtly aggressive behavior and avoiding open conflict with members of the majority, may cause later combat training to be less effective.

(Remember this was written in 1953.)

On intelligence an intriguing picture emerged. The mean IQ of fighters worked out as 91; for non-fighters it was 78. But at the same time the mean IQ of the whole group of combat soldiers was 85, that is, 15 points below the average. In other words, the less intelligent men in the army are those who are sent to fight and among them the more intelligent make better fighters – though they are still well below average.

The personality differences which distinguish good fighters from poor ones also showed a remarkable consistency. Five main factors were isolated. The most important was leadership, which consisted, according to the personality scales used, of poise, spontaneity, extraversion, freedom from anxiety and independence. Next in importance was a 'masculinity' factor – outdoor adventurousness is probably a better way of describing this. Third, came intelligence, already discussed. The fourth factor was, interestingly enough, a sense of humour. The good fighter preferred sharp wit and biting sarcasm rather than the flatter types of humour such as simple joke-telling. The fifth factor isolated by the tests was 'emotional stability': the good fighter was more stable, less anxious, less prone to depressions than the poor fighter.

Other systematic differences were found between fighters and non-fighters. For example, they appeared to have quite different patterns in their early lives. More of the fighters' families had owned their own business; more often the non-fighters had to leave school early to help support their families. More people were financially dependent on the non-fighter while he was in the service. Fighters were more likely to describe their home life as harmonious than non-fighters. In contrast, among 40 per cent of the families of non-fighters the father had actually died before the soldier was eighteen, twice as many as among successful fighters. The HumRRO investigators add:

The absence of the father may possibly have meant the lack of a male figure or model with whom the subject could identify and thus establish a masculine role. It might be recalled here that some of the largest differences in the tests differentiating fighters and non-fighters were scales measuring masculinity, in which the non-fighters tended to score in the less masculine direction.

Build was important. Fighters were less ectomorphic and more mesomorphic than the non-fighters and they were, on average, an inch taller and eight pounds heavier.[2] In the HumRRO study, fighters took part more often in sports at school than non-fighters and were more likely to prefer body-contact games – football, hockey or boxing rather than tennis, say, or athletics. Non-fighters tended to date more during their early and middle teens. The first sexual experience with a female occurred, on average, a year earlier for the non-fighters (at 14 years 3 months compared to 15 years 8 months for the fighters). But the non-fighters got married a year later than the fighters, perhaps because they could not afford it any earlier. Finally, in politics the non-fighters were likely to regard themselves as more liberal than the fighters.

Some of these findings are more surprising than others. That they should be so consistent is, however, an important point; and the fact that they could be measured *before* the soldiers underwent the experience of combat was regarded as a significant advance, even if at that stage no verification in real combat was possible (we shall see in a moment how simulated combat was used for verification purposes years later).

Other studies have shown that the fighter is much faster at psychomotor activities than the non-fighter. He does mirror-drawing tests and pencil mazes quicker and he also has faster reaction times. On sociometric questions fighters were preferred more than non-fighters though the differences were not as great as might be expected. Sociometric tests, used widely by many military academies, were shown by these

2. Build also matters for seamen (7). Four hundred men on an aircraft carrier were questioned by Eric Gunderson, particularly about which of their personal characteristics they were dissatisfied with. There were three main areas of dissatisfaction: size, weight and intelligence. The more they were dissatisfied, the worse the men were in combat.

studies to be more useful for screening out non-fighters than for, say, deciding which of the fighters would make good leaders (8).

Birth order also makes a difference. The best account of this is given by Yehuda Amir, an Israeli psychologist at Bar-Ilan University, and her colleague, Shlomo Sharon, from Tel-Aviv (9). In 1968 they published results of a survey of nearly 5000 candidates for officer status in the Israeli Defence Forces between 1961 and 1966. On the basis of much research (by others) into the effects of birth position on personality, they predicted that first borns would be more anxious than later borns – mainly because of their lonelier life at home – and that this would lead them to try to enter positions in the army that exposed them less to contact with the enemy. Though this theory seems a bit thin on the face of it, Amir and Sharon did indeed find that first borns in the Israeli Defence Forces were less likely to volunteer for combat and that even when they were allocated to combat duties, patrols with first borns in them were 'somehow' less likely to encounter terrorists. More than a decade earlier Paul Torrance, in a study of jet aces in the Korean War, had found much the same – that first-born aces engaged the enemy less often than later borns and reported more anxiety over flying (10). The general picture is also supported by naval studies (11). Robert Helmreich, one of the psychologists who spent some years studying the behaviour of divers in Sealab, a naval underwater psychological laboratory, found that first-born divers were more frightened of being underwater than later borns and also that they did not carry out their military tasks – laying mines, fixing nets or listening devices – either as quickly or as reliably as later borns. (Danger does seem to be a factor: studies of first borns as children show that they are less likely to take part in dangerous sports.)[3]

Later studies, including some in simulated combat in the Arctic and on manoeuvres in Germany, looked at Marines and signal corps, as well as the groups so far mentioned. Much the same results were obtained (the racial differences continued – and continued to be top secret) though there were refinements: 'sense of duty' and 'stamina' were added as qualities which identified a fighter (12). By the mid- to late-fifties, however, the research was paying off in terms of the way men

3. Incidentally, the US Army, in its eagerness not to let any good fighters slip through the selection net, even looked at how well military criminals compared with men with clean records. Some 247 men with dishonourable discharges were studied but, though possibly aggressive in the barracks, this group of men performed rather worse than the rest in combat. Armies lose little by keeping their criminals in the stockade (13).

were allocated to combat status. For example, on the basis of the data collected by HumRRO on the front-line in Korea, any group of 1000 soldiers contained roughly 184 outstanding fighters, 753 adequate fighters and 81 non-fighters. No one at that stage knew how many outstanding or adequate fighters were being kept in reserve, support or supply positions. By 1958 (had there been a war on), using the selection techniques based on the findings above, any 1000 soldiers would then have consisted of 250 top fighters, 700 adequate fighters and 50 non-fighters – to the army a significant improvement (14). And, in fact, all the more marked when considered in terms of a US Army of one million enlisted men. The US Army's Personnel Research Branch calculated that the following changes had been brought about by the psychological research: in an army of one million, if 32 per cent were assigned to combat specialities, the gain from the introduction of psychological aptitude tests was as follows:

	Top fighters	*Adequate fighters*	*Non-fighters*
Using basic information only (physical and mental profile; just IQ)	58 880	235 200	25 920
Using the PRB's and HumRRO's methods	80 000	224 000	16 000
Difference	+21 120	−11 200	−9920

Just as important the extra fighters would have gone where they were clearly needed. According to the PRB about half of the increased top fighters would have gone to the infantry, the rest being divided among artillery, armour and engineers.

GAMES WITH GUNS: TRAINING TECHNIQUES

Better selection of good fighters is not, however, the only way to improve the efficiency of an army or navy. Another way is through training – providing a young soldier with the opportunity and the incentive to learn to fight better. Military psychologists have identified two crucial aspects of combat training which need to be taken account of if the men are to benefit properly from what they are taught about battle.

Motivation

In a system like the army, particularly the career army, little thought is given usually to rewards, to the positive side of discipline as it were, except for awards for sheer bravery. Probably, few opportunities are thought to exist. Yet psychological learning principles show quite clearly that positive reward is far more effective in producing desirable behaviour than punishment is in eliminating undesirable behaviour. The US Army Training Center, Human Research Unit, at Fort Ord in California, therefore decided in the mid-1960s to see if it could improve combat performance by experimenting with the positive rewards open to the army.

The initial approach was to give certain incentives during basic combat training and see what effects this had. No less than forty-three incentives were open to the commanders, some of which will only sound like incentives to men who have been in the army. These ranged from being exempted guard-duty, being allowed to sleep late once every week, being first in the chowline for a week, being given a month's supply of Brasso, to a letter of commendation being sent to a soldiers' parents or being allowed to take meals at the NCO club for one day. The results showed plainly that some types of reward work and some do not. The rewards were divided into three types: recognition (such as receiving a letter of appreciation from the CO), material rewards (like a cash benefit, or some other object), or autonomy or freedom (extra leave or choice of next assignment). For some men recognition was the most effective reward, whilst for others autonomy was. No one wanted material reward. Judicious use of the minimal sanctions available, therefore, can make the men learn their basic combat skills better. This was not a striking experiment, nor were the results particularly newsworthy. But it could well be the most important experiment in this book (and probably the cheapest too) (15).

Simulation

In combat, the noise, the danger, and the general hazards are unfamiliar to most and a unique experience, one for which preparation can probably never be complete. By definition the first engagement can never be pre-empted. But simulation of battle is perhaps the most hopeful way of equipping a soldier to cope with at least some of the problems he will eventually face. Psychologically, there are three important aspects to it.

Some army schools are now introducing more realistic elements into basic training. One example is the use of the 'quick-kill' method of firing a combat rifle. Normally, men learn first to fire their guns at stationary targets, sometimes at bull's-eyes, sometimes at the silhouette of a soldier. Even in this latter case, however, the situation rarely resembles that which is encountered in battle. Rarely is there time to find a tree on which to rest your arm, or even time to take proper aim. Hardly ever will the target be stationary. The quick-kill method requires the soldier, in training as in battle, to react quickly – so quickly that he will often have to fire his gun from the hip, without taking proper aim. The targets, silhouettes of part of a person, pop up unexpectedly and the soldier has to react adequately in the small time available (16). Tracers also help in providing the trainee with feedback on how well he is doing. (Incidentally, in the US Army – and it may be true for other armies though probably not for the British – another advantage of quick-kill training is the incentive it provides for men already familiar with guns when they join up. In the US Army, according to one survey, no less than 62 per cent of the recruits had experience with guns before they were recruited.)

The next area of simulation is probably the most familiar to many of us: the field exercise, in which a mock battle is fought. Here the sounds and inconvenience of battle can be reproduced fairly faithfully even though, save for one or two celebrated cases, blank ammunition is used. Mock battles, therefore, require the use of umpires to rule on who has been killed or wounded. Basically, this method has always been felt to be a better training for the commander, because general tractics can be more faithfully reproduced, than for the enlisted man, because the individual combat situations cannot be so well duplicated. However, there are two recent advances which make mock battles that much more real. One is the product of psychologists; the other of physicists – though, since its effects are primarily psychological, it will be reported here.

Using research on pop-up targets and some details available from various wars, psychologists have been able to compile pretty good data about the range at which the average soldier can hit various targets. They have now devised targets which are of such a size, colour and shape that, if they can be seen, they can be presumed to have been hit. For example, an infantryman on manoeuvres may now wear a number on his helmet; if it can be read by another infantryman (the 'enemy'), the first man is judged to have been killed. Similarly, a tank will have a slightly larger number which can be read by the ordinary infantryman at much greater distances but not put out of action because the infantryman has not the means to do it. On the other hand, if the anti-tank battery can read it, then the tank *is* out. The battery tells the number to the umpires and, if the number is correct, the tank is instructed off the mock battlefield (17).

This is a fairly cheap method and is widely used nowadays. Eventually, however, it may be replaced by a more sophisticated method that is rather more realistic albeit considerably more expensive. It is currently only in use at Fort Ord and has been called 'the war game to end all war games'. Basically, it is a mock battle in which all the weapons are fitted out to fire weak lasers. The weapons are in radio contact with a central computer, as is a small receiver in the helmet of every soldier taking part. Each weapon's laser beam is coded according to the fire power it would normally have (so a rifle cannot knock out a tank). If a man fires his weapon at the enemy and actually hits a special receptor sewn into the 'enemy's' uniform this is registered in the computer which, in turn, radios a message to the victim telling him he has been 'killed' and must leave the battlefield. Just to make sure, the computer also cuts off his ammunition – his laser beam. Though this is all pure physics so far, the technique does mean that simulation of battle conditions is now much better for the infantryman, and not just for the tactician, and so future soldiers should be much better prepared than in the past at knowing what camouflaging they can risk, for example, and what they can not. This judgement is, of course, an important psychological ingredient in war, and the laser system seems likely to teach this better than other types of simulation (18).

Apart from the use of simulation techniques in basic training, there is the most important use of simulation in studies of stress. This can take two forms. It can be used as a screening technique to assess who is likely to break down in battle. Some of the work in this area is extremely controversial in that it exposes soldiers to unnecessary and even harmful stress situations. Such work is dealt with in detail in chapter 9. The other use of stress simulation is more specific – that is, in training men to cope practically with emergency situations. Such an example is the Nimrod flight simulator used at the RAF marine reconnaissance training base at Kinross in Scotland. The simulator is used to train pilots to convert to Nimrods and to teach the men emergency procedures.

Faulty dial readings and flight characteristics can be deliberately introduced into the simulator, in complete safety, and the performance of the crew monitored. Actual battle conditions are not reproduced but emergencies stemming from, say, being hit on the wing so that an engine is set on fire *can* be simulated. Furthermore, emergency can be piled on emergency so that the situation, in terms of having to behave quickly and accurately, does become stressful in a *real* sense if not in a battle sense. The RAF, at least in this context, takes the view that its prime responsibility is to train men to *deal* with stress rather than to screen out those who cannot take it. They therefore present their pilots with all sorts of contingencies so often that their responses can be carried out almost without thinking. Crews often report, after they have met a real-life emergency, that they 'just carried out the correct drill' and had completed it before they realized they were not in the simulator (19).

Armies, we can now see, have acquired the wherewithal to select the best fighters, to motivate these men so that they learn their basic combat training with the minimum amount of fuss and time and, in certain circumstances, to prepare them for the terrible stresses battle is bound to entail. There are also specific things the individual soldier has to be able to do if he is to survive at the front. At its most basic he has to be able to aim his gun properly, to navigate his way across unfamiliar territory, to go on patrols using what camouflage the surrounding countryside offers, to distinguish friendly tanks and aircraft from enemy ones and so forth. The US Army identified in one study, for example, forty-one essential skills for the infantryman. These include such things as demolitions, human maintenance, hand-to-hand combat and use of indirect supporting fire. But only recently have they become legitimate fields of interest for the military psychologist – since in fact thirteen of the skills were identified as most in need of improvement. These included: firing from a fox-hole at a moving target; distribution of fire at suspected enemy positions; knowing when to throw a grenade; and the use of cover. Already, the psychologists have produced some intriguing developments and it is to these that we now turn.

MARKSMANSHIP

From time to time, armies get worried that rifle marksmanship is not what it could – or should – be. In 1968, for example, during the

Vietnam War, the US Army ordered a series of no less than twenty-one experiments to explore the 'significant training problem in the area of marksmanship' (20). These provide a useful starting point for the discussion of aiming because the experiments were done under both day- and night-time conditions. They thus serve as a good introduction to the way gun shooting is researched.[4]

During the daytime, and against single targets, it was found that when the target was more than 50 metres away the semi-automatic mode of fire produced more hits than the automatic; at about 50 metres there was little difference between the two, but inside 25 metres automatic firing was better. Time to the first hit is regarded as the best measure of aiming ability (see Figure 1).

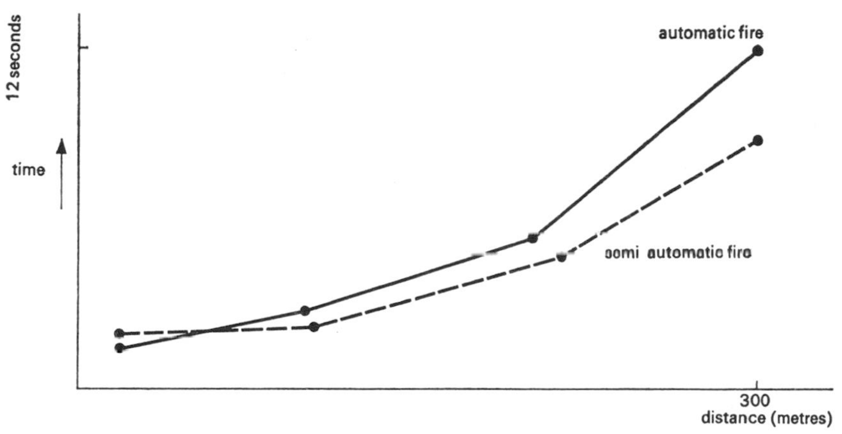

Figure 1 Mode of fire as a function of time to first hit.

However, in real war, so as to give the enemy as little information as possible about your own position, the number of trigger pulls to first hit are very important. Here the effect of mode of firing is even more important (see Figure 2).

In other words, automatic fire may kill the enemy quicker, but only at greater distances (200 metres plus) is this not offset by the risk of giving away your own position.

4. There are those – psychologists among them – who claim that an army of marksmen is not what is needed. Instead, they say, armies – especially in war-time – should mass produce men who can use a weapon *fairly* well. They have been responsible for the training programmes which use pop-up silhouette targets, and concentrate heavily on showing soldiers what targets actually look like in a war. While pop-up targets are widely used now, most armies still seem bent on producing super marksmen.

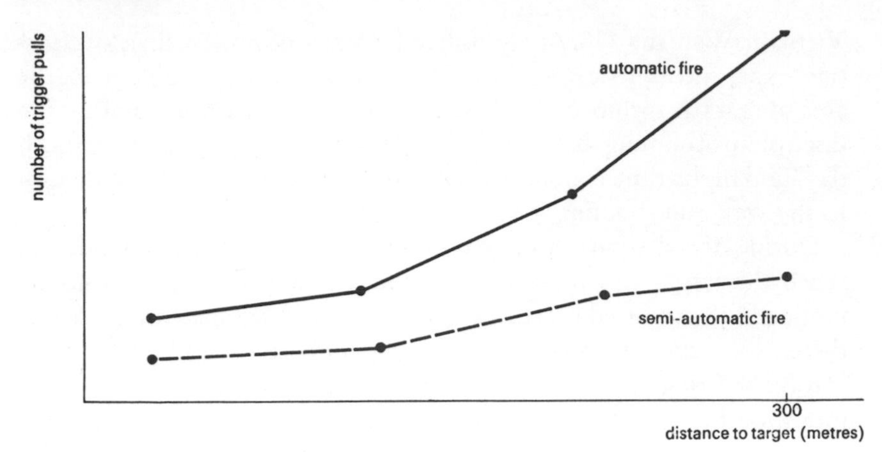

Figure 2 Mode of fire as a function of trigger pulls to first hit.

Quick-kill shooting has caused some controversy in military circles, as many people have been worried about the accuracy of fire with this method.[5] Figure 3 shows that, as the distance to the target increases, quick-kill gets relatively worse.

Kneeling positions, generally, give the most rapid time to first hit.

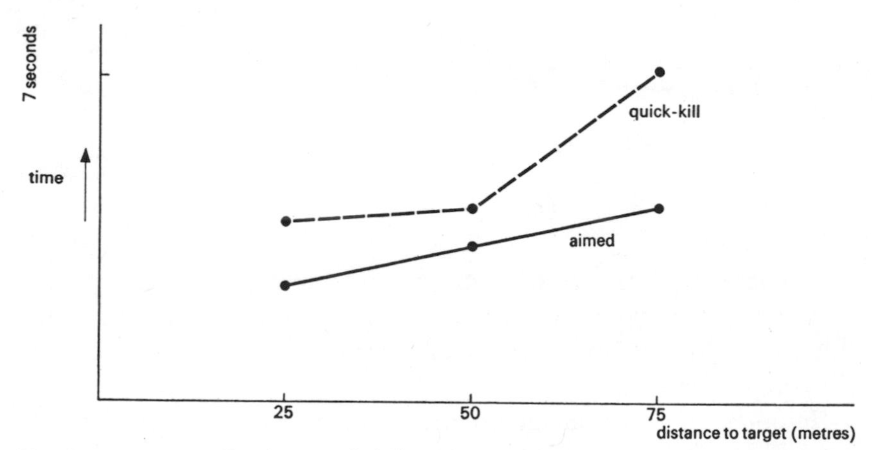

Figure 3 Time to first hit: quick-kill versus aimed fire.

5. The British Army has developed a different method for dealing with quick-kill situations. Its infantry use a night-sight known as the Tri-Lux, with a larger aperture which may be used for rapid aimed fire at close targets in the daytime. Major D. Stopford, in 1969, concluded that the British system was better than the US system, so a HumRRO research team set out to verify this. They found that under 25 metres aimed fire *was* superior to quick-kill but that this was true of aimed fire in general and not just the British Tri-Lux sight.

At the more distant targets, the prone position is the next best, followed by standing. These results also apply if the number of trigger pulls to first hit is the criterion. If the soldier lies down it is as well to remember that he is a better shot if he aligns his body *with* the rifle *not* at an angle to it. All of these positions, however, are normal only during offensives. During defensive actions the soldier will usually fire from a fox-hole or a bunker. The main thing to remember here is that men in these defensive positions are faster and more accurate marksmen.

The two criteria for determining the best carry position for the rifle are comfort and speed onto the target. When there is no threat then clearly comfort should come first. It is when there *is* immediate threat that maximum readiness counts. In these conditions US and British practices differed, so HumRRO compared them. In the British ready position, the butt of the weapon is placed high on the shoulder pocket so that when the gun is raised, a minimum head movement is required of the marksman. For a right-handed individual, the right hand is on the pistol grip, the left hand is on the stock beyond the carrying handle, and the weapon is slanted down and to the left across the body. This position was less comfortable than the US position – the gun held the same way but level across the body, at roughly waist height – but it was shown to be faster to first hit. The two positions were equal in the number of trigger pulls to the first hit (21).

At night things are rather different for the rifleman. Then, automatic fire, using the three-round burst, is superior to semi-automatic fire, both in the total hits achieved and in hits per trigger pull. And incidentally, it takes no longer to fire a three-round burst at night than to fire a single round in the semi-automatic mode. Nevertheless, automatic fire will always use up far more ammunition, and this is a factor which often operates against its use.

Night sights have improved marksmanship significantly in the dark (see Figure 4).

At night the general relationship between kneeling and the prone position is maintained, provided that some sort of night sight is used. This need only be a strip of white tape along the barrel to help in aiming. But without night sights the prone position is best in the dark. The use of tracers for night firing produces a considerable improvement in both speed and accuracy but it has unfortunate side-effects. If someone trained in the use of tracers has to shoot at night *without* them, then he is nowhere near as good as men trained without tracers.

Finally, one of the applications of learning theory to the psychology

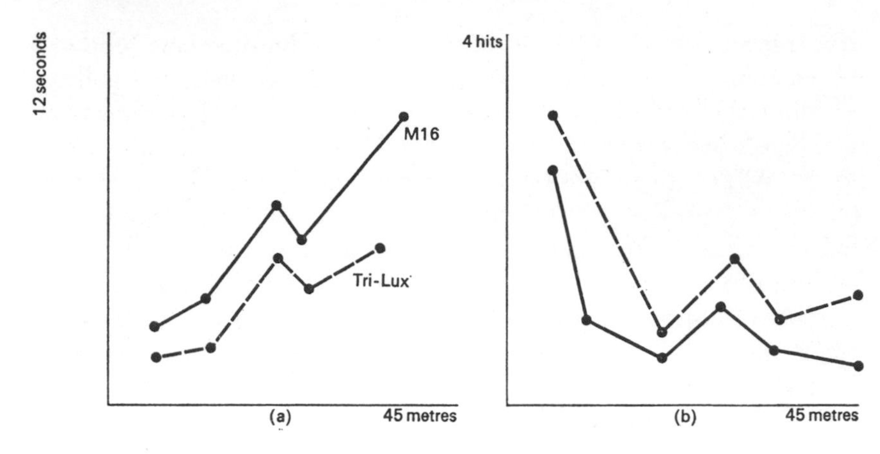

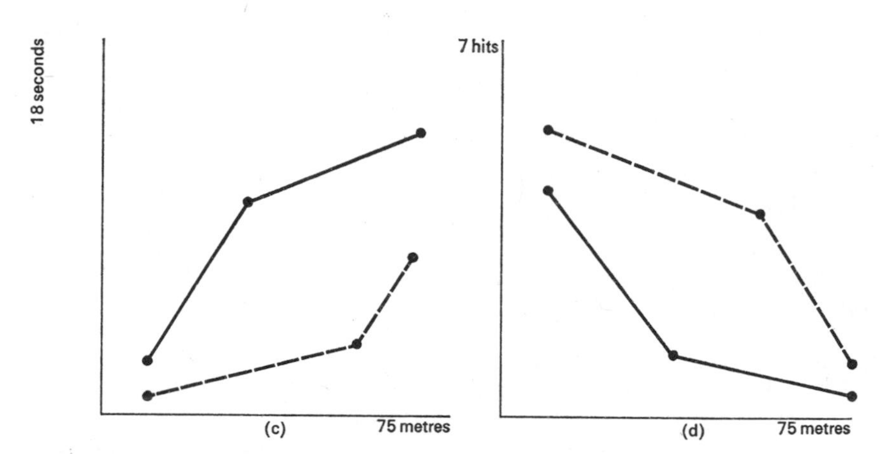

Figure 4 Effect of night sights: (a) time to first hit at night (starlight); (b) number of hits under half-moon illumination; (c) time to first hit (full moon); (d) total hits (full moon).

of aiming is that there must be knowledge of results for learning to occur and errors to be corrected. Soldiers should be told more than they had hit or missed the target. Psychologists at the weapons department of the US Army Infantry School at Fort Benning, Georgia, adapted basic combat rifles to fire tracers and even, on occasions, white tape. This proved particularly effective in teaching men the subtle 'feel' of a rifle – say, on the hip. They became proficient much quicker than with the old 'hit or miss' methods (22). With night firing, of course, lack of feedback is a particular problem. The HumRRO psychologists introduced daytime practice of night-firing techniques so that the soldiers could see their errors. Sure enough, they found that men trained in night-firing techniques – three-round burst, with night sights – during the day performed considerably better in the dark than those trained on the same techniques, for the same length of time, at night (23).

A new course, constructed on the basis of some of the results reported here (for example, daytime training for night firing, changes in firing position, change from quick-kill to aimed firing), was compared in its efficiency to the old-established course. Figure 5 shows how effective the new programme was.[6]

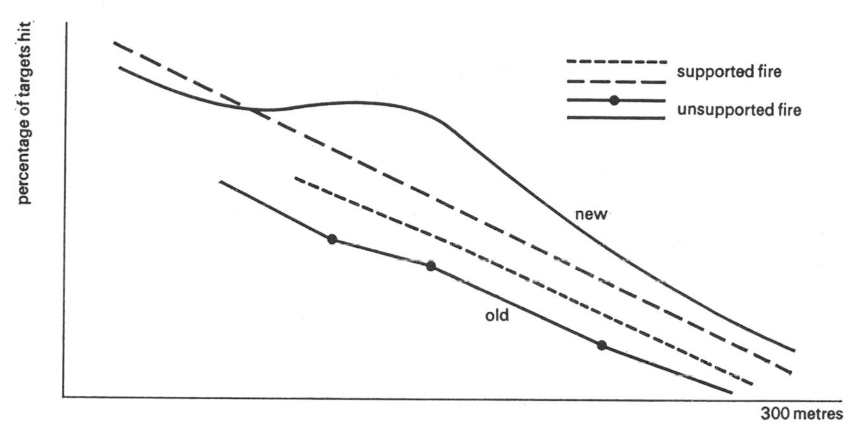

Figure 5 Comparison of hit probabilities: old and new programmes.

Of course, there are major differences between firing a rifle on an ordinary training range and firing the same gun in combat. Albert Prince, at HumRRO, studied the effects of danger on rifle firing by having marksmen shoot on a range at a time when a series of dynamite charges were exploding increasingly closer to him. Three successive bull's-eyes would stop the explosions. Prior to the experiment all the marksmen were given a psychological test to measure their general level of anxiety. Prince had the idea that people who are normally highly anxious might be less affected by stress than those less so; hence, they might be exceptionally suitable as marksmen in a war. In fact, it

6. Other improvements introduced by psychologists include personalized stocks – for greater comfort – and various devices to stop riflemen from flinching. It was found that 38 per cent of shooting errors stem from flinching.

did not work out like that. Everyone was affected to more or less the same extent by stress (the dynamite). In all cases accuracy of fire dropped off significantly as the explosions got closer. The chronically anxious were in any case much worse shots than the less anxious. This latter group, therefore, might be expected to perform best under fire in real war. Not the answers Prince expected, but no less useful for all that (24).

A classified study, carried out at the Human Engineering Laboratories of the US Army at the Aberdeen Proving Ground in Maryland, developed a different method to investigate the effects of stress on rifle firing. The experiment, under the direction of R. R. Kramer, found that stress – in this case simulated artillery fire – had 'a considerable effect on firing – more on accuracy than on rate of fire – but that some styles of shooting were more resilient than others. Firing from the shoulder was affected more than others in terms of rate but not in terms of accuracy . . . therefore, particularly at night, when too much extraneous fire could give away your position, firing from the shoulder should be encouraged' (25).[7]

There are other, less-obvious types of battle stress: for example, the use of night-sights may impair dark adaptation. And night-sights can, in some cases, produce chronic headaches in troops so that the allocation of who does what on night ops needs to be carefully watched by the commanding officer.

Altitude, also, exerts an influence on marksmanship. W. O. Evans, from the US Army Medical Research and Nutrition Laboratory, reported an experiment in which troops had been rapidly taken from sea level to Pikes Peak, in the US Mid-West, which is 14 110 feet above sea level, to see how well the troops performed on certain military tasks. One of these was the ability to fire the handgun (26). As Figure 6 shows, there was a 35 per cent decrement in both the accuracy and the speed of firing during the first four days. And it took two weeks before the soldiers regained their full sea-level performance.

In the same field, a still more intriguing experiment was carried out by R. G. Smith, at Edgewood Arsenal, in Maryland, on the effects of drugs on aiming (27). Many 'chemical agents' (they are nowhere named in the report) affect visual acuity and can have an affect on military skills. The men in the experiment had to use their rifles against

7. Missile aiming is particularly susceptible to 'combat degradation'. Various attempts have been made to train missile operators to be resistant to stress, or to design controls that are themselves less susceptible to the conditions of war (28).

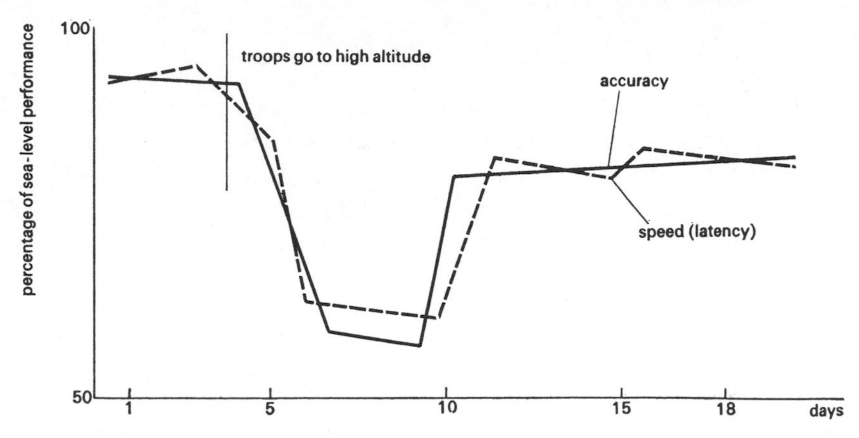

Figure 6 Effect of altitude change on marksmanship.

eight fixed-silhouette targets at 200 metres. They then had to repeat this five hours after being given a chemical agent and again after twenty-four hours (see Figure 7). The experiment was repeated for rifle loading, grenade throwing and rate of firing – all with much the same results.[8]

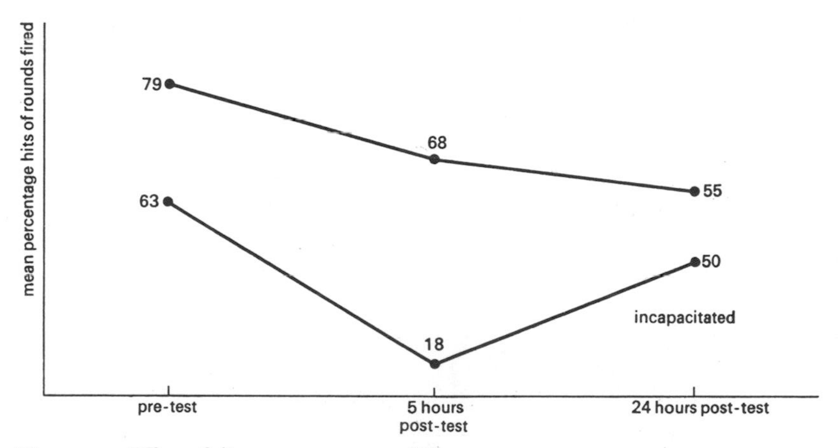

Figure 7 Effect of drugs on accuracy of fire.

8. It is not clear from the account I have seen whether the 'aim' of the experiment was to see how soon after an enemy drug attack the infantry would be ready to get going again or how long enemy troops would remain incapacitated after such an attack by US forces. It is possible, however, that as early as 1966 the army had LSD problems and was hard at it, seeing the effect of LSD on the military capabilities of its men. The revelations, in 1975, that the American security services had used LSD experimentally in the 1950s and 1960s, adds to the suspicion that the chemical agent in these studies was indeed LSD (29).

THE SOLDIER'S ATTITUDE TO HIS WEAPONS

S. L. A. Marshall's work on weapons usage in Korea has shown that the M1 was a popular weapon with the soldiers, that the BAR was the mainstay, but the carbine was seen as worthless. Much controversy in the field of personal weapons has centred on the value, or otherwise, of the bayonet. A major study of this was carried out in the US Army in 1969 when it was found that officers were more likely to carry a bayonet than their men and that in Vietnam carrying a bayonet was rare. Most people thought its use was mainly in special operations (raids, ambushes, etc.) rather than offence or prepared defence where, after all, automatic weapons preclude close combat. Most said they thought they could handle it as well as needed but less than 50 per cent said that they thought the bayonet added to the aggressiveness of their units (30).

However, when the men did have to use their knife in hand-to-hand combat, twice as many men preferred a sheath knife to the bayonet. On the other hand, more than 80 per cent reported that the bayonet was indispensable for riot control and disaster relief. The main reason was its psychological effect, though 13 per cent thought that it enabled control without gun-fire (31).

The survey showed that the men in fact would have much preferred a modified bayonet that could double as a survival knife. The shape preferred and the existing bayonet is shown in Figure 8.

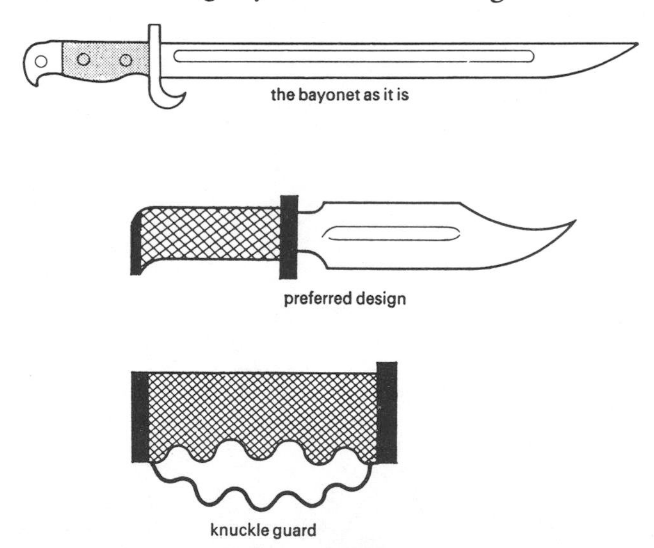

Figure 8 Existing bayonet and new design based on soldiers' preferences.

When asked whether they felt confident of their ability to face an enemy soldier in bayonet combat, 70 per cent said they were. However, when the men were asked to rank the combat skills in the order of most training needed, bayonet training came out next to top (see Table 3).

Table 3 Rankings of the need for extra training in combat skills

1	Marksmanship
2	Bayonet training
3	Knife fighting
4	Judo
5	Karate

3 Artillery

TOO LITTLE, TOO LATE: GROWTH OF RESEARCH PROBLEMS

For the armed services, 'the control of gunfire' with large weapons – anti-aircraft batteries, aircraft guns, and now, of course, missiles – is a far greater problem than rifle marksmanship. This is particularly so because the weapons themselves are complex and frequently need more than one person to operate them (or they only operate effectively in sequence with other guns). And partly it is because the consequences of a miss with these weapons is both more expensive and more dangerous. It should come as no surprise, therefore, to find that the study of marksmanship with these sophisticated pieces of equipment has a much longer history than the study of rifle marksmanship. (And even with the advent of heat-seeking SAM missiles, it is likely to remain one of the most crucial military skills for some time to come.)

The problems of aiming complex equipment first became apparent in the Second World War. In 1942 the US Navy gave priority to training anti-aircraft gunners, and the techniques developed proved so successful that they were incorporated into other training programmes – for example, engineering crews.

For aircraft gunners, in the early stages of the war when machine guns were still in use, aiming was roughly comparable to firing a pistol from one moving car at another. However, the development of computing machine-gun sights eliminated the guess-work, but complicated the gunners' task in other ways. For example, in some designs the gunners' right and left hands had to perform different operations at different speeds (1).

In 1942 and 1943, with aircraft dominating the scene, requests were made for research on naval gun crews, height-finder operators, range finders, radar operators and anti-aircraft batteries in general. Field artillery made frequent and serious errors. With a target 5000 yards away range estimation could be out by as much as 500 yards, sometimes with the result that artillery shelled its own troops. In May 1944, a study on the sources of error in field fire was begun (2). The B-29 plane was also introduced in combat in 1944. Dissatisfaction of the gunners with their combat equipment and the rising threat of Japanese suicide attacks led to a request for study of the B-29 gunsight in June of that year. Faced with these problems and repeated requests for, research, the military psychologists did eventually get started. Although many of the projects did produce practical results, they were often too late. Because the requests were unusual, they took time to process and the war was over. Too many times the psychological aspect of a problem was not immediately recognized, and no doubt many people paid the price of this neglect. (Part-way through the war a special committee was set up to look through the blue-prints for all new equipment and to iron out any psychological problems which could be foreseen.) It serves as a useful reminder that psychology has always been more important in military affairs than many people have been prepared to give it credit for.

Eventually, three important psychological problems linked to gunnery were isolated:

(a) The relationship of a gunner to the target and to his gun changed in the mechanized system – it became indirect. The representation of a target by a dial pointer or an oscilloscope blip did not necessarily look like the target. In such presentations the actual movements of the target might be changed in apparent magnitude, direction and rate.

(b) The number of unit tasks was increased. Frequently several gunners had to cooperate. Successful performance required that all units be carried out adequately at the same time. Failure resulted from deficiencies in any single unit.

(c) Mechanization required that delicate complex equipment be maintained and calibrated.

These three factors combined to prevent the individual gunner from knowing whether his duties were well or poorly done. Most important of all, knowledge of results was distorted, delayed or even prevented altogether. So the problems were identified; only towards the end of the war and after did solutions begin to show themselves.

IMPROVED AIMING

The major problem during the war, before the advent of radar, was to know how accurate gunners were and where and why they made the errors they did. To begin with a 'checksight' was introduced. This was a second sight on the gun aligned with the gunner's own, through which his instructor, an experienced gunner, could look and assess the trainee's errors. This was fairly reliable but had the drawback that a trainee could not be better than his instructor. So the trainee's gunsight was then fitted with a photosensitive cell pointing in the same direction as the sight. The target planes in training were fitted with a bright light so that when the sight was correctly aimed at the light, the light activated a buzzer through the photosensitive tube. The success of this method led directly to other means of error-recording for other gun systems – for example, in aircraft (3).

Such basic advances in the simple measurement of performance also enabled psychologists to study other – less obvious – bad habits which affect aiming. One problem in combat is that all gunners tend to open fire too soon – before the enemy is in range. Research by D. D. Wickens showed that this was also true of untrained gunners not in combat (4). The tendency, therefore, seemed to have nothing to do with fear or danger, or wanting to get at the enemy before he could hit back, but was a perceptual problem. This being the case, merely giving gunners true knowledge of their performance decreased their errors considerably.

Triggering patterns

Other studies, with B-29 gunners particularly, were made into triggering behaviour. It was found, for instance, that the gunners were just as likely to fire when off target as when on, clearly a waste of ammunition (5). So unexpected was this result that further studies were carried out. Sample triggering patterns were obtained from various gunners and found to vary markedly (see Figure 9). Some gunners triggered frequently, some infrequently: but what did emerge from the data, as is apparent from Figure 9, is that the ratio of time spent firing to time spent not firing was relatively constant for any one gunner but differed from man to man. In other words, each gunner fires in his own, mainly unconscious, rhythm. It seemed to the investigators that the attention of the gunner was so occupied by the task of

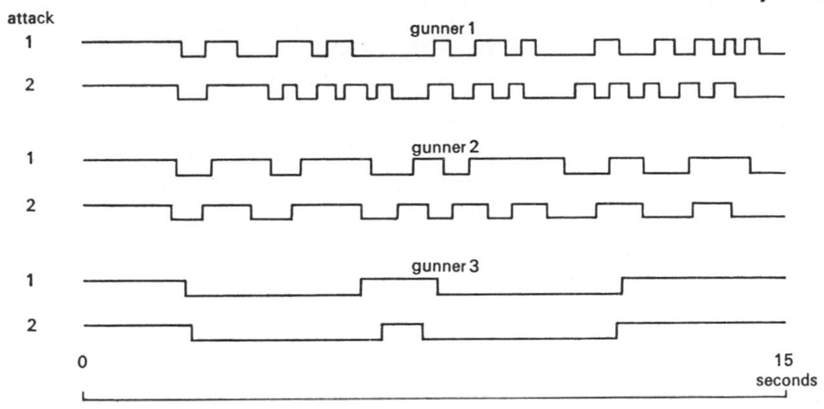

Figure 9 Triggering patterns for three gunners.

tracking and ranging onto the target that he could not make a clear and discriminating choice of the moment at which to fire. The whole procedure was so complex that it taxed the limits of human ability.

The operational doctrine was therefore changed. The gunner was henceforth trained to press the trigger *continuously* once firing began. If, for reasons of ammunition shortage, it became necessary to fire in bursts, then an interval time-device was installed to permit intermittent fire *despite* continuous triggering. In a twenty-three-day study up to 22 per cent improvement in the number of hits was recorded.

Use of tracers

Since the Second World War new techniques and new equipment have required the practices developed by psychologists to be up-dated. One such example is the increase in recent years of low-altitude penetration and attack by high-speed aircraft in which the plane's angular velocity is exceptionally high, making it a difficult target. As a result, all US direct-fire, forward-area weapons (light anti-aircraft guns and infantry-type weapons) now use tracer ammunition as a primary or auxiliary technique for adjusting fire. Tracers are no longer new – but there is still much disagreement about their value, even among military commanders. HumRRO was therefore called in to try to settle the disagreements.

The office reviewed the use of tracer ammunition since the Second World War for the Joint Chiefs of Staff. The survey was implicitly sceptical about the value of tracers, making the point that their principal use in air-defence seems to be limited to establishing the initial aim point of the weapon – and several psychological factors that adversely affect human use of tracers were outlined.

First, humans have a natural limit to their depth perception, stemming from the fact that their eyes are on average only 65 mm apart, and this can lead to inaccurate judgements about the tracer's location in space.

Second, there are certain visual illusions in tracer streams which gunners have great difficulty in overcoming. This is mainly because the line of fire of tracers is different to that of real bullets. As the incendiary powder burns away so the projectile loses mass, and this changes its trajectory. The tracer is thus affected by gravity differently to the normal bullet (see Figure 10a). Furthermore, the sweep of the gun in tracking gives the trajectory the appearance of a lateral curve which can be more confusing to the gunner than no tracers at all (see Figure 10b). In addition, the angle of approach of the tracer onto the target accelerates on closing to such an extent that there is another

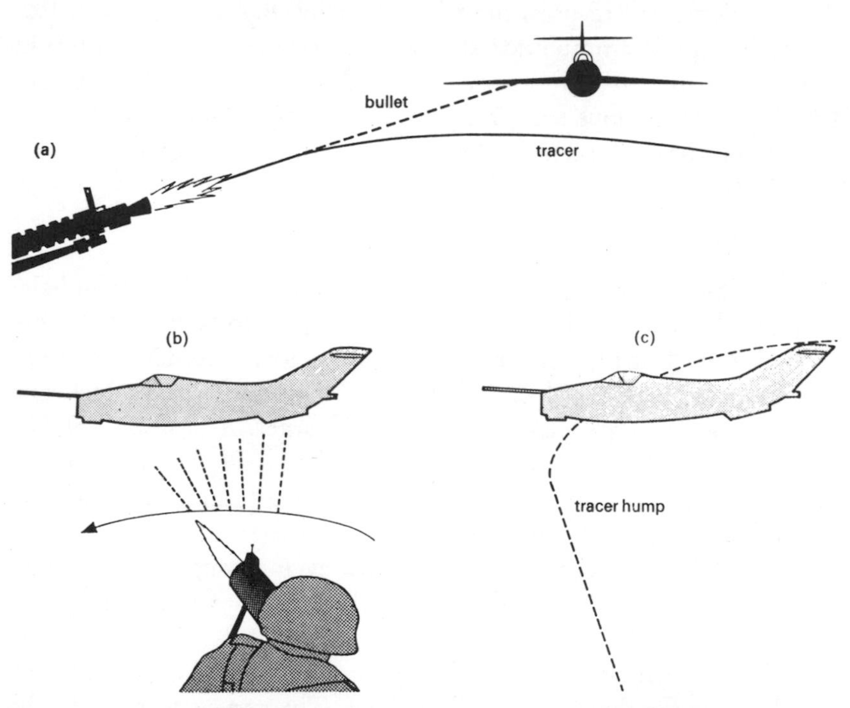

Figure 10 Tracers: (a) trajectory compared to normal bullet; (b) apparent lateral curve caused by sweep of gun; (c) trajectory when approaching target showing tracer hump.

lateral curve, so sharp this time that the phrase 'tracer hump' was coined to describe it. In other words, a tracer which at one moment appears to be heading towards its target appears the next moment, because of the speed of the target, to fall hopelessly behind (see Figure 10c).

Third, the rate of firing may in any case exceed the ability of the observer to process the information.

To these psychological factors has to be added the mechanical problem of actually turning a weapon at the speed required in forward-area positions where angular velocities and accelerations of the target are very high. In fact, it appears that the speeds are so high that many weapons simply cannot follow them. The HumRRO survey also noted that the British Army, which was very attached to tracers in the Second World War, discontinued them a long time ago, except for some training purposes, no doubt because low-level flights made them more trouble than they were worth (6).

The most positive finding about tracer use was the most unexpected: enemy pilots were more afraid of tracer fire than of fire they could not see. Second World War evidence did seem to show that pilots would give up at farther ranges if resisted by tracer fire. On balance though the HumRRO survey found against tracers and they do now seem on the way out.[1]

Infantry versus aircraft

Despite the problems of high angular velocity, the improvements in marksmanship brought about by the psychologists encouraged the infantry to train its men to engage low-flying aircraft with small arms. However, because of the very specialized nature of this type of warfare, the army did not want to risk expensive aircraft at low levels and so, when it officially asked HumRRO to develop 'a training technique in the engagement of aircraft with small arms', it

1. A new training device, developed at NASA as a spin-off from the Apollo programme, makes use of tracers. It is basically two electronic screens plus two simulated cockpits – one the gunner's, the other the pilot's. The screens simulate tracer firing at the aircraft and enables the pilot to take evasive action. The tracers are actually represented on two screens, one view being the gunner's, the other the pilot's. The tracers obey the gravity rules and the illusion rules. The device enables the gunner to see the same tracers from the pilot's viewpoint to see whether they really do fall away quickly as they appear to from his own standpoint. The device is in its early stages yet and no one knows how successful it is likely to be (7).

also stipulated that the training, if at all possible, was to be done *without* the use of real planes or even drones. A tall order – but what HumRRO produced was not merely effective: it cost virtually nothing and just shows what can be done with a little ingenuity (8). The whole project is of special interest because it is an example of a military situation, once thought obsolete, being re-created as the nature of war changes. For many ground forces, training in the basic skills required for direct fire against aircraft ceased after the Second World War both because of the advent of high-speed jets and because automatic guns and missiles were judged (rightly) as more effective. However, the extensive use of helicopters and the reduced speeds of jets when flying at very low altitudes have re-created the low-level air threat. (Helicopters travel at 150 knots, reconnaissance aircraft, manned or drone, fly at 75–300 knots and jet attacks take place at 400–500 knots against ground targets.)

Two basic problems face the infantry in tackling aircraft. These are range estimation and lead estimation – the amount by which the rifleman has to fire in front of the aircraft to allow for the distance it will have travelled by the time the bullet reaches it. The HumRRO research dealt with both aspects of this (the two are interrelated, of course, because the farther away the aircraft, the more lead there has to be). To give some idea of the extent of the problem Table 4 shows the lead distances for aircraft of different speeds at various ranges.

Table 4 Lead distance as a function of speed and range

	Lead distance in aircraft lengths		
Gunner-to-target range (metres)	*100-knot target (UH1-A helicopter 12·2 metres long)*	*200-knot target (OV-1A Mohawk 12·5 metres long)*	*300-knot target (F-100 Super-Sabre 14·6 metres long)*
205	1	2·5	3
220	1·5	2·5	3·5
250	1·5	3	3·5
300	2	3·5	4·5
400	2·5	5	6
500	3	6·5	8

Special air-defence sights have been developed on a number of occasions to help the gunner compute lead length. The Japanese used one in the Second World War and the North Vietnamese reportedly had a simple bamboo rear sight, attached to the side of the rifle stock. The sight has three holes: the hole nearest the stock is used for firing at helicopters; the middle one for fixed-wing targets and the outside hole for high-performance aircraft. HumRRO developed an American version of these based on much the same principle (9).

Even with sights, however, a method of firing is required. Three have been developed. The first is where the gunner estimates the amount of lead required and keeps his gun moving ahead of the aircraft firing the whole time. This is known as the *changing-lead technique*. In the second, the gunner selects an arbitrary lead, but one that is well ahead of the aircraft and fires into it continuously. If he is accurate then the plane must fly into his bullets. This is known as the *fly-through technique*. The third method is where several guns adopt the fly-through techniques together, all firing at some arbitrary point ahead of the aircraft. Once again the aircraft should fly through this hail of ammunition. This is the *pattern-of-fire* technique. Although these three techniques of aerial fire have been used under different conditions, it has to be said that even now the relative efficiency of the three methods is not known. The most generally favoured method is to use changing-lead against slower targets (up to 200 knots) and pattern-of-fire against higher performance aircraft.

HumRRO's achievement was really the development of a series of gadgets – in some cases cardboard models – which helped trainees learn to allow the correct amount of lead. The first one is shown in Figure 11. When the trainee estimates correctly the coach's pointer is aiming at the target. Using this method, speeds of up to 125 knots can be trained for. When the men subsequently changed to M14 rifles (using 7·62 mm ball ammunition) they averaged a hit rate of approximately 4 per cent.

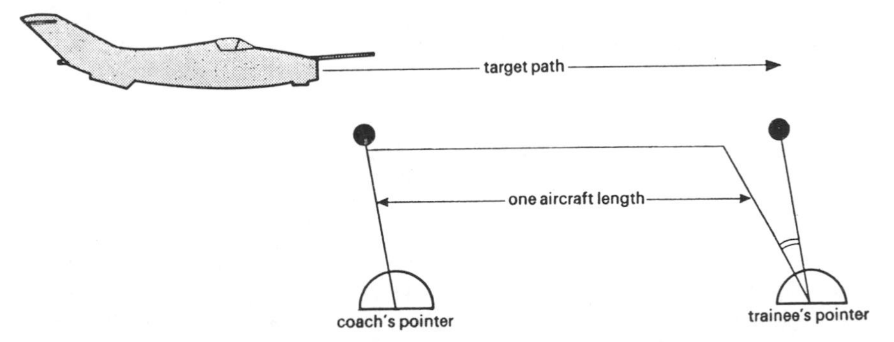

Figure 11 Device to train lead estimation.

This compares favourably with the hit proportions obtained in live firing tests by the US Army Combat Developments Command (10).

For faster aircraft, at greater ranges, simple gimmicks were used to help the men calculate distances (see Figure 12). Trainees moved towards the target until they estimated, using the method shown in Figure 12, that they were 350 metres from it. Their actual distance was then measured using a tape measure. The training continued until they could estimate the range to within 6 per cent (roughly one half of one step).

Other cheap methods were developed. One-foot models of MiGs preceded by silhouettes were sent down wires at scale speeds. Provided

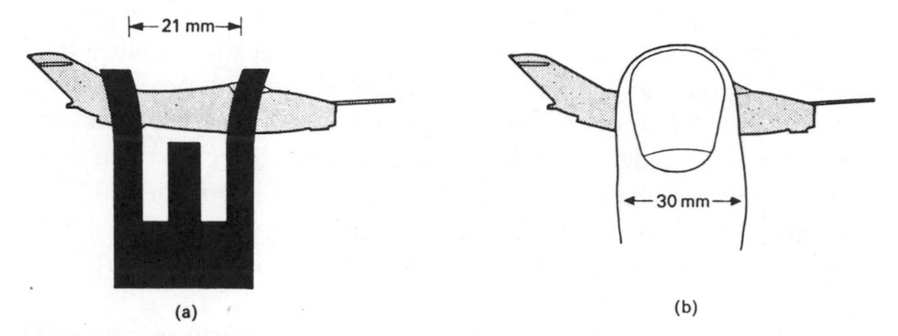

Figure 12 Gimmicks for distance calculation: (a) front sight picture of Russian MiG 21 at 350 metres (aircraft length 13 metres); (b) index finger–target relationship at 350 metres (index finger held at arm's length) for Russian MiG 21.

the trainees aimed correctly at the silhouettes, they should have hit the target. For faster speeds the targets were pulled by jeep. The men then transferred to a service weapon – the M14 rifle. They had to fire at a 20-foot target sleeve towed 5000 feet behind a Navy aircraft at approximately 150 knots (11).

Men on the course were compared with another contingent of men with similar military experience, marksmanship ability, age and so on, but who had not had the HumRRO training. The experimental group achieved an average of 2·3 per cent hits whereas the control group managed only 0·7 per cent hits. A significant improvement after a very cheap training. It was decided therefore that the reduced-scale system could be effective in developing in men a skill for engaging aerial targets.

SEEING WITHOUT BEING SEEN: RECONNAISSANCE, SURVEILLANCE AND CAMOUFLAGE

Survival in battle will often depend on which side first sees, recognizes or accurately locates its targets. By the same token, the more one can keep one's own side hidden the healthier life will be. Target detection and camouflage therefore may be regarded as different sides of the same coin.

Consider first the wide variability in targets. Air defence units, for instance, have to be able to tell the difference between friendly and enemy aircraft from their shape or sound. 'Image interpreters' pore over aerial photographs looking for regular blobs which – among a mass of irregular marks – indicate a man-made target. Point men in infantry platoons need to be able to spot disturbances in the vegetation of a trail that indicates the presence of a mine ahead. Helicopter rescue squads need to cover large areas of search for small objects, like people in the sea. And radar operators have perhaps the most difficult job of all – selecting from a batch of dots on a screen which are real targets and which are irrelevant 'noise'. In training the soldier it is usually impossible to give him practice in every situation he is likely to face. Consequently, some attempt at generalization of targets has to be made (12).

Target characteristics

The ease with which a target is seen or heard will depend to a great extent on the surroundings. Making allowances for this, however, it is now known that certain types of target are more readily perceived than others – a useful matter both in search manoeuvres for offensive forces or survivors and for the camouflage tactics of defensive forces.

Size and shape. Naturally, the bigger the target the better it will be seen. But for the military man the crucial point is the distance at which certain sizes can be distinguished. The results of several experiments now seem to suggest that, distance for distance and area for area, circular targets are more readily seen than non-circular ones. So, whatever the merits of the old wagon trains in forming themselves into circles for protection at night, in a modern war that would be fatal. If the main danger is expected from the air then it follows that tanks, lorries, armoured vehicles and so on should disperse themselves randomly about the place. On the other hand, if they need protection against local attack then a compromise may have to be reached. It is then worth knowing that for shapes the increasing order of 'camouflageability' is: geometric forms (squares, triangles), then long thin rectangles and finally, most difficult to see, multi-legged forms (though avoiding straight lines, where possible). Objects also get more difficult to see as the length-to-width ratio increases, other things being equal (13).

Distance. Distance as already noted, is intimately related to the size of a target. But distance is often subject to illusions, so its vagaries must be clearly understood by soldiers such as the artillery gunner who have to estimate range. Several points have emerged from the research. For example, if other cues are missing (such as intervening trees), objects on the horizon are usually judged closer and larger than they really are compared, say, with similar objects up in the air. Similarly, experiments have shown that mountains are usually perceived nearer than they actually are – and this is particularly true of the foothills where, of course, many targets are likely to be hidden. Distance also interacts with elevation. For example, detection of targets is fair at 5° elevation, poor at 45° elevation and continuously better through to 90°. At 75° targets can be detected at 3·5 times the distance they can at 45° and identified at 2·5 times the distance. Research at Fort Benning in Georgia has shown that with each increase in range from 50–300 yards human targets in scrub need to be 25 per cent farther away from each other for the distance between them to be perceived without error. This is very useful knowledge for a commander who wishes to conceal the size of his unit for as long as possible. Detecting humans at night is a particularly difficult task – if you shoot and it wasn't the enemy, you may have given your position away. The Fort Benning research showed that under a full moon the average soldier is correct in identifying human figures half the time when they are roughly 80–90 yards away. With no moon that comes down to 27–30 yards. This thus gives the defender reassurance of his judgement if he can assess range well and offers some guidance for the advancing soldier (14).

Contour. Contour occasionally acts independently of shape. Research at the US Army Human Resources Center shows that when targets are camouflaged people tend to see sides before angles and so conceive of the targets, first, as symmetrical, equal-sided figures (15). Clearly this can give rise to errors and should be explained to the men in training. The research on contour, however, is not as conclusive as that in other areas: in some circumstances simple figures are easier to see than complex ones, though in other situations this arrangement is reversed. By and large it appears that right angles are easier to see than either obtuse or acute ones and apart from the pure circle, straight lines are more readily perceived than curved ones.

Number. Estimating the number of troops or ships in a given area can be a crucial task. It is also one on which individuals vary a great deal. Fortunately, these differences seem to follow fairly standard patterns so it is possible to allow for them. Basically, people covering a large surface area will tend to *over*-estimate the number of a particular target, whereas those surveying a small area will tend to *under*-estimate the number. Therefore, two people can survey an area, one looking at the whole picture, the other at just a small area, and their two estimates can then be averaged; or one can work out the 'true' figure from the known tendencies of people to under- or over-estimate; these have been calculated and can be looked up in tables (16).

Colour. The effect of colour depends largely on the degree of contrast between an object and the background. (This is the whole point of camouflage, after all.) Where colour on its own *can* count is in twilight – when the light begins to fade objects that retain their colour longest stand out. Research by the American Optical Society shows that full colour vision begins in conditions approximating full moonlight. Below these light levels, purple stands out the most and yellow and green merge best into the surrounding gloom (17).

Movement. Movement often gives things away, of course. Joseph Wulfeck, in the US Armed Forces Committee on Vision (18), was able to make a number of conclusions about the perception of movement: movement is easier to perceive when there is background than when there is not; things look slower in peripheral vision than in central vision; there is a breakdown of contour with movement across the retina; the best background for identifying acceleration is random texture; identification is poorest when starting and end points are not seen. A further report by Robert Gottsdanker (19) also found that deceleration is perceived better than acceleration. Probably the most important practical point to come out of this is that, given that movement *will* attract attention, the unit or vehicle or person doing the moving should aim to make identification as difficult as possible. It appears that the best way to do this is to try to ensure that you are only seen in motion – not accelerating or decelerating but in full flight.

Camouflage. Research into hiding targets – camouflage – is progressing quite quickly. For example, psychologists at the US Army Tropic Test Center, have worked out the distances at which men and other targets can and cannot be seen in a jungle environment (a drop of 14 per cent detectability for every 10 feet over 40) (20). More intriguing still, at the US Army Combat Development Command a special paint is being developed with 'high diffusivity': this minimizes glare and reflection and is good at disguising vehicles (21). The Army Weapons Command at Rock Island, Illinois, has produced a paper entitled 'A biochemical approach to camouflage for army application' (22). This is a survey of the use of camouflage by animals and particularly explores 'chameleon paints', which change colour with the light, and texture paints, which reproduce the fur of animals so as to blend equipment better into deserts or jungles. The Advanced Research Projects Agency (ARPA), the source of so many way-out military ideas, has also developed in idea ofr a reversible camouflaged uniform (23). This has four colours in the normal freeform on one side, but on the inside it is either black, for night use, or else is designed not to look like a uniform at all but, in the case actually mentioned in the documentation, to look like a Vietnamese peasant's outfit.[2] Sounds a bit risky.

Friend or foe? Identifying targets

Spotting a target is one thing. If, however, you are to take retaliatory action you need to do rather more than spot it. In the case of tanks or aeroplanes, for example, you need to be able to identify it as the enemy and then to know how far away it is, in which direction it is moving and, to a lesser extent, what type it is. For aeroplanes it is also necessary to know the angle of elevation of the plane. Psychologists have tried to provide trainee gunners and other target spotters with systematic and inventive means of identifying the enemy.

2. The Army's Natick labs have even produced a world camouflage map, which shows commanders what camouflage is needed where (24).

The low-altitude air assault doctrines developed during the Vietnam War showed up certain psychological problems. The main difficulty was with forward-area soldiers whose job it is to spot the aircraft as far away as possible. It was found that they *can* be effective but there *are* problems. These are: most people untrained in range estimation tend to grossly over-estimate the distance – this reveals the gunner's position too early; dark-painted aircraft are recognized sooner than aluminium-skinned aircraft; if the gunner has a rapid-fire weapon with a low-capacity magazine and opens fire too soon, he may be out of ammunition by the time the aircraft *is* in range; most errors are made on the incoming flights, where most people have a tendency to over-estimate; they tend to be much more accurate on the outgoing flight – but of course it may then be too late.

Another operational problem of some importance is that although the amount of light does not affect range estimation, the angle of elevation of the plane does. Low-flying aircraft tend to be *less* accurately estimated (possibly because they *look* larger) than the higher flying aircraft (the experiment which showed this dealt with aircraft at 75 feet and 400 feet – producing target elevations of 9° and 55°) (25).

Two important problems have been isolated by HumRRO. One practical difficulty in researching this field, however, is cost. Many of the experiments – and the actual training, of course – need real aircraft to fly as targets. An F-100 jet fighter and a H-23 helicopter – the usual 'targets' in the tests – are very expensive. So the HumRRO team has tried to see if scale models are equally effective in training. The initial results show that the scale ranges are not merely as good as, but in some cases better than, real targets for training range estimation. This is, however, confined to the outgoing aircraft; for incoming targets the scale ranges are much worse than the real thing (26). A later experiment has shown that this can be overcome. The simple expedient of having the gunner look at a target through a bunched fist, so that cues from the ground are excluded, provides a cheap and effective way of eliminating their effects. For ranges up to 2500 metres it was found that after twenty trials only, using the bunched-fist procedure, the average error in the estimation of targets was – 14 metres (that is, an *under*-estimation of 14 metres). For other training methods it was +127 metres (that is, an *over*-estimation). It was also found that the underlying bias to over-estimate the range of approaching targets can be much reduced by this training. Hum-

RRO has concluded that gunners can be trained to estimate aircraft range under field conditions *without* using live aircraft as targets (27).

The basic task of simply seeing an aircraft and estimating its range, therefore, contains some hidden problems – though with a little research and a bit of training these can be sorted out. It is not always as straightforward as this though. For one thing in battle no one can guarantee perfect visual conditions. Working in the desert near Tonopah, Nevada, another HumRRO team therefore compared visual search for a target with auditory tracking techniques and also looked at how the performance of the naked eye compared with binoculars (28). The targets in this experiment were F-4C fighter jets and B-52 bombers. Each flew at low altitudes but at tactical speeds and at ranges up to 3300 metres. In one direction, the gunners' views were unobstructed but in the other direction there were low hills providing a background against which the aircraft had to be seen. It was found that the average gunner can see a plane, whatever type it is, at about 500 metres before he can hear it. Using listening only, untrained people tend to consistently estimate the plane as ahead of where it really is. On the other hand, it appears that people using this method are consistent in their error and so, when visibility is bad, tracking by auditory techniques is possible if allowance is made for an individual's average mistakes. It was also found, interestingly, that the farther away from the flight path the observers were the more accurately they spotted the targets (29).

But the most important results probably concern the use of binoculars and identification of the plane's constituent parts. After all, this is what helps identify the plane as friend or foe. Binoculars do *not* help gunners see the targets any quicker, whether the target is against a clear sky or against a background of low hills. In fact, when the hill background is near (4000–6000 metres) the aircraft are actually better seen *without* binoculars. Once the aircraft has been spotted – located – but is too far away to be identified with the naked eye, binoculars are of use (30).

The features of the planes, whether they are fighters or bombers for example, can be identified in a set order. This, say the researchers, is not only important for identifying friend or foe but is also a useful guide in estimating range. The point at which certain features of a plane can be recognized is a set distance which soldiers can be taught. Table 5 shows some examples.

Table 5 Sequence detection of structural components

	Unaided vision				*Binoculars*			
	Fighters		*Bombers*		*Fighters*		*Bombers*	
	F-4C	*A-6*	*B-52*	*B-58*	*F-4C*	*A-6*	*B-52*	*B-58*
200 metres away from the flight path								
Fuselage	1	1	3	2	3	2	5	3
Wing	2	2	1	1	1	1	1	1
Vertical stabilizer	3	3	4	4	4	3	3	4
Wing extension	4	4	—	—	2	4	—	—
Canopy	5	5	6	5	6	5	7	5
Nose	6	6	5	6	7	7	4	6
Horizontal stabilizer	7	7	7	—	7	7	6	—
Air intake	8	8	—	—	5	6	—	—
Engine pod	—	—	2	3	—	—	2	2
1400 metres away from the flight path								
Fuselage	2	2	2	3	4	3	4	4
Wing	3	3	1	1	2	2	1	1
Vertical stabilizer	1	1	4	2	1	1	3	3
Wing extension	4	6	—	—	3	5	—	—
Canopy	5	5	6	5	6	6	6	5
Nose	6	4	5	6	5	4	5	6
Horizontal stabilizer	7	7	7	—	7	8	7	—
Air intake	8	—	—	—	8	7	—	—
Engine pod	—	—	3	4	—	—	2	2

The first structures, with or without glasses, were identified on the fighters at about 4000 metres (the planes had been mostly *detected* at between 5000 and 6000 metres). The second structure was identified at between 3600 metres and 3800 metres and so on. With training this could prove a very effective way of estimating range.

There has also been some experimentation on the effects of prompting the identification of targets – that is, by having someone guess the target just before he can positively identify it. It appears that, with ships at least, target identification can be improved by this method in comparison with confirmation, – that is, by telling someone he has got it right *after* he has responded. But the evidence is contradictory (31).

Forward-area observers, who do most of the target identification, tend to exaggerate the performance of any particular crew. Research at George Washington University has shown that where the forward observer is accurate, then his decisions mean that the target is engaged earlier. If, however, the observer is not so good, the chief behind him takes relatively longer to make up his mind and the team, as a whole, is slower than it would be if it did not have a forward observer at all (32).

Crew chiefs appear to fall into one of two types – cautious and accurate and the less cautious and less accurate. The less cautious chiefs did not often wait for their forward observer's opinion (communicated by field telephone). The cautious ones, however, did wait and here an accurate forward observer is especially helpful, since he ensures quicker reactions without a loss in accuracy (33).

Also found to be important were the directions from which an aircraft was flying. When the aircraft flew directly overhead, most chiefs did wait to hear what the forward observer had to say. But again, when the aircraft was offset from the crew and observer, say flying parallel to them, the chiefs usually did not wait, but went ahead on their own account. It seems reasonable, therefore, that some chiefs should be made to use forward observers, because the practice will improve their accuracy some of the time. Research is now going on to select the chiefs who need such help.

Time: the loneliness of the long-distance vigil

The effect of time on target detection is one of the most complicated problems of all. Under some circumstances, as you would expect, the longer someone has to look for a target the more likely he is to find it. Or, put another way, an observer should spot more targets in a longer time. Once he has spotted the target, an observer is also more likely to identify movement or assess velocity the longer he has to look. But this is true only up to a point; the relationship is not straightforward. For example, the observer or look-out will not necessarily spot double the amount of targets in double the time.

Boredom affects the rate of target detection particularly with the radar operator or night look-out. This is the well-known problem of *vigilance,* identified early on in the Second World War when, after the invention of radar which promised to solve so many military problems, it was soon discovered that radar operators were still missing many targets on the screen even though their lives depended on not doing so. The surprisingly low level of performance of look-outs, aerial observers and sonar operators became at one time practically the most critical military manpower problem. However, with the trend towards automated systems in recent years, vigilance research has shifted and now most attention is directed at the problem of man as a monitor of automatic target-recording equipment.

The essential defining characteristic of the vigilance phenomenon is a systematic change in the probability of detection of a signal over time. In general it is errors of omission, rather than of commission, which are of most interest to the military since weapons are irrelevant against targets which have not been detected. The main trend with long vigils is that detection ability gets worse as time goes by. Typically, this decrement starts very soon after the vigil has started and becomes greater as the vigil or watch continues, with the greatest drop occurring during the first thirty minutes. Figure 13, taken from one of the early British military studies of vigilance, shows a typical result.

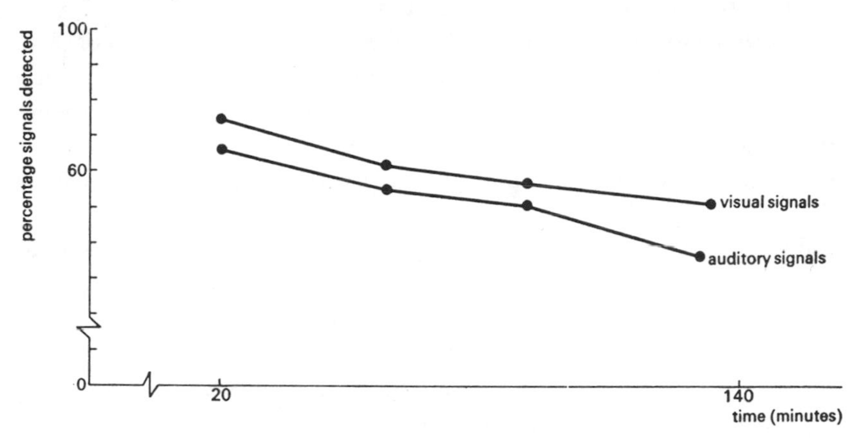

Figure 13 Decrement in signal detection over time for visual and auditory signals.

In some cases the circumstances of the display affect detection (there was up to a 46 per cent drop in one experiment); in still others, there are sizeable differences between individuals. (Some people – an important minority for the military – do not show a performance decrement during prolonged vigils.) The factors affecting vigilance can be conveniently split into three: those concerned with the job itself – watching a screen or the sky; those affecting the surrounding environment in which the task takes place – is it day or night?; and those affecting the individual who is doing the monitoring (34).

The factors of the task. In general, the longer a vigil lasts the greater the drop in performance. But if the monitor expects a short vigil he will perform better than if he expects a longer watch. The only exception to this is the 'end effect' – towards the end of a watch, say the last half-hour, detection improves. Vertical display is the best arrangement for detection. It is commonly observed that in-coming target pips are ordinarily not detected until they are well in towards the middle of the screen. In an effort to minimize this effect, certain search techniques have been developed – they are discussed below. A recent innovation at the US Navy Engineering Psychology Branch enables targets to change colour as they move across the screen. This change is more likely to bring a target to the attention of the operator than if the target stays the same and *is* a practical possibility (35); in Britain a screen has been developed which lights up in alternating bands. Both devices increase the number of targets detected.

There is also an interaction of detection with the distribution of pips. For example, in one experiment it was found that in the quadrant of a screen which displayed pips at the rate of 72 per hour, 71 per cent were detected, whereas in another quadrant on the *same* display at the *same* time and where the rate of pips was 12 per hour only 63 per cent were detected (36).

Several studies have now shown that the more frequent the signals the higher the percentage of them that are spotted, though it has also been found that more false targets are spotted when there is a high rate of pips (37). This has led to the technique, discussed in more detail at the end of this section, of introducing artificial signals into a screen in order to improve the detection rate of real signals – it sounds strange but it actually works.

To begin with, it was found that detection rates were similar whether the targets were presented visually, auditorily or tactually. This led to the belief that whatever caused the decrement was located in the main part of the brain and was not associated specifically with the eyes or ears. Though this still remains plausible, more recent studies have shown that visual monitoring is better than auditory monitoring which in turn is better than tactual monitoring. The introduction of deliberate rest periods consistently improves detection, look-outs being not only more accurate after a rest period, but quicker, too (38). Any form of rest helps – even being allowed to fidget improves performance. The most dramatic effects of rest were probably those obtained in the early British studies by J. F. Mackworth who found that if he gave a rest period of thirty minutes after every thirty minutes of work there was no decrement at all – and an improvement in some cases (39).

Environmental factors. Noise has been found, by Donald Broadbent in Britain, to improve some people's performance on a visual vigil. Extreme heat (e.g. 97° F) and cold (7° F) affect detection adversely, particularly the combination of heat and humidity. From this, and from the other studies discussed above, it makes sense for ship's captains to have shorter radar watches in the tropics or in polar regions than in more temperate zones.

Isolation is also important, although of course the precise meaning of that word varies. In extreme circumstances it can mean the absence of *any* sensory stimulation. In others, merely that stimulation is limited (as at night or in cloud). In still another sense it may mean the complete absence of other people. It is probably fair to say that the occasions on which a target would have to be detected in conditions of sensory deprivation are extremely rare. The more important form of isolation from a military point of view is probably where the look-out is alone and the surroundings are unchanging (say on a ship's bridge at night, or in the radar room in a submarine). In these circumstances, it seems that groups are better at target detection than lone individuals. On audio-visual sonar displays, for example, two individuals working together usually spot 11–20 per cent more targets than lone individuals, and three people working together detect 6–15 per cent more targets than two people. Groups of more than three, however, do not boost detection any more. If someone has to maintain a vigil with others around who are doing something else (which is likely to be the normal military situation) then this does not appear to affect the number of targets detected. The only exception is when persons in authority are present. People do significantly better in the presence of an authority figure than in his absence (40).

That may be comforting to senior officers – until it is realized that it only puts them on a par with drugs. There seems to be fairly wide acceptance that benzedine sulphate reduces the vigilance decrement. One of N. H. Mackworth's studies in 1950 shows that benzedine sulphate could actually *eliminate* the decrement with no reported side effects. Several recent studies have confirmed Mackworth's original results but though dramatic it has not proved easy to find a direct practical spin-off from the drug research. It is probably not really open to the military to give their target detectors benzedine in order to improve their performance. Instead the theoretical implications of this research are probably of greater, long-term interest. Alcohol, incidentally, does not produce this effect.

The effects of diet are far less dramatic than those of drugs but possibly more important since they can be manipulated. Two results particularly are worth noting: balanced meals or high fat diets produce better vigilance than high carbohydrate meals; breakfast *is* important. Men who have no breakfast or coffee only do significantly worse on target detection than men who did have breakfast, however light. A lesson for commanders and quartermasters.

Lack of sleep for short periods of time seems to have no effect. On longer tasks sleeplessness does begin to take its toll – but even here it is not very great. Loss of sleep does not appear to be a major variable in vigilance performance.

Individual differences. It appears that introverts are better on vigils than extroverts and the less active (non-fidgety) individuals also appear to do well. (Motor activity in general is negatively correlated with success in target detection.) Intelligence matters, of course. Men with low scores on intelligence tests like the Armed Forces Qualifying Test do worse on target detection than those with high scores and show a greater decrement. Age and sex exert their influence as well. Those between twenty and twenty-nine do better than those between sixty and sixty-nine *but there was also a greater performance decrement among the younger group*. And, incidentally, although most studies – the military ones especially – have used men, females appear to be superior to males at least 'during the last half hour of a two-hour watch' (41).

Apart from these fairly conventional differences, people also appear to differ in their observation style. Some people tend to observe at all times, whereas others only really concentrate when they think a signal is about due. By and large the continuous style leads to better detection, and this may be trainable (42).

The physiology of the body affects matters. Particularly important is posture. The response of the frontalis muscle, over the eye (and in which there are large individual differences), affects response latency. Autonomic activity – like skin conductance or blood pressure – is also related: the more it increases, the worse the target detection gets. It has been suggested that changes in skin conductance could be used to activate devices which would boost the alertness of a drowsy observer (43).

One recent textbook devoted to the subject of vigilance contained just under 600 references, so I can make no claim to have more than scratched the surface of the topic. I have, however, given the main points of vigilance research as they affect the military. And on the basis of what we have discussed so far, we can at least sketch out several human engineering principles which should ensure that look-outs and radar operators perform at their best.

Improving vigilance. Some techniques are obvious: signals should be as large in magnitude as is reasonably possible; they should be made to persist until seen (if possible); the area in which the signal appears should be as restricted as possible. Although real signal frequency cannot often be controlled, according to a HumRRO report (44), 'it is desirable to maintain signal frequency at a minimum of twenty signals per hour. *If necessary this should be accomplished by introducing artificial signals to which the operator must respond.*' Also, wherever, and however possible, the operator should be given knowledge of results; where possible, he should be given *anticipatory* information (a sound buzzer for example). It also helps to improve the man's understanding of a target; that is, to point out the differences in brightness (for example) between targets and 'noise'. Also, in a busy area accuracy may be crucial; so tell him this and he will identify fewer targets but make fewer mistakes. If a man is asked to 'look for targets', his performance drops in the period following an identification. If, however, he is asked to search a particular area on the screen (or reconnaissance photograph), it does not. Noise, temperature, humidity, illumination, etc., should be automatically maintained at optimal levels. The system should be so designed that operators do not work in isolation from other individuals; or from higher ranks. Where possible, the system should be so designed that individual watches do not exceed thirty minutes.

Owing to the development of highly complex computers capable of distinguishing many features simultaneously, vigilance may not be the problem in the future that it was in the past in air defence systems. Human vigilance tasks will still occur, however, in high density systems such as the crowded air space found in Vietnam. There is a future here, HumRRO believes, in selection research. Selection should take into account the individual differences already discussed. The four that are considered the most promising for research are: visual acuity; intelligence; initial detection thresholds; and observing style.

Finally, there is quite a bit that can be done in the *design* of the vigilance set-up. Much research needs to be done along the lines of the optimum arrangements for detection; would working in pairs help?

Which people do better in pairs? How does rest affect pairs? What is the best rest period? How often should it be?

Danger in the dark: night operations

Combat in the dark poses innumerable problems and a whole series of experiments have been given over to night operations. A HumRRO survey found that there were eight basic skills needed at night: land navigation; target location and identification; the ability to avoid detection by the enemy (mainly through noise discipline); communication; the ability to install equipment by touch; engaging targets (how to fire without giving away your position); rifle fire (aiming in the dark). The basic problem, if you like, is not to train soldiers to 'see' at night but to adapt their combat behaviour to the dark (45). Mainly, the experiments that have been completed have been on target acquisition and on aiming and it is these we shall examine here.

Tank crews featured prominently in the early experiments: the US Army Human Research Unit started with a brief to develop methods of training tank crews to be more effective at night both in terms of combat skills and in knowing how best to illuminate a given situation either when attacking or defending. The first study, completed in 1958, attempted to find out the ranges at which approaching tank-size vehicles could normally be detected on a starlit night (a) when the observers had to look towards a tank searchlight and (b) when there was no searchlight turned on. For the vehicles used (M48 tanks or two-and-a-half ton trucks) detection and recognition range proved almost identical. The searchlight, however, did not blind or dazzle the observers as much as was expected. In fact, recognition ranges were greatest when the approaching vehicle was travelling down the brightest part of the light beam. On average, identification could be made 260 yards earlier than when the searchlight was not on. Most important of all perhaps, because it was so unexpected, was the finding that observers *inside* the beam of an incoming searchlight could see tanks approaching from *outside* the beam though when the observer was outside the beam his recognition range for other tanks was greater. Individuals differed by as much as 200 yards in their recognition of the approaching tanks at night. This is a large difference and consequently the possibility of selecting men for this crucial task is now being explored (46).

Tactically, therefore, the following possibilities arise:

Offence. Troops advancing should remain along the edge of a searchlight beam and use the stray light along that edge for their own purposes; for large distances (say, 1500 yards) the light should be shone straight at the enemy position (provided that it is known); as the distance to the enemy gets shorter (say, 700 yards) the light should be switched to an oblique angle, *not* pointing at the enemy though with one's own troops still outside the beam, on the far side away from the enemy. This is shown in Figure 14. Provided it is followed, the enemy will have least chance to spot advancing troops.

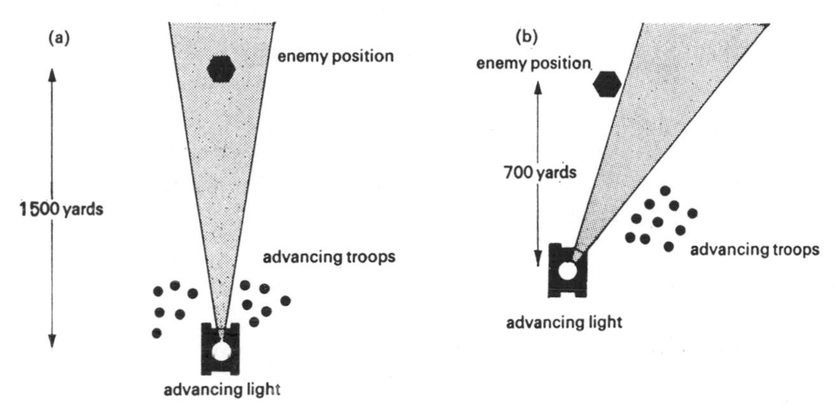

Figure 14 Relative position of searchlight and advancing troops: (a) when enemy position is 1500 yards away; (b) when enemy position is 700 yards away.

Defence. Naturally in a defensive position this procedure is reversed – special attention should be paid to those areas just outside the light.

Of course, keeping the light on exposes the tank to enemy fire. So the effects of turning the lights on and off at intervals have been explored. (This itself poses problems – what is the optimum period for the light to be on but not to become a target? and what is the best time in which the tank commander can spot what he is looking for? The former aspect of this has apparently been determined though is at the moment secret. Details about the second aspect are available.) Observers were stationed at the searchlight source and at 10, 20, 40, 80 and 160 yards from the light along a line approximately at right angles to the axis of the beam. There were three types of combat target to locate – a jeep, a tank and an armoured personnel carrier, at each of four distances (655, 780, 900 and 1055 yards). The light was turned on for two minutes for each presentation. For all presentations observers who were to the side of the searchlight made more detections than those at the searchlight source. For the first thirty seconds of light the naked eye was the most effective in detecting targets, but after one minute binoculars were more effective. As in studies with aircraft, once the target had been detected, identification was easier with binoculars than the naked eye. Identification performance at the light source was particularly inferior when the targets were further away (47).

Tactically, this means of course that target observers should stand away from a tank-mounted searchlight. Roughly speaking, there is a 50 per cent chance that an observer using binoculars will detect a target within thirty seconds if it is at 900 yards (for a tank) and 750 yards for a jeep. It is a short step from this result to the calculation for the time a searchlight should be switched on. Figure 15 shows that a law of diminishing returns operates. In the first minute there is a 55 per cent probability of detection which only increases to about 62 per cent in the second minute. This gives some idea of the upper limit of the flash needed. For about thirty seconds there is roughly only a 33 per cent probability of detection. From a detection point of view alone, therefore, the optimum time appears to be not less than twenty seconds and probably not more than a minute. This would of course be tempered by the risk to the tank from keeping its light on for so long (48).

Searchlights, of course, are not the only things which can give away the position of a tank. If it is firing then the flashes from its gun may serve the same give-away function. It is, however, difficult to locate

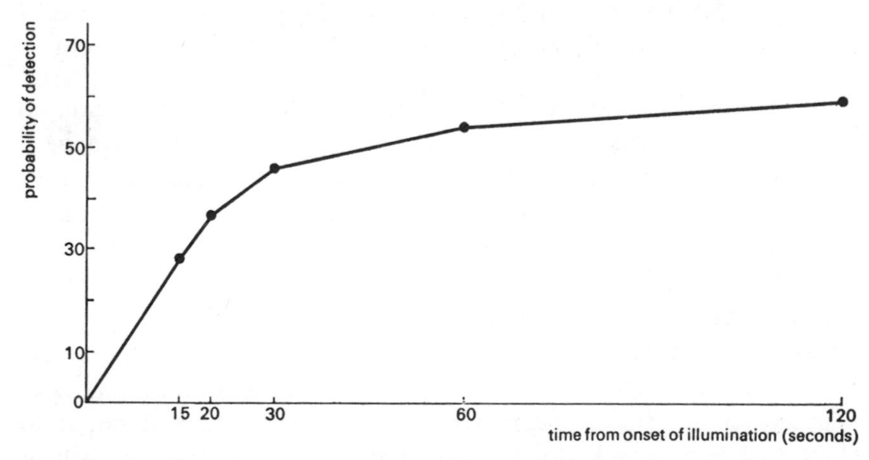

Figure 15 Probability of target detection using searchlight for up to 2 minutes (per cent).

gun-flashes and no training is normally given for it. Thirty-one gunners were tested for their ability to lay the main gun of a M48A2 tank against forty flash positions and also to locate the same flashes in a periscope field of view. Performances were judged in terms of (a) hits on a target 2 metres square at 500 and 1000 metres and (b) hits on the same target by fragmentation of a high explosive round. One interesting point to emerge was that inexperienced gunners were every bit as good, if not better, than experienced gunners at the localization. In the test of laying the main gun against flashes, accuracy was very poor: 12 per cent and 5 per cent direct hits at 500 metres and 1000 metres respectively. Practice produced no appreciable improvement in these particular tasks. But the researchers, though their results in this case were disappointing, do say that if a reticle could be used which retained the image of the flash on the sight, important improvements could follow (49). Research has already been done by the US Army Human Research Unit to determine the optimum design of this equipment.

Interference with dark adaptation gives rise to temporary blindness. This cannot be avoided when it is caused by enemy flashes. Dark adaptation is also interrupted when a soldier has to use an image intensifier – a night-vision device developed in the 1960s. These devices present the user with a green image whose brightness is above the normal photopic threshold. A soldier may use these sights on and off for as much as thirty minutes in a watch, and have to perform other tasks as well, so their effect on dark adaptation is of some concern. Experimenters (at Fort Knox) therefore examined the effects of the interruption of dark adaption on two military tasks: (a) walking parallel to a ground-mounted guideline, keeping as far to one side of it as possible (these guidelines are used to alert and guide soldiers who are moving across country at night through or near hazardous areas); (b) firing an M14 rifle at stationary silhouette targets (50).

Various intervals of dark adaptation interruption were introduced into the experiment – from zero minutes up to ten minutes. Time to first round, duration of fire and target hits were all measured. Interruption of dark adaptation in one eye meant that the men had to be 10 per cent closer to the guideline than if no interruption had taken place; in both eyes, they had to be 20 per cent closer. Between two and three minutes were needed to re-adapt. Time to first round was increased by two or three minutes and duration of fire by half a minute or more. Target hits were not affected.

This means, then, that on guideline-following it is best to allow the men time to recover; usually in cross-country movement there will be time to do this. Firing presents more of a problem. Probably the best way round it is to have one man using the image intensifier and another firing at the target – so avoiding a delay of several minutes which could easily be fatal. Another way round this is to use the *non*-shooting eye for the sight so that the shooting eye remains dark-adapted. But it appears that the non-shooting eye can interfere with shooting so this may not be a satisfactory answer.[3]

Target detection when aiming missiles. Many missiles have high intensity light sources in their aft position, either due to the sustaining power, side thrusts or tracking flares. These do not interfere with target detection during the day, but at night this glow can disrupt vision markedly and even prevent the missile aimer from seeing his target. Two experiments were carried out at the Aberdeen Proving Ground in 1964–5 to see what limits there are to this (51).

The target – a truck covered with drab camouflage – was either unilluminated or lit by a 6-volt lantern 6 feet away. It was viewed from 400 metres away. Between the target and the observer, to simulate a missile, lights were stationed a quarter, a half and three quarters of the way along. Once the observer had acquired the target in his binoculars, one of the intervening lights was turned on for fifteen seconds. The observer had to say whether or not the target was obscured while the light was on and then had to 're-acquire' the target as soon as the light went off if he had lost it in the meantime. The experiment showed that about half the targets were lost for a critical period with a bright light and about a quarter with a dim light. Providing the missile aimer with a lantern to shine on the target to offset the flash of the missile, did not help either. The army concluded that flashes greater than 100 candlepower are unacceptable, and this was the guidance given to missile manufacturers.

As with other areas of research, however, night operations has its futuristic elements and it is perhaps fitting to end with brief mention of one such project beginning at the US Army's Behavioral Science

3. In some cases there are racial differences in these results. 'The night vision of colored and white soldiers' is the title of a restricted paper at Fort Bragg showing that whites are slightly better than blacks at some night-time military tasks.

Research Laboratory (BESRL). Dr A. Hyman, from that organization's Performance Enhancement Laboratory, is working on the use of covert responses in military situations. What this means is that automatic and skeletal responses – stomach contraction, resistance in the skin, some muscle changes, which are beyond conscious control but may react to change before a man is aware of them – will be tapped and measured by 'listening' devices fitted to muscles or skin. When they broach a critical level, a buzzer or something similar will in effect warn the soldier that incipient danger has been detected. This technique should not only be valuable in night operations but could also increase the ability to detect far-off sounds and to distinguish the noise of real targets from the wind, the chatter of the jungle and so forth. But it is a long way off, if it ever comes to fruition at all (52).

4 Special Skills

Special military skills – bomb disposal, for example, or reconnaissance or code-breaking – are required by many soldiers over and above the basic abilities we have just been considering. They require expensive training and so a great deal of thought and research has been devoted to an assessment of whether men can be selected for training in such specialities, thus minimizing wastage.

HOW TO BE STEALTHY: DETECTING MINES AND BOOBYTRAPS

The detection of anti-personnel mines, anti-tank bombs and boobytraps is a more complex skill than most, partly arising from the conditions in which it is carried out. It has been treated separately by military psychologists.

The importance of these contraptions in modern wars has perhaps not been paid enough attention. But a study by George Magnon in Vietnam showed that, in 1967, no less than 33 per cent of the casualities were sustained from contact with mines and boobytraps (1). Another study by the Picatinny Arsenal, involving twenty-one tank crews, showed that surface-laid anti-tank mines placed as a barrier resulted in sufficient delay in two military operations to enable anti-tank weapons to be brought to bear on the assaulting force (2). All this, together with the observed fact in combat that some men are much more adept at spotting boobytraps than others, encouraged the US Army Mobility Equipment Research and Development Center to see whether this ability could be spotted in advance with psychological tests and the men singled out for special training that would make them even better mine detectors (3).

Preliminary research in Vietnam, which involved interviewing expert and inexpert detectors, showed that two psychological factors seemed linked to success: the ability to use concepts and the ability to visualize spatial relationships. The relationships, however, were only moderate and unreliable to the extent that, presumably, some of the more inexpert mine detectors had been killed and so were unavailable for testing and interview. A study was therefore begun at Fort Benning to try to isolate exactly what mine detection is, what psychological capabilities it involves, and how these can be identified and brought out in training.

From what was already known about the psychological factors which affect visual discrimination and search and from a job analysis of point men (the men in squads who go on ahead, looking for enemy personnel and for boobytraps), a list of twenty-four individual characteristics was drawn up as possibly relevant (see Table 6).

Of these, it was decided that the characteristics with stars were capable of being tested.

Table 6 Individual characteristics of possible relevance to mine-detection ability

Physical	*Personality*
Age*	Dogmatism*
Sex	Individual motivation
Visual acuity*	Anxiety*
	Perceptual style*
Personal	*Native aptitudes*
Cultural affiliation	For discriminating environmental change*
Smoking habit*	Separating figure from ground*
Religion	Extracting relevant information*
Speed of movement	For integrating information
	For making deductions from partial information*
Mental	
Intelligence	For integrating special knowledge
Acquired skills	*Acquired knowledge*
Background experience	Of mines*
Formal training	Of use of mines, etc.*
Combat experience	Of cues associated with mines, etc.*

These tests were given to 104 men of the 197th Infantry Brigade none of whom had been in combat. They then had to cross three routes – a wooded track, an open area, and a dirt road – each of which was 'mined and boobytrapped'. An example of the route they had to cross is given overleaf in Table 7.

On the first two routes the devices were concealed so that they were

Table 7 Test route for mine detection

Type of device and how activated	*Location*	*Above or below ground*	*On or off trail*
	←— 3 metres —→		
Small Schumine – P	●	B	On
Small grenade – TW	●	A	On
Small AP mine – P	●	B	On
Large 105 mm Rd – CD	●	A	Off
Medium AP mine – P	●	B	On
Small grenade – TW	●	A	On
Small Schumine – P	●	B	On
Large Claymore – CD	●	A	Off
Small grenade – TW	●	A	On
Medium AP mine – P	●	B	On
Small grenade – TW	●	A	On
Small AP Mine – P	●	B	On
Large 105 mm Rd – CD	●	A	Off
Medium AP mine – P	●	B	On
Large DH-10 – CD	●	A	Off

P = Pressure; TW = trip wire; CD = command detonated (i.e. enemy fire from off the trail); Rd = round

moderately difficult to detect; cues were present and were the type that might be noted after a device had been in place for a short time. These cues were variations in colour, camouflage, vegetation, soil, size, shape and texture. Errors built in consisted of inadequate camouflage, failure to renew it, continued use of the same camouflage technique, disturbed soil, disturbed vegetation, mine or boobytrap partially exposed, triggering device exposed, anticipation by tactical conditions. The dirt and gravel road was 'mined' with anti-tank devices which were completely buried. The situation simulated that which an infantryman would face during a road-clearing operation.

After all the tests, and before they actually started the course, the men were briefed on observation methods used by experienced men and on the cues that might indicate a mine or boobytrap. They were then told to assume that (a) they were in a tactical situation, acting as a point man for their small reconnaissance patrol; (b) their operations area was known to contain various types of mine; and (c) their mission was to locate these devices visually so that a path could be cleared through the area. The men could not use sticks or prods and were only allowed to bend as far as the waist. They were watched and their performance rated by experienced observers.

Two types of result emerged. One was the personal characteristics of the men who were good and of those who were not so good. The other was a more general finding about which mines are easier to detect and so forth. It was found that the best predictions of actual performance were given by a group of eight of the tests. Accuracy of prediction was about 70 per cent, a considerable improvement over chance. The highly proficient detector moves slower than the poor detector, naturally puts more effort into the search, has a better educational level, is more likely to have taken part in many different school and leisure activities, is more open minded, and has better visual acuity. Table 8 shows the fairly obvious fact that large mines or boobytraps and those above ground are easier to detect than small ones and those underground. But probably of more interest are the actual percentages of detection, which give a good indication of the levels of risk involved. (In the section on stress it will be seen how important accurate statistics about risk are in order to control fear.)

Table 8 Percentage of mines detected and average distance at which detected

	Per cent	*Distance (feet)*
DM-10 (Russian Claymore mine)	88·9	17·1
Hand grenade/trip wire boobytrap	77·5	2·9
105 mm round	68·8	11·9
M18AI anti-personnel mine	67·1	14·1
Schumine	52·9	1·9
M16 AP mine	46·9	2·6
M25 AP mine	45·6	2·2

One interesting variation was that on the anti-tank mined road some smaller mines were actually easier to detect than the larger ones. An important factor was the *position* of the mine on the track. Those in or near the middle of the road were detected more easily (about 80 per cent probability) than those at the sides (about 60 per cent probability).

Another point, not spelled out by the army, but built into the research, is the Vietnamese habit when placing mines. From my own analysis of the available information, it appears that about 60 per cent of mines were placed at or near the centre of the track, and 40 per cent at or near the edges. Pressure appears to be used on 66 per cent of occasions, trip wires on the remaining 33 per cent.

It also emerged that a search speed of 12–13 metres a minute resulted in a 66 per cent rise in detections over a search speed of 28 metres per minute. This may bring home to point men the amount of time they need for their job. The search procedure that was most effective was an area search (looking ahead and off the trail first), followed by looking right and left as the gaze is brought in and, thirdly, looking very carefully to the immediate front where the next footfall is going to be.

Analysis of the basic cues used by successful detectors also helps camouflage experts. The study showed that variations in colour were most frequently used by the detectors (29·6 per cent), followed by variations in texture (21·8 per cent), exposed triggering devices (16·9 per cent), and variations in shape (14 per cent). Devices detected more frequently by variations in colour were the Schumine, M25 anti-personnel mine, M16 anti-personnel mine and the Claymore. The initial cue was usually the contrast of the device's colour with its background. Devices detected more frequently by shape were the 105 mm artillery round and the DH-10 (the VC Claymore). Although these devices were camouflaged, the trainees were apparently able to detect them by the characteristic shape of their exposed portions. The thin wire stretched across the trail at varying heights had a high detection rate (77·5 per cent) but a rather short average detection distance of 2·9 feet. The detection rate was much lower for trip wires below knee height. While not listed as the most frequent cue for any one device, texture was the second most important cue overall. Often, reflected light was seen by the men, especially a glint from the sun. Finally, detection rate was highest for the road course, next the wooded and lowest for the open country course.

BOMB DISPOSAL

Psychologists at Fort Benning have recently been refining their personality tests which help select men good at bomb detection; but it is probably one of the few areas of military psychology where the US experts can learn something from their British counterparts. For obvious reasons, the British have had to become rather good at selecting men for a different, but closely related, kind of stealth: the defusing of bombs and boobytraps.

The involvement of psychiatrists and psychologists in the selection of British bomb disposal operators began in the spring of 1974. A psychiatrist flew to Belfast to join the bomb disposal squads at work in the province. For a few weeks he lived with the squads to try to analyse the types of stress which the men face, in order to screen out recruits who were not suitable. The move was made necessary by changes in the type of men available for the bomb squads. At that time most of the army's disposal officers were fairly old and had been recruited in a more peaceable age, before Ulster grew so ugly in 1969. Their jobs were straightforward, inspecting the explosives at quarries and so forth. The upswing in bombings in Ulster, however, meant that younger, less experienced men had to be recruited and some of them stood the strain less well than the older men (between 1969 and 1974, the British Army had two, possibly three deaths of disposal officers that could have been avoided).

As a result of the time the psychiatrist spent in Ulster, and other studies, three tests were introduced, designed to measure in all ten psychological characteristics. One test was the clinical analysis questionnaire – about fifty questions relating to a man's sleeping habits, behaviour at school as a boy, his attitudes to his home, to the mistakes he has made in his life and so on. Also measured in this test is something known as 'suicidal disgust': individuals who resent certain things that have happened in their lives often harbour unconscious feelings of suicide and they are clearly unsuitable for bomb disposal. The second test used is the dynamic personality inventory, which measures willingness to pay attention to detail and a factor, coded Pf, which is a measure of fascination with fire, winds and storms; people who exhibit this are ruled out. And third is the 16Pf test which measures among other things the preference in people to work with objects rather than with others; these are believed to be more suitable than most for disposal work. Other qualities looked for in the testing are the ability to carry out delicate operations with one's fingers and the ability to distinguish an object from its background. This can be crucial in detecting the odd screw out of place which will signal that a bomb is boobytrapped. It is a difficult area, with no room for mistakes. But in general the tests are thought to have been successful in ruling out people who are not quiet, careful or stealthy (4).

THE VIEW FROM THE GODS: AERIAL RECONNAISSANCE

Reconnaissance nowadays usually involves photographic (or radar) reconnaissance from low-flight helicopters, high-flying aircraft or space satellites. In practice the photographs are by no means as clear as the military would wish them to be. Added to that is the difficulty of having to view even familiar objects from unfamiliar positions (that is, from overhead). By 1974, no fewer than thirty-four *series* of studies had looked into this problem, masterminded by psychologists from the US Army's Behavioral Science Research Laboratory (BESRL) and carried out in conjunction with scientists from various other organizations (5).

I can do no more than give the main results of this work. Specialist readers who are interested are referred to BESRL's Technical Research Report 1160, *Summary of BESRL Surveillance Research*, which contains the most complete account of what has been achieved so far (6).

There are four forms of aerial reconnaissance: one is high-level flight (45 000 feet or so) with drone aircraft; and three using helicopters: tree-top flight at top speed; tree-top flight with the occasional 'pop-up' to, say 200 feet, to get a better view; and tree-top flight with the occasional stop and dismount on high ground. The three methods using helicopters were compared in the 1960s in a field exercise in Bavaria, Southern Germany. The performance of a platoon of reconnaissance vehicles was also looked at (7).

More than thirty runs were made by the helicopters in each type of flight. Each area recced contained two target complexes – two stationary armoured vehicles, and a row of five on the move with two or three security vehicles travelling ahead. It was found that the helicopters acquired 60 per cent of the available targets regardless of tactics but that, allowing for the number of rounds fired against the helicopter from a range when it could have been hit, the low-dismount flight was best. Furthermore, in this method the helicopter was heard before being seen only 23 per cent of the time. The low-dismount tactic was slower – it took thirty-five minutes to do what was done in twenty minutes by the low/pop-up tactic and ten minutes in the top-speed flight tactic – but the minimized danger was considered to offset this time disadvantage. The vehicle platoon took much longer but was seen only 3 per cent of the time. Since the stationary targets fired three times as many rounds at the reconnaissance patrols as did the moving ones it was concluded that helicopters are most suitable against moving targets but that ground patrols should be considered where targets are likely to be stationary and time is not important.

The ambiguities which arise from high-altitude reconnaissance photographs pose different problems. One is that there is so much material – so much in fact that it threatens to overwhelm the image interpreters in time of war. (Image interpreters are the backroom boys whose job it is to sift through the photos looking for military and other intelligence. In the US Army they are trained at the intelligence school at Fort Holabird in Maryland. In most of the experiments which are described below, the subjects were all recent graduates of the Fort Holabird image-interpretation course.) One of the first experiments undertaken by BESRL was to see how the scanning of the photographs could be speeded up. Two methods were developed from experimentation (8). One involves going over the photos twice. First, very rapidly indeed, rating their probability as to how much valuable military intelligence they are thought to contain; then those photos that achieve a certain probability are gone over again in more detail. The second method is to scan the photographs systematically, say in geometrical patterns. Both these methods improve the number of targets detected; however, they also give rise to the false identification of targets – something which can be very costly in military terms. A second series of experiments has therefore explored the development of 'keys' – drawings or photographs of both correct targets and those commonly mistaken for them – in order to prevent errors. It has been found that if drawings *and* photographs are given to the image interpreter, with both oblique and plan views, he is much more likely to spot them in a reconnaissance photograph than if he has only one kind (see Figure 16). Table 9 shows the proportion of errors made by image interpreters scanning photographs brought back from two reconnaissance missions in Vietnam.

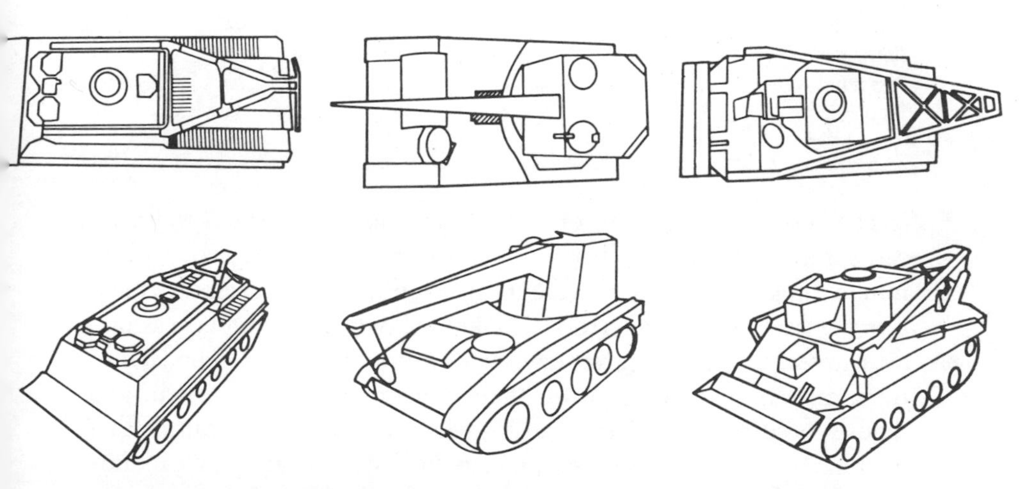

Figure 16 Examples of schematic large-scale vertical and oblique views of targets used together for aerial reconnaissance scanning.

Table 9 Percentage errors in scanning aerial reconnaissance photographs

Target type	*Mission 5536*	*Mission 6358*
Weapons position	42	29
Vehicles	16	6
Sampans	13	29
Supply points	10	19
Personnel	10	2
Tunnel entrances	4	1
Mine fields	0	10
Bunkers	6	6
Other objects producing false alarms		
Graves	34	29
Craters	22	3
Bush and trees (ground)	20	−6
Brush and flotsam (rivers)	0	25
Straw	12	26
Wells	12	1

The proportion of omissions which the image interpreters made is shown in Table 10.

Table 10 Percentage omission by mission and target type

Targets omitted	*Mission 5536*	*Mission 6358*
Weapons positions	85	93
Sampans	88	92
Personnel	99	100
Trenches	88	85
Road blocks	(none to be seen)	67

The production and use of error keys, showing graves, wells, ox carts, mud puddles and other features commonly mistaken for targets, resulted in a 39 per cent increase in the number of right responses and a 26 per cent decrease in errors. The keys also drew attention to associated features which suggest they are not military targets – the lack of track marks surrounding them, their non-strategic emplacement and regular occurrence, and the paths which skirt them, indicating that they have been there for some time. A final way to attempt to eliminate errors is to attach 'costs' to them and give trainees an error score which, if exceeded, means that they have failed to qualify. But this is not a simple scale: tanks and missiles are important omissions, supply facilities the least serious regardless of tactical position, and minefields are more important in offensive situations, less so when a unit is on the defensive. Team interpretation is also considerably better than individuals working alone.

The US Army's surveillance aircraft can now be fitted with special reconnaissance equipment – side-looking airborne radar (SLAR) – which the pilot can view in flight. The problem with this, however, is just how the pilot or co-pilot can detect targets in the relatively brief time that a particular stretch of ground is on the screen (about half an inch of imagery is produced each minute at a scale of 1:500 000). Exciting though the equipment may be to some, when BESRL began its research only 18·5 per cent of available targets were actually being spotted, making the device virtually useless. According to the latest documentation, BESRL's training techniques have improved this somewhat but the accurate location of targets is still not good enough for operational purposes (reckoned in the US Army to be to within 100 metres if artillery fire is to be directed onto a target) (9). This appears to be a good example of equipment outstripping the ability of soldiers to use it.

THE BIONIC PILOT

Given the cost of modern military aircraft, and the wide variety of tactical activities which are more suited to them than any other military machine, the pilot now rates as one of the most valuable and highly skilled fighters in war. All countries lay great emphasis on the selection and training of their pilots, not only because of this crucial role but also because it is extremely expensive if pilots leave the service before they have fulfilled several years of duty.

Selection research has mainly looked at the way a wide variety of psychological tests are linked to pilot performance and to whether he stays in the service. It is a narrow field of research but there have been many studies. Suffice to say that a large number of results show that the good pilot is not the anxious type, is high on psychomotor adaption (i.e. uses his arms and legs well), has a good mechanical sense, is introverted but can, when he wants, get on well with other people. By and large men identified on these personality tests are less likely than most to suffer from motion sickness.

Not all good pilots are good in combat, however: and the links between a pilot's ability to withstand stress and his physiological response patterns is now being investigated. Only now has modern technology enabled the various psychophysiological instruments to become small enough to fit into cockpits without discomfort during combat, which is why advances have not been made sooner. Professor Lionel Haward, of the University of Surrey, is one of the foremost researchers in this area. He has looked particularly at ejection seats and the psychology involved in preparing for this hazardous task: as it is likely to be the most stressful part of flying it should be a good measure of pilots' responses (10). He has also used a very interesting research technique – for, besides studying men in mock-up ejector seats (in which they just shoot up a tower), he has also hypnotized a number of Royal Navy pilots and instructed them to image that they were in the process of ejecting. Previous research has shown that this can be a good way of simulating the psychophysiological reactions which actually occur on real, stressfull occasions. From his research Dr Haward has been able to show that good pilots have stable patterns of physiological response whereas poor pilots do not. He has even been able to predict which pilots would 'freeze' when the time came to eject: it actually happened with some men under hypnosis.

Since 1966 Soviet work with pilots has developed techniques to improve combat performance. These include such devices as mnemonics to remember flight zones, simple diagrams on how to direct their attention round the cockpit and so-called Pavlovian techniques for handling fear. These include 'inner dialogue' – that is, the pilot talks to himself, telling himself that he *can* complete his mission, and avoids unpleasant thoughts by deliberately concentrating on memories of successul engagements, intercepts and so on (11).

Neither hypnosis nor inner dialogue, however, even begins to compare with the latest pilot research at the Advanced Projects Research Agency in Washington. The ARPA breakthrough is, literally, to link weapons directly to brains. A computer, linked to the human brain by wires and electrodes planted beneath the skull, has been programmed to recognize the pattern of electrical activity in the cortex when a pilot merely thinks of a particular word. The ultimate plan is to have a small computer fitted into the fighter plane of the future, just behind the flight deck. As the pilot dons his helmet in the cockpit, fine needles will project into his scalp (though he will not be able to feel them). The needles will be connected by wires to the computer. As he approaches his target he will merely have to think 'fire!' and the computer will recognize this thought and relay a triggering instruction to the plane's rockets or bombs under the wings. In the case of rockets it need not end there. As he watches the rocket zoom towards the target area, the pilot will eventually be able to direct it simply by thinking 'left', 'left again,' 'down a bit,' 'dive!' The computer can stay in radio touch with the rocket so that the pilot's instructions get through right until the fatal impact (12).[1]

The reason for developing the mechanism is mainly because, as the Russian pilots have found, there are so many things to do these days on the flight deck of a modern combat aircraft: the two movements needed in guiding the plane, especially at low altitude, *and* a rocket may be antagonistic. Non-fighter planes – for example, big jet troop carriers (the US Army now has Boeing 747s as flying garrisons) – now have so many dials on the flight console that some accidents have been attributed to pilots missing a crucial reading. With a brain-reading computer, the centre of the console would contain a single television screen. The pilot would merely need to think 'altimeter' or 'air speed indicator' or 'fuel gauge' and the computer, via his helmet, would recognize the instruction and the appropriate dial would appear on the screen.

NAVIGATION AND NAP-OF-THE-EARTH FLIGHT

The navigational problems facing the modern pilot arise from the need for an immediate military response over any type of terrain and in any area of the world; because of the enemy's expected dispersion in depth, military aircraft may be required to penetrate a significant distance into enemy-dominated areas at low level; helicopters – in general most suitable for this work (e.g. OH-6, AH-1G, AH-56A) – are getting faster so that speeds of up to 250 knots are eventually envisaged.

In low-level or 'nap-of-the-earth' flight a pilot flies low following the contours (natural and man-made) of the land. He pre-plans a broad corridor, based on known features, which has an axis pointing towards his target. Then, within this corridor, he weaves this way and that,

1. It will also be easier from now on to load the missiles. The US Navy has developed an exoskeleton – the 'Handiman' – which may actually be worn as a 'mechanical garment'. This device, using a principle known as 'force feedback', links into the pressure exerted in someone's muscles and dramatically amplifies the wearer's strength by a factor of twenty-five to one. That is, when the wearer lifts 25 lbs it 'feels' as though he is lifting 1lb. Weights of up to 1000 lbs are possible with the exoskeleton and, fitted to a real man, make it ideal for cargo handling, salvage or rescue operations. A real bionic soldier (13).

taking advantage of tree lines, dams, etc. The idea is to take the enemy by surprise (hopefully to get under his radar). He cannot see you coming and the angular velocity over a given defensive position is such that it is much harder for him to hit you even when he does catch sight of you. It is, therefore, a valuable technique. But it is also a very hazardous one. It completely changes many navigational skills. Several units in Vietnam, for example, reported difficulty in achieving and maintaining the required proficiency in navigation which nap-of-the-earth flight demands.

As is usual in these situations, HumRRO was called in. The US Aviation Human Research Unit at Fort Rucker wanted to see what changes in equipment and training procedure were necessary to improve low-level navigation. The major part of the research, declassified in 1971, was tried and tested and modified in the light of US Army experience in South East Asia. The work-unit set up by HumRRO to explore the problem was code-named 'Lowentry' (14).

There are four basic problems involved in nap-of-the-earth flight:

(a) From a low-flying aircraft the appearance of the ground is altered. Visibility is limited to the horizon profile which in vegetated areas may be 50 metres or less. Objects – natural and man-made – appear in oblique relief form and not in planimetric form (i.e. as they would appear on a map, as if seen from directly above). This creates problems of coordinating map symbols with objects seen in relief. Low flight, of course, leads to rapid angular velocity, one result of which is that flat features – rivers, roads and so on – flash past at very fast apparent speeds. In addition, the relationships between features are much more difficult to observe: distances between them must, therefore, be inferred from the time taken to get to them. One answer to this is to fly at ground speeds that are in multiples of 60 so that mental calculations, based on so many miles a minute, become easy (15).

(b) Probably the most important single problem is that low-level flight reduces the amount of time the pilot can devote to navigation. At higher altitudes the aircraft can usually be trimmed for hands-off flying and a full minute or so can be spent looking at the map and calculating position, bearing or course. Studies for the Canadian Defense Research Board by R. E. F. Lewis show how much this state of affairs is changed in low-level flight (16). Lewis's studies were carried out with helicopter pilots flying at minimum clearance of 25 feet above obstacles. He found that 27 per cent of the flying time was spent in

Table 11 Duration and percentage of glances at map during low-level flight

Duration (in seconds)	*Percentage of glances*
0·5	5
0·5–1	30
1–1·5	30
1·5–2	20
2–2·5	10
2·5–3	2
3–3·5	1
3·5–4	2

looking at the map but the duration of the glances is shown in Table 11. In other words, four seconds was the maximum time the pilots were willing to look inside the cockpit and in practice their glances seldom exceeded 2·5 seconds. Navigational aids must therefore deal with this sort of problem. But in any case Lewis concluded that aircraft flying in this way 'will sooner or later hit wires, trees and birds'. His conclusion was supported by the fact that in twenty-five low-level test hours there were six potential wire-strike incidents, five of which were only averted by warning from the investigator–observer who had familiarized himself with such obstructions on the routes.

One attempt to get round the general problem was developed by the short-lived 'Tiger' training course at Fort Sill, Oklahoma. There, a 'look–fly' procedure was adopted in which the pilot stopped his helicopter, studied his map for the next flight segment and flew on from memory (17).

(c) *Maps and charts.* The average distance over which military aviation missions are conducted is increasing so much that under present conditions manouevring over 2000 or more kilometres in three to four days is a possibility in some types of tactical warfare. This has significant implications for the pilot's map requirements, not least among them being the interrelationship between psychological factors and logistics. Put simply this means that he may not be able to stow the bulk of maps required at the special scales needed for low-level flight. The perceptual problems of maps, which cause this bulk, are more crucial now than ever before.

There are three ways in which maps can be modernized to meet the new requirements: by changing map scale; by new encodement format; and by using photographs as maps.

Map scale. Current military scales are 1:25 000; 1:50 000 and 1:100 000 for local flights plus 1:250 000 for en route flight. (For high-level flights 1:500 000 is adequate.) On the other hand, one study of Marine helicopter pilots indicated the highest preference for map scale was 1:150 000, or about two nautical miles per inch. This map scale is non-existent in some armies (e.g. the US) but HumRRO concluded that it may in fact represent an optimum combination for providing maximum information with minimum handling problems. (To give some idea of the problems of bulk, the scale 1:100 000, which is regarded as just about feasible, needs forty sheets, weighing 7 lbs, to cover an area of 200 × 150 nautical miles, say roughly the size of Scotland.) (18)

In a separate study the effects of map scale on position location were investigated (19). It was found that error is related to map scale – that is, the larger the scale the smaller the error. On the basis of this it was concluded that a scale of 1:100 000 was adequate, in that average error would be about 200 metres and a 400 per cent drop in bulk would be achieved over the 1:50 000 scale standard tactical map. This study also confirmed that intersections of roads and streams were the most favoured features for pilots to fix their position by and that since these features are carried on 1:250 000 scale maps they are suitable for en route use, though not for approach.

Encodement format. The encodement format of maps – that is the symbols for churches, bridges and so on – is based on the assumption that the time available for map study and interpretation is unlimited. Yet investigations have shown that with the standard tactical map it usually takes *a minute or more* to obtain from it a mental conception of the terrain configuration. This is far too long for the nap-of-the-earth pilot. At low levels, flat surfaces simply cannot be seen. Lakes, rivers, some roads and so on are therefore of limited usefulness in low-level navigation.

Also, at night these features are even less used. In a study at Fort Benning helicopter pilots/navigators were asked the features they made most use of in navigation (20). Table 12 shows the way they adapt their behaviour at night.

Even when features can be seen, it is often for too short a time for navigation to make use of them. The low-level map therefore needs to make more use of objects, natural or man-made, with a considerable vertical dimension. Combining these two findings, HumRRO came

Table 12 How night-time affects navigation

Most helpful features	*By day*	*By night*
Roads, railways	42	39
Bridges	20	7
Lakes, dams	7	0
Large buildings, towers	0	9
Contours	4	2

up with maps which show objects as little drawings in relief: the symbol on the map actually resembles the appearance of a feature from the aircraft. It is, however, difficult to represent, say, a tall building on a map as it would appear from all directions. A compromise was reached and now nap-of-the-earth maps, at least, show a drawing of a church-like building for churches for example (21).

Pilots themselves have well-developed preferences for certain types of feature to be used as map references. For instance, they prefer pictorial representation of cultural and drainage features (roads, railways, rivers, buildings) rather than of relief or land cover (trees, mountains, valleys, etc.). But they also think that it is important to have everything in relation to other features, i.e. a prominent church must be represented on the map as part of a village, not just as standing on its own (22).

Shadow maps – in which the shadows cast by natural features when the sun is due south, say – were thought at one time to be a useful and quick way of helping the pilot assess his direction. However, they are now thought to be highly dangerous for low-level flight because, among other things, valleys appear as peaks and peaks as valleys.

Photographs as maps. One method that has promise is the use of aerial photographs as maps, especially those that have been touched up by a cartographer. The correspondence of photo to ground feature is much easier to make than between map and feature and experiments are under way to see if corridors of photos help navigation (23). Overhead photos do not seem to pose quite the same translation problems as do maps: the human brain seems able to understand what an object – seen from above in the photo – will look like from the side much better than the features on a conventional map. Experiments at the American Institutes for Research, with 600 lieutenants, found that a photograph, with roads, rivers, railways and so forth, touched up in two colours produced very efficient maps indeed (24).

(d) There are, in addition to these perceptual difficulties, certain psychophysiological factors associated with low-level flight. Vertigo is a serious hazard at low levels since it is more likely to result in an accident. The greater need to concentrate on the actual flying at low levels probably cuts vertigo in comparison with high-level flight but the greater stress associated no doubt offsets this advantage. The greater degree of turbulence at very low levels can also make some manipulations impossible. It has been estimated that on 5 per cent to 20 per cent of missions, plotting on maps, manipulating hand calculators and writing may not be possible (25). But the most serious psychophysiological effect of low-level flight is the way it accelerates the rate of fatigue. The danger and complexity of low-level flight can lead to fairly simple, primitive reactions: freezing, fleeing or fighting. There is evidence that at higher speeds delays or failures in responding to enemy actions may be more common in low-level than in high-level flight (26). A survey by Brant Clark and colleagues, at the US Naval School of Aviation Medicine, gave the name of 'fascination' to a series of pilot reactions in low-level flight. In such a state the pilot may become so involved in a large number of tasks that he neglects some critical aspects; he may become engrossed in a single aspect and neglect all the rest; he may simply daydream; he may have an unrealistic attitude to a potentially dangerous move or area; he may be attracted, almost magnetically, to a target; he may feel that events are happening very slowly.

Research is currently underway to see whether certain personality factors are related to low-level navigation performance.[2]

By and large, low-level navigation is a hard skill to acquire and the pride that many military pilots take in it is not misplaced. On the other

2. The Personnel Research Branch of the US Army has looked into the general question of whether some people make better navigators than others. Starting from the commonsense view that some people are good at remembering places, a team of four psychologists reviewed the literature and interviewed many civilian psychologists working in the field. Their preliminary finding was that a test of about 200 questions would be needed to separate out the good navigators from the poorer ones, plus a few visual psychological tests such as sorting out the following sentence: 'Eig htee neull stedm on froi mamis fa nytrum it.' Encouraging results were obtained from a small study of officer cadets, but the project appears not to have progressed beyond this point. (27)

hand, many of them are nowhere near as good as they say, or think, they are. The Vietnam experience, above all else, has shown this. At low levels the pilot functions, at most, at only 90 per cent of his performance at high levels and in some this may go as low as 10 per cent. The improvement of performance, however, will come with more psychologically designed equipment, rather than with different training techniques. The basic requirement appears to be to keep the procedures simple.

NIGHT LANDINGS

One of the main areas of night operations research apart from target acquisition and aiming has been the difficulties of landing on aircraft carriers in the dark. There are four times as many accidents at night as by day – one every 2500 landings; so Clyde Brictson and Joseph Wulfeck looked at twenty-one experienced Navy pilots who were qualified to land on aircraft carriers in the F4 Phantom. Using radar they collected landing details both during the day and at night. They found that the chief error was in altitude estimation – at night pilots approached the deck lower than they should have and often had to make last-second adjustments to get on the deck. This resulted in more long landings (catching the fourth cable rather than any of the first three) or missing altogether and having to go round again. So strongly did the research suggest that pilots be given help at night with their altitude readings that the study was repeated with 1000 landings in the Gulf of Tonkin under real battle conditions (day and night strikes against heavily defended targets in North Vietnam). Results were the same and now pilots may be talked in from the carriers with special attention being given to their altitude (28).

CODE-BREAKING

One final special skill which we shall consider briefly is the ability to decode secret messages. This is, of course, one of the most secret of research fields, but it appears that psychological tests to select cryptographers started during the Second World War and are based on the fairly simple notion that many intelligence tests are themselves decoding exercises – spotting the hidden patterns in series of designs or words which are deliberately arranged to mislead. During the war, for example, a cryptographic aptitude test was devised which consisted mainly of items similar to those in the normal intelligence test. The aim was to upgrade the standard of intake at Chanute Field in Illinois, where code-breakers were taught on a course that was, apparently, never short of applicants (29). The other area of research in codes has been in how long they take to break. On the principle that any code can be broken given time, the psychologists have tried to classify the principles of encodement according to the length of time normally taken to decipher a code. The idea of course is that rapidly perishable information need not be coded quite so thoroughly as more important and more durable intelligence.

THE 25 PER CENT PAY-OFF

Many of the studies reported in the last three chapters are very specific and in themselves could never be expected to change the course of a war. Measuring the fractions of a second during which a pilot glances at his map while navigating, the number of inches at which a point man spots the average boobytrap, or knowing that in a jungle the enemy is 14 per cent more visible at 40 feet than at 50, are not the sort of military intelligence that would stop a Wingate or a Westmoreland in his tank tracks. Nevertheless, taking together all the studies covered it is worth pointing out that the *average* improvement in military performance which results is no less than 25 per cent. And since, as we have seen, the research covers many of the combat skills, then to achieve such an improvement through manpower alone would require somewhere in the region of 200 000 more men in the US forces and 80 000 in Britain. Such a calculation should prevent us from underestimating the effectiveness of research into the psychology of combat. No doubt researchers are more prone to publish figures of experiments which work than of those which do not – so 25 per cent is possibly an exaggeration. On the other hand, this figure excludes the effects of programmed instruction on soldier training. In one study, the reduction in training time achieved by the use of programmed instruction averaged 40 per cent (30). Neither does this general figure include the research that has tried to improve the operations of groupings of soldiers – pairs, squads, platoons – which have often been even more effective in boosting combat performance than experiments with individuals. It is to these that we now turn.

5 Soldiers in Groups

Important as it is for armies to make sure that the individuals in the front line are the best there are, in modern war just as in the past soldiers fight in groups – squads, platoons, companies, regiments. The way these groups hang together through the stresses of battle is even more important, most soldiers will tell you, than the quality of the individuals who go to make up the units. The good squad is more than the sum of its parts. The single most important psychological factor affecting the fighting ability of a squad or platoon is without doubt the quality of its leaders. Chapter 7 is given over entirely to the psychological problems involved in leadership. But leadership apart, there are many other aspects of military group life that affect combat ability – from the way close friends are disposed about the line to the way aggressive soldiers relate to more mild-mannered individuals.

Some of the earliest studies of military groups were carried out in the western desert during the Second World War, when it was found that the extraordinary mobility which the open swathes of desert offered affected soldiers quite differently from the more constricted warfare elsewhere in Europe (even the rates of psychiatric breakdown varied) (1). But more systematic study did not begin, as with so much else in military psychology, until the Korean conflict. And by then, the emphasis was not on the larger grouping – company or regiment, say – but the platoon, the rifle squad or the aircrew, groups of men ranging from about half a dozen up to about forty. Since even in larger units men form themselves into groups of friends and acquaintances of only a handful, it is this size that will be our main concern now.

THE KOREAN LABORATORY: EXAMINING SUCCESSFUL SQUADS

The Korean War became quite a psychological laboratory. The Personnel Research Branch of the US Army, the Operations Research

Office of Johns Hopkins University, and the Human Resources Research Office all sent psychologists to Korea to interview soldiers either actually at the front line or immediately they had returned from it, to try to find out what differentiated the units which found the enemy and killed him from those which appeared unsuccessful on both counts.

Several early studies found that effective squads shared a better 'emotional climate' than ineffective ones. This was not, however, a simple matter of the more the men liked each other the better the unit was. For instance, Rodney Clark, a HumRRO psychologist, found that a soldier in the effective squads would choose one colleague to spend the night in a bunker with (a dependable member of their rifle squad) but would choose entirely different people as likeable or helpful – the types he would go on leave with (2). This confirms other studies which show that there are several distinct functions – in terms of human relationships – which have to be successfully carried out if the squad is to be effective. What matters, however, is not that the dependable soldiers are different from the likeable, but that both psychological functions should be present. There are no problems if the man a soldier wants to spend the night in a bunker with is not only dependable but also likeable; the trouble starts when there is *no one* in the platoon whom the soldier feels is likeable. The important point to take in here is that non-combat-related factors *are* important to the effectiveness of a squad on the battle field.

Psychological duties

Five crucial non-combat-related 'psychological duties' were identified by Clark in his Korean research and have been confirmed since. Some of them are normally the prerogative of the senior members of the squad (and are discussed in more detail as aspects of leadership in chapter 7). But all have to be carried out by someone if the squad is to be effective. They are:

Managing the squad: this involves most of the formal aspects of leadership – supervising supplies, channels of communication with HQ, etc.

Defining rules and procedures for acceptable behaviour: this involves mainly verbal behaviour – initiating discussions, encouraging the men to speak up on anything they need, and saying precisely what is expected of each man.

Performing as a model: it is important for senior members to behave in such a way that other squad members want to behave like them.

Teaching: the teacher has to be skilled in some attributes, or know something considered important by his squad mates *and* he has to be able to demonstrate what he knows in such a way that the others understand it; all squads need men in a position to teach the less experienced.

Sustaining with emotional support: almost, in Clark's words, 'a process of therapy'. In Korea when squads were properly functioning, personal differences between men came out into the open without festering for ages, and were resolved. The squad members developed more confidence in one another, were closer and more harmonious. The men who encouraged this behaviour, however, were – according to Clark – more than listeners or sympathizers. 'They seemed to have the knack of recognizing when an individual's desires were in conflict with the welfare of the group and to be able to take the initiative in helping such an individual adjust his conflicting efforts to harmonize with the squad's goals.'

Table 13 shows how often the formal leader of the units failed to meet all five of the requirements of the squads – and how often other squad members performed these functions.

Table 13 Performance of psychological duties in sixty-nine combat squads (2)

	Function performed by:		
Function	*Squad leader*	*Assistant squad leader*	*Other squad member*
Managing	64	37	1
Defining	35	19	14
Modelling	13	8	8
Teaching	14	6	9
Sustaining	11	7	11

Note that on three of the functions the lower ranks contributed as much as or more than assistant leaders and that it was on the less formal roles of defining rules and providing emotional support that the lower ranks were most active. Since these are the areas where the leaders – usually the NCOs – most often fall down, it follows that men possessing these qualities should be inserted into squads performing ineffectively.

Clark's research has also led to an analysis of the way an effective squad develops. To begin with, when the men first come together, teaching is the most effective activity in developing cohesion, the first objective. This implies that good teachers and men with properly defined skills who can act as models are the most important initial members of a squad: the leader should seek to bring them on first. Next, the defining of acceptable behaviour becomes important – this creates loyalty to the squad. Third, the sustaining role becomes paramount: this makes it possible for group goals to emerge which everyone accepts. Then, and only then, can extra modelling and teaching take place. But this is not the teaching of ordinary military skills so much as behaviour designed to inculcate combat aggressiveness into the squad. This comes last of all (3).

Tests for determining successful squads

By the time this kind of analysis was finished the Korean War was over and the men were back in the USA. With the knowledge gained on the front line the peacetime aim became to see whether quick, simple, and above all cheap psychological tests could distinguish in advance successful squads from unsuccessful ones. A whole series of Personnel Research Branch studies were devoted to this and, eventually, came up with a field problem carefully designed to include manoeuvres and actions which show up well the differences between good and poor squads. These differences were then related to a wide number of *group* psychological tests.

The field exercise developed contained reconnaissance, attack and defence elements, in the course of which the squad was attacked by surprise, had to take a prisoner and had to split up into smaller units with the leader able to go with only one of these. Blank ammunition was used to enable umpires, hidden along the course, to rate the performance of the squads and the ratings were compared with the personality tests. Three types of result were obtained (4).

First, it was found that effective squads consisted predominantly of men with a certain psychological make-up. Squads with men who were rated high on paranoia did poorly. So did squads with men who scored high on nervousness (not necessarily the same as behaving nervously). Men who were sociable were found to be good for squad combat effectiveness – the more there were in a squad the better it was. It helped, too, to have some older, conventionally 'masculine' men in the squad. Another factor was psychological homogeneity among the men. For example, where the men were similar in their levels of aspiration (for life in general and in the army in particular) and where they had similar levels of confidence in their physical prowess, the units were better in combat.

Second, the mutual respect which the men had for one another also mattered. Men were given a questionnaire which tapped the extent to which they wanted to share activities with other squad members. These were either non-military activities, garrison activities (not related to combat) and combat duties. The more isolates a squad had – that is, the more men who were not chosen by anyone for shared activities – then the worse it was at fighting. (An additional finding was that the more a leader distributed his choices among his men – that is, the fewer favourites he had – then the better was the squad.) The third factor was garrison behaviour. Men who kept clean quarters, clean weapons, reported promptly for duty, continued to maintain discipline when the leader was absent, also performed above average on the firing line. Possibly this reflected a general motivational factor.

Perhaps the single most important new finding which came out of this series of research was that the effective squad is one which is *psychologically homogeneous*. This meant, therefore, that, by the late fifties, psychological testing of the potential combat soldier was able to do two things. First, to separate out the good fighting individual from the poor one. And second, to put the good one in with other fighters he would get on with, so forging a unit with a proper cutting edge. The tests were not perfect, but, between them, were able to produce roughly a 15 per cent improvement in squad effectiveness – no mean achievement.

SQUAD SIZE, STOOGES AND SHOOTING

Group processes in combat units

Since the end of the 1950s the selection methods for fighting units have been refined and modified in a number of ways. Psychologists have now shifted their attention to other ways in which they can improve the fighting abilities of armies by tampering with the group processes of the small unit, be it rifle squad, bomber crew or special forces dropped behind enemy lines. Partly this has been done through theoretical studies, into what – for example – makes people change when they are in groups: why do some become more aggressive or more resilient (5)? But, more practically, it has been done by looking at the relations between people in groups – especially the communication and coordination of soldiers in fighting units. This has led to some definite changes in the way small combat units are now organized.

Briefly, though, the more theoretical studies first.

At West Point, the US military academy, it has been shown in a number of experiments that people in groups get more worked up than people alone (6). Given the opportunity, they are far more cruel and aggressive. In itself this may not be so surprising – that is why riots happen, after all. But the corollary *is* more surprising: it appears from the experiments that though anger is 'contagious', its relief is not. In other words groups get worked up more quickly than individuals – but calm down more slowly. This is not necessarily a good thing in combat. Combat aggressiveness is generally a cool efficiency rather than 'fighting mad'. The research also showed that when one member of a group sees colleagues fighting a common enemy, the action may calm the fighter's ardour but not that of the man who is witnessing the fighting. In other words the men who, through lack of skill or opportunity, have not done their fair share of killing (say) will remain more worked up than those who have. It makes sense, therefore, for squad leaders to make sure everyone gets a fair crack at the whip if internal dissension is to be kept to a minimum.

Identification with a group has specific effects. Arnold Buss has shown that men will actually stand more physical pain (electric shock) when they feel they are members of a tightly knit group than when they are alone (7). Competition actually increases pain tolerance. Clearly in combat, with its noise and other threats, these results are relevant. Buss also found that group identification can be heightened by stressing competition between groups that are different – share different customs, for example. It makes sense, on this basis, therefore, for commanders to organize their fighting units so that very different groups are close together for campaign purposes. In Vietnam the presence of Australian troops, with very different military customs to the US troops, injected useful competition that at times was a great help to morale and fighting efficiency.[1]

1. The Australians, it seems, stood out in Vietnam. They had longer hair, looser uniforms, the familiar wide-brimmed hats. They earned substantially less than US GIs, so could rarely frequent the more expensive Vietnamese bars and restaurants used by the Americans. They were thus rather isolationist, the more so since they were a token

Clay George has found that groups need their own codes – their own words for familiar things, their own idosyncratic ways of doing routine procedures (8). Thus, if several ways for doing something or several words can be thought up to describe everyday objects, morale rises, group solidarity coalesces and the unit becomes generally more effective.

Squad size

Military groupings have evolved over the centuries; the conventional groupings of squad, platoon, regiment or division must be fairly stable or they would not have survived. Nevertheless, formal psychological research into warfare is only a recent innovation and it may be that the size of a fighting unit could influence military efficiency: for instance, conformity within a group may be related to size. This is one thing that Harold Gerard has been investigating. He used groups of people ranging from two to eight, and gave his subjects tests of conformity (9). In these tests the subjects were shown a stimulus – say a line – which they had to guess the length of. The experiment was manipulated so that the subject always made his choice after the other 'stooges' – up to seven in number. These stooges progressively made bigger errors so it followed that the larger the error to which a subject agreed the more conforming he was.

Gerard's results showed that although, in a very general way, the larger the group the greater the conformity it exerted, it was also true that the greatest amount of conformity occurred in groups with *five* members. No doubt squads are organized in eights for logistic reasons

---

force and they knew it. Their base camp was spotless in comparison with the Americans'; they were more self-sufficient, doing themselves many of the chores that the Americans employed Vietnamese for. This made their security better. The distinction between the officers and men (inherited from the British Army) was more marked than in the US Army, and this made officer life pleasanter for the Australians.

All this, according to Peter Bourne (10), affected their psychology. They were very self-conscious, critical and competitive. Rumours grew up that, because their units went for days without meeting any VC, the guerrillas were terrified of them and preferred the 'softer' Americans. They also took immense pride in their health record, adhering strictly to the regulations. The number going down with malaria was much less than the Americans. However, the Australians, unlike the Americans, had an uncertain time in Vietnam, and this made them critical of their command. One negative effect was that psychiatric breakdown rates were higher in the Australians than in the Americans, especially among those who, despite the general isolationism, came into frequent contact with the Americans. Numbers were small, however, and the discrepancy not statistically significant.

but it should perhaps be remembered that groups of five will tend to be psychologically the most 'solid' and especially useful therefore in fluid guerrilla situations. David Chester of the US Army's Personnel Research Branch found in Korea that morale was highest when men were trained in groups of four and if these four were kept together at all times (11).

Killing by coordination

Perhaps the most remarkable piece of research on coordination within military groups was that carried out by Clay George when he developed a method that taught men in a rifle squad how to control their fire so as to improve their kill ratio (12). The specific reasons for the research arose out of the difficulties facing a rifle squad in a jungle environment such as you might find in Vietnam or Africa. Here, with powerful rifle fire available, leadership is difficult when the squad is dispersed. In a conventional war, for example, the fire position might be that shown in Figure 17. Here, no real coordination is necessary.

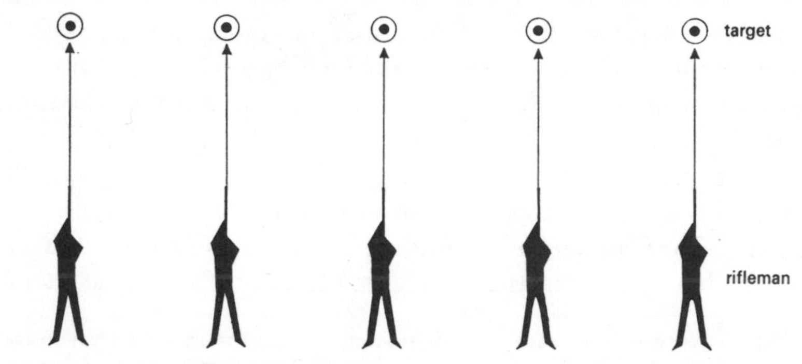

Figure 17 Conventional firing pattern.

On the other hand, with modern weapons one remaining man in an enemy squad will probably possess enough fire power to still kill all the remaining men in the opposing squad even if they outnumber him by four to one, provided he has a tactical advantage – say, of cover. George therefore devised the following procedure to train men to cope with this type of situation.

Men in squads of four or five had to fire at fleeting pop-up targets in front of them but, once they hit their target, that target would not reappear until all targets had been hit – and the man could not score any more hits of his own during this time. At this point, therefore, the man who had just hit his own target had to turn his attention to those of his fellow squad members. George found that there was wide variation in the men he tested in their readiness to coordinate their fire with their colleagues. Some would fire on the targets when it was not necessary or even dangerous as it wasted ammunition because the target was already 'dead'.

He therefore developed two methods which, he found, could help remedy this. In one, the men were trained only to fire at their own targets and at those immediately adjacent to them. In the other – his most imaginative innovation – George redistributed the ammunition available in the squad. The wildshooters were never put at the end of the line, where they might feel exposed, and they were given less ammunition than the rest of the squad – so that they had to husband it and make sure every shot told. The two arrangements are shown in Figure 18.

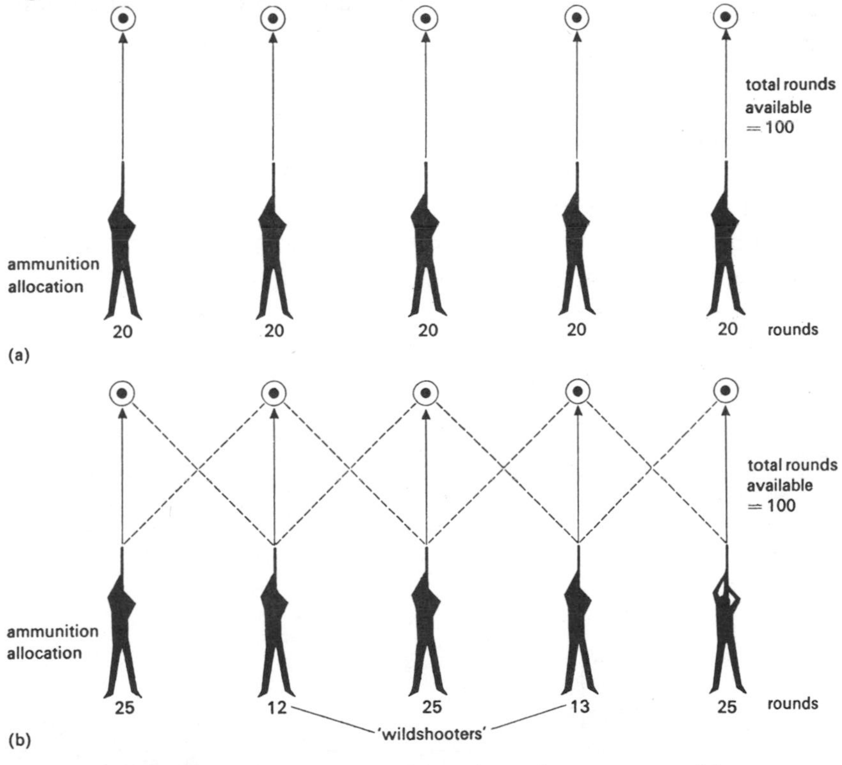

Figure 18 Allocation of ammunition: (a) traditional arrangement; (b) experimental arrangement.

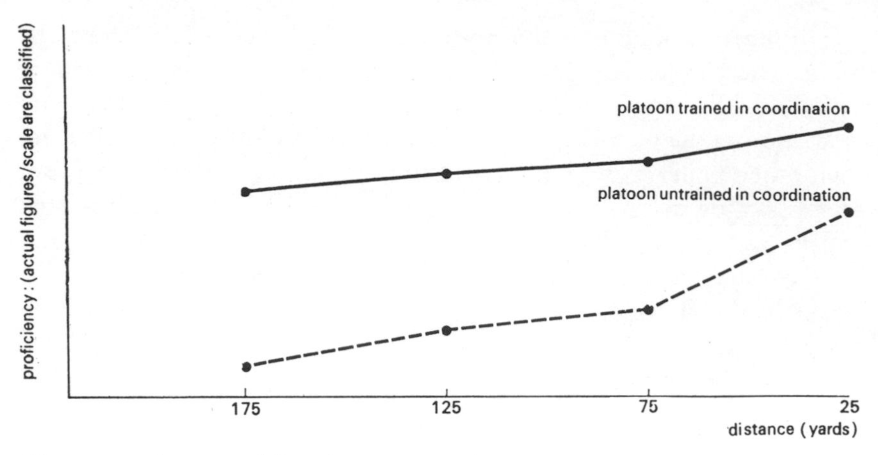

Figure 19 Accuracy of fire of troops trained in coordination compared with untrained troops.

Next, to make training even more realistic, George arranged the pop-up targets so that if they were missed for a certain amount of time at one distance they reappeared *nearer*. Into this situation he put two groups of men. One group was given one-and-a-half day's practice of individual firing, the other group was given the same period of training in coordination. The results are shown in Figure 19. The difference is self-evident.

Another by-product of the coordination training was the psychological effect it had on the platoon members' feeling for one another. Training for coordination clearly helped self-esteem within the squads

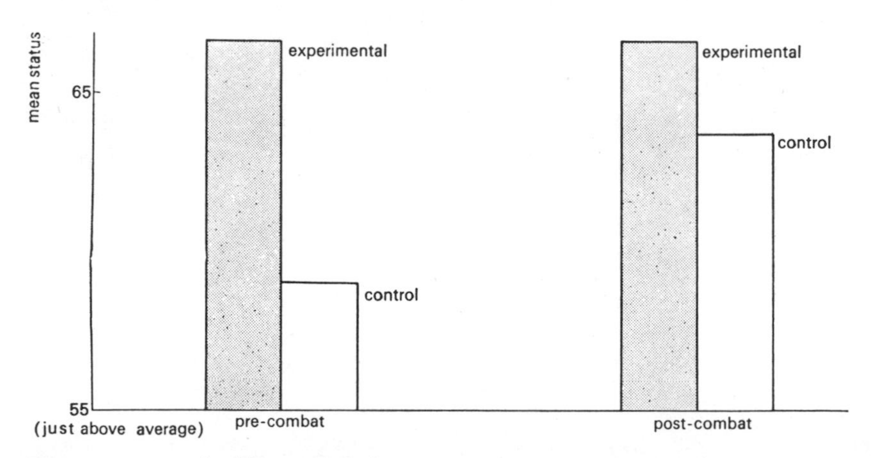

Figure 20 Rated self-esteem of platoons trained and untrained in coordination.

as Figure 20 shows. This was measured by the usual sociometric techniques. However, this effect is a temporary one – or most of it is. After simulated combat, the self-esteem within the control group rises towards that for the platoon trained in coordination. But it can do no harm to have platoons start with this high self-esteem.

George's experiments have grown increasingly sophisticated (13). He has investigated the effect of coordination in emergencies. For this he introduced casualties and weapon failure into his simulated combat. The teams were told to advance until brought under simulated machine-gun fire; then they had to go to ground and provide covering fire while another (simulated) team moved against the target area. The platoons had 840 rounds to use against fleeting targets.

In a preliminary conference the men were instructed in fire distribution and team coordination. Once the firing began, weapons malfunctioned according to a previously arranged plan and casualties also occurred. The platoons were rated on their coordination behaviour – for example, alternating attention between a man's most immediate target and that of someone else's, shouting the whereabouts of fresh targets, noticing when a particular weapon was out of action. Again, no absolute figures are given – being classified – but the relation between coordination and kill ratio is shown in Figure 21.

In his final experiment in this series, George studied eight full-strength squads (14). Four of these, the control squads, were given a conference on the rifle squad in attack, a day's practice with blank ammunition and a morning with live rounds. The other four, the experimental squads, were given the same tuition plus a conference on

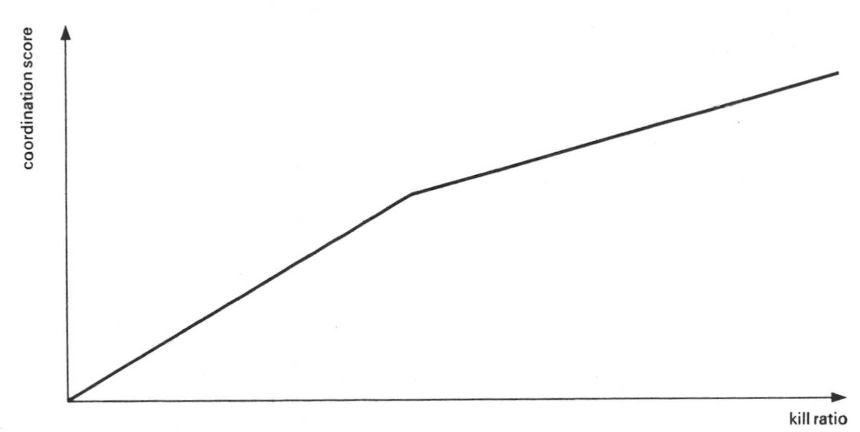

Figure 21 Accuracy of fire in relation to coordination training.

squad coordination, and their firing sessions included emergencies like those described in the previous experiment. The experimental squads were encouraged to criticize their own performance at the end of the training. On the afternoon of the second day the two types of squad were taken to a densely wooded area they had never seen before and which, they were told, contained well-camouflaged targets. Their job was to gain as many effective hits in a set amount of time. Georges' results are presented in Figure 22.

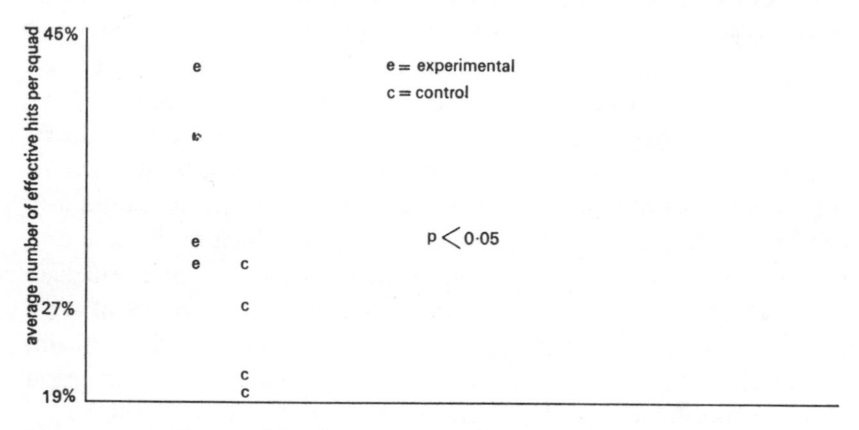

Figure 22 Accuracy of fire for troops trained in coordination and untrained troops in field test.

George's most important conclusion is that, to some extent, training in coordination can replace leader control. Given how important leadership is rated in the armed forces this is a significant departure. But, says George, you must make three things clear to the troops: how effective coordination is – and this can only be done by showing them; the importance of practice; the need for feedback on their performance, including feedback on how they performed in emergencies.

The other psychologist who has concerned herself with this area of research is Adie McRae. She has looked at combat-related skills complementary to those investigated by George – for example, coordination in intelligence-gathering and interpretation, and tactical decision-making. To this end she has been trying to devise tasks which provide experience for combat groups in coordination and, at the same time, has been looking at which interaction is the most effective.

The research aspects of her experiments are more interesting than the training side. For example, she used a maze test – an exercise in which a problem has to be solved which demands the cooperation of all the members of the group simply because no one member has enough information himself to solve the problem (15). She sat each man in a small cubicle within hearing of his three squad mates but out of sight. In front of him were two switches which only he could see or manipulate and a light which everyone could see. Each switch had two positions and when one of the eight switches corresponded to the master switch the light went on. The task was to switch on the light as quickly as possible and with as few mistakes as possible.

McRae found that the squads were very bad at first but quickly learned that coordination helped. What was more interesting, however, was that coordination could be split into at least two types. One type McRae called organizational. These were shouts or calls by the men to organize the squad in some way; they were aimed mainly at the squad in general. The second was a more specific interaction – like specifying a particular action for oneself or someone else, requesting or giving information and so on. McRae found that the more a squad interacted on the organizational level, the quicker they solved the problem – but they did so *without* a drop in mistakes – these continued to be high. With the more specific type of interaction, however, errors went down and time taken did also – the squads solved the problems more quickly and without making as many mistakes. She also noted a tendency for squads to dwell on organizational interaction to begin with. But the effective squads transferred quickly to more specific interactions whereas the ineffective ones did not make this transition.

McRae also found that the intelligence of a soldier, or his status within a group, affected his readiness to interact with his fellow squad members. Very often the reason a squad did not progress from the organizational phase to the specific one was related to personality problems within the team. This method, then, besides giving men practical experience in coordination which improved squad effectiveness, could also be developed as a way of testing whether the balance of a squad needs to be changed before it goes into combat. Further research in this area is proceeding.

L. T. Alexander studied air crews at a coastal location, and found that they could be taught to work together if their jobs were organized in certain ways. The air crews' job is to maintain a round-the-clock surveillance of the off-coast skies of America, on the look-out for hostile aircraft. This surveillance is maintained by radar. Every blip on the screen has to be checked with current flight plans and even identified blips are checked every two minutes to ensure they are on course. All 'unknown' blips trigger the scrambling of interceptor aircraft. A final function of the air defence centre is to 'tell off' blips at the edge of the area to adjacent air defence centres.

During training, crews are deficient both in individual skills and in team-work. It proved easier to rectify individual mistakes than team mistakes, which is why Alexander was asked to tackle the problem. His difficulty was to define the air-defence task in terms of behaviour that could be measured and involved more than just one member of the crew.

Alexander in fact identified four functions in the task: (a) input – that is the detection and correct location of an aircraft; (b) the determination of its course; (c) the correct following of that course and making sure the aircraft maintains it; (d) the appropriate tactical action – telling off or scrambling. By simulation the input could be controlled so that, on all four tasks, performance could be measured. Alexander then took four thirteen-man crews and studied their progress over two months (16). Two groups were debriefed – told of their performance – after each session whereas the other two merely had the opportunity to practise. The results were remarkable. The debriefed crews increased their detection rate by up to 62 per cent whereas the controls stayed virtually the same (and in one or two tasks actually got worse). Initially crews were missing approximately 50 per cent of all aircraft, so this kind of testing is very important. Alexander found that in the debriefing sessions, by merely making the measures of performance available, each specialist began to learn the strengths and weaknesses of *other* specialists, giving more objective analyses of their own performance. This reduced error and facilitated cooperation; people gave and accepted criticism more easily, and *acted* upon it. An important result, probably with applications wider than this one study.

In some training sessions with infantry similar tasks are done both with and without leaders. Though military law is fairly strict on who is in command in any given situation, it is now generally conceded that, in guerrilla wars especially, men may have to operate in 'informal' groupings and they should be trained to recognize what tasks can be done by an individual and which ones cannot.

Finally in this section we can take a look at some naval research – into the reduction of hostility within groups to enhance team performance (17). The idea of this research is to make crews more effective by reducing the jealousies and destructive competitiveness which may drain them of productive energy – without, of course, reducing the healthy competition for promotion and so forth. The basic idea is that effective squads are those where self-esteem among the members is both high and spread among them in sufficient quantity for no one to feel left out. Albert Pepitone has been studying the way competition within a squad should be structured, what rewards and punishments should be given in order to foster competition without it spilling over into counter-productive hostility. Experiments have been carried out in which some squad members have to risk themselves or their self-esteem on behalf of their fellow squad members. An outline of the study has been presented to the relevant military authority but detailed results have not yet been published.

COMMUNICATIONS AND THE TACTICS OF TOUCH

Communication is clearly essential to the successful functioning of a combat squad. Recognizing this, the US Army, at least, has conducted several studies which examine the ways this communication can be manipulated during fighting and the extent to which this makes squads better fighting units.

First, we shall look at the effect of the distribution of squad radios, since this is the single most important technical development, even from a psychological point of view, in recent years; then at the experiments into a new type of communication – through the skin.

Walkie-talkies

With advances in radio technology, walkie-talkies are now so small that it is possible, virtually, to equip every member of the army with one (as is happening in most police forces around the world). The question arises as to whether the increased communication offered by this development is worth it – is the ability to communicate of paramount importance on the front line or does it get in the way of effective fighting? A walkie-talkie may after all divert attention from important sound cues, and mere possession of a walkie-talkie may in itself attract enemy attention to a soldier.

A simulated tactical problem was developed for the purposes of the experiment (at HumRRO), based upon the kinds of tasks that might normally be performed by an infantry rifle squad in the type of decentralized operations common in Vietnam. Three typical combat situations were simulated: (a) a daylight search and destroy operation, ending with an assault; (b) a night raid; (c) a night defence. There was simulated sniper fire, non-walking casualties, and streams, forests and ditches to be negotiated, etc. In each phase special events were imposed so that certain communications became necessary and these served as a basis for measurement as to how effective the squad was at coping with its mission. The squad was assumed to be acting as part of a platoon which, in turn, was acting as part of a company in a decentralized operation (18).

During the experiment, every type of radio variation was tried – from having no one with a radio, or with just the platoon or squad leaders having one, to a situation where everyone had a two-way walkie-talkie.

Six research staff acted as judges in the exercise and the time taken for each tactical development to get back to the squad or platoon commander and for him to act on it was recorded, as was whether or not anyone was shot. The time taken to spot a boobytrap, report this back to the squad leader and for the squad leader to issue a relevant instruction is an example of the type of measure taken in the experiment. The squads were also rated by the observers on several other factors – how quiet they kept, the extent to which they remained in visual contact with the rest of the squad, their skill in following designated procedures, the coordination of movement between fire teams, when moving forward, say, and keeping the various leaders well and promptly informed of all relevant developments. Each of the thirty-two test squads consisted of ten men drawn from the 197th Infantry Brigade, a number of whom had combat experience in Vietnam.

The greatest differences occurred in the extent to which squads took up new trails, reacted to sniper fire, reported back on enemy fire, and made adequate arrangements for night defence. But on *all* factors – spotting boobytraps, defending a landing zone, conducting a successful assault, organizing the evacuation of casualties – the distribution of the radios had a consistent effect on the amount of time these operations took. It was shown, without much doubt, that too many radios in a squad can be harmful to the safety and effectiveness of that unit. The most effective form of radio distribution is to have the platoon commander and the squad commander each with a two-way radio. It makes little difference if the remaining members have radio *receivers* except at night when it can, under certain circumstances, be advantageous for everyone to have one. But to equip all the men with transmitters as well as receivers was catastrophic. Everyone talked to everyone at once, everyone was a commander, people talked more than was usual and more than was necessary. And the squad commander was under a special burden in that he had to carry two instruments – since his communications with his platoon leader had to be on a different frequency to his communications with his men. It is rare for psychological experiments to be so conclusive.

Modern and available as they are, walkie-talkies may not always be the complete answer to field-communication problems in wars of the future. In a heavy artillery battle, for example, it may be impossible for someone to make himself heard; and, probably even more important, in the classic jungle guerrilla fight silence is often crucial. But how can a squad commander communicate with his men when he cannot see them, and cannot use the walkie-talkie for fear of giving away his, and his men's, positions? One answer may be 'tactual communication'.

Tactual communication

Tactual communication was developed by a specialist unit at Fort Benning with the code name 'COMTAC', standing for 'Tactual communication as a medium for increasing control in small-unit operations' (19). The idea is to fit out the men with electrodes attached to their bodies and through which the commander can give them coded instructions by means of small electric shocks – a sort of morse code – to tell them what to do. Work in this direction is in fact fairly advanced.

The first problem was to find out what was the best form in which to give the shock – so that the soldier could feel it but not feel pain. Early results showed that you had to steer clear of the more hairy parts of the body – so legs and forearms were excluded. An anodal pulse of 0·5 milliseconds proved to be the best configuration. The ideal place to have the electrodes is somewhere on the body where there is a high ratio between pain and touch thresholds and these have been found to be, in decreasing order of effectiveness: upper arm, abdomen, chest, back.

Next the number, arrangement and spacing of the electrodes were considered. For two electrodes it did not matter on which area of the body they were placed, but as more electrodes were used it became clear that the abdomen was the best place, in terms of accuracy of recognition of messages being coded. As to spacing, the maximum accuracy was obtained when the electrodes were approximately two and a half inches apart. Further research showed that soldiers could more accurately distinguish patterns of pulses in arrangements that went from right to left rather than up or down and that not more than two or three electrodes should be used, otherwise the men get the codes confused.

The electrodes have now been tested in all sorts of field conditions. For example, the effects of noise on recognition accuracy has been studied, as has the effect of sweating, which is likely to happen when the soldier is frightened. Other studies have examined the effects of body position and extreme cold. In one or two cases these special conditions have affected pattern recognition marginally, but it is now generally accepted that, under most conditions, the properly trained man can recognize at least 95 per cent of the coded messages sent him in this way.

SPECIALIST CREWS: GUNS, SONARS AND BLOCKADES

We shall end this chapter with a look at specialist combat groups whose role centres around some sophisticated and expensive piece of equipment – a helicopter, say, or a submarine – and where the men have to work in close coordination under stress. To illustrate the problems faced, and the methods of study used to find solutions, we shall look at three types of squad or crew: the anti-aircraft battery, the anti-submarine helicopter, and the patrol gunboat.

Defending New York: anti-aircraft batteries

Every artillery man knows that there is a difference between the potential killing power of a radar gun system and the system's actual performance when men are operating it. The vital point is the size of that difference. In the early spring of 1953 HumRRO research officers discussed with military staff how a project might be initiated to improve the anti-aircraft batteries' abilities, then giving cause for concern (20). As a result, six measures of critical activity in an AA battery were isolated: the range of radar pick-ups; the firing range scores; radar maintenance; artillery maintenance; the rating of the commander; and 'adverse personnel actions'. 'Strike' missions were flown against these batteries at 15 000 feet and the batteries had to follow them, distinguish them from radar clutter or noise, estimate altitude, start range, azimuth, and forward the information so that the gunner could 'fire'. The trackings of the crew were later compared with the actual course of the plane and the same was done with the actual angle of the gun when it was 'fired'. Maintenance behaviour was measured in a fairly straightforward way. Adverse personnel actions were based on such things as court martials, the battery punishment book and so forth. The AA units used were those actually in place to defend New York, San Francisco and Seattle. They had the M33 radar and were fitted with either 90 mm or 120 mm guns.

The most important finding of this early experiment was that range of radar pick-up is *not* related to firing score. In other words, the crews which accurately spotted the planes a long way away were not necessarily the ones which shot them down. Early identification was related to adequate maintenance of the radar but not to either adverse personnel actions or artillery maintenance.

Most important of all, perhaps, firing accuracy did not appear to relate to *any* of the measurable characteristics in the experiment. As a result, HumRRO next tried a rather more sophisticated approach to this bit of the problem (21). Tom Myers and Francis Palmer looked first at thirty AA crews at Fort Ord in California. At that time the US Air Force was itself developing a crew dimensions description questionnaire (CDDQ) which had found some thirteen dimensions of crew behaviour related to general effectiveness. Myers and Palmer used just four of these, which they thought would be suitable for AA crews: harmony, intimacy, procedural clarity (the extent to which each man's job was spelled out) and stratification hierarchy. Independently of this exercise, information was collected on the ability of each crew to pick up targets accurately on their radar screens. Two variables were found to be related to crew effectiveness. First, where the leader was seen as separate from the rest of the men the crews did better; and second, crew intimacy was related to effectiveness. However, this was a curvilinear relationship, crews with a moderate degree of intimacy performed best at radar spotting. Those who were very intimate with one another or not intimate at all performed much less well.

The intimacy finding, a surprising one at first sight, was later confirmed in a further experiment. Myers and Palmer next looked at sociometric ratings and group productivity among the same radar crews. Each crew consists of between eight and thirteen people – the crew itself being part of a larger unit of about 100 men. Crews consist of three status types – the range platoon sergeant, the radar mechanic, and the chief radar operator, plus the subordinate whose main job was to operate the equipment. As in the previous experiment the success of the various batteries in picking up aircraft and locking onto them was the criterion. One hundred and four test missions were run in the experiment.

Interpersonal relationships were measured by the simple sociometric device of asking each crew member which three persons he would most prefer to go 'on a pass' with. The most important finding of this part of the experiment, and that which fitted closely with the last finding, was that the greater the tendency for a crew to choose itself as members to go on a pass with, the *less* effective it was as a crew in spotting aircraft. This was true for all crew members but especially so for high ranking men. If they chose members of their own outfits then their crews were particularly bad at aircraft spotting (22).

At this point it is perhaps worth emphasizing the size of the differences which human factors can produce in this field. In these current experiments, for example, the top third of the batteries had an average range of spotting aircraft that was 30 000 yards (roughly 20 miles) greater than the poorest third – 40 000 yards as opposed to 10 000. Personal qualities are therefore very important. The final stage of the experiment we have just been describing was therefore to isolate those personal qualities related to successful batteries.

Leaders – the range platoon sergeant, radar mechanic and chief radar operator – in radar crews that performed well were found to differ from leaders in poorer crews in several ways:

Personality: when crew leaders were more assertive in social situations, or placed higher value on perseverance and efficiency as goals, crew performance benefited. Other personality dimensions, however, like enthusiasm, preference for dealing with people rather than objects, emotional maturity and authoritarianism, did not discriminate leaders of good and poor crews.

Intelligence: leaders of good crews were more intelligent than leaders of poor crews. However, the average intelligence of subordinate members of good crews was no higher than for their equivalents in poor crews.

For non-ranking members of these crews, some aspects of personality also mattered: the tendency for non-ranking members of good crews to value perseverance and efficiency as goals in themselves differentiated them from their equals in poor crews even more thoroughly than it did with their leaders. They were also likely to show greater independence of action than subordinate members in poor crews. Many other personality variables did not differentiate between good and poor crews.

Another finding of interest, however, was that members of good crews were more often chosen as off-duty companions by other battery personnel than were members of poor crews. Myers and Palmer conclude that 'batteries whose crews perform best are cohesive at the battery level (of 100 men or so) rather than at the crew level (of a dozen or so). One explanatory hypothesis would be that when crews associate more with members of the larger unit, a greater understanding may be gained of the roles played by the several sections in the battery.'

Sinking subs by betting on bombs: helicopter crews

The US Navy have employed similar techniques in the study of what makes an effective helicopter crew when it comes to spotting, and dropping bombs on, enemy submarines under water. Experienced flight teams were used in the experiment – twelve crews from Naval Air Squadrons 3 and 7 stationed at Norfolk, Virginia (there are eleven such squadrons in the US Navy). Each crew consisted of three men – a pilot, navigator/co-pilot and sonar operator (23).

The crews had to fly missions, track a sub and try to put it out of action. The time each crew took to do this and the accuracy of its bombings were recorded. This criterion was compared with the interactions of the crew members throughout their mission. The kinds of interactions that were looked at were: the number of messages transmitted by each crew member; messages indicating a willingness to accept risks; corroborated messages; interpolative or extrapolative messages; intuitive messages; repeated messages; and so on. A typical tactical situation is shown in Figure 23 overleaf.

The naval investigators found that four types of communication were associated with 'miss distance' – that is, with accuracy in bombing (and therefore, by implication, with actually finding the enemy submarine in the first place). First, the interaction of the successful helicopter crew is characterized by what is termed a 'probability structure'. This is jargon for crews who were willing to talk in terms of probabilities – to speculate beyond the information they had in an interpretative, extra-

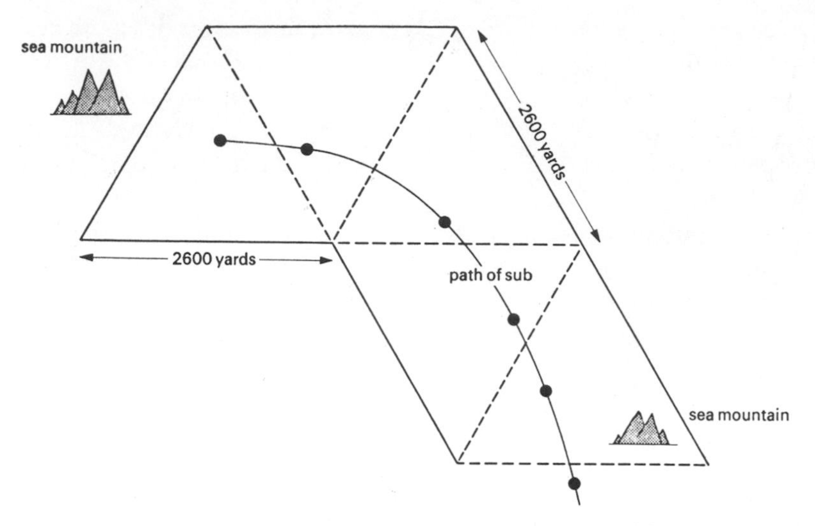

Figure 23 Sample tactical problem: the pilot is told to patrol between the solid lines, proceeding always in triangular directions (dotted lines). He does not know the route of the sub, nor is he aware of the 'sea mountains' which he must distinguish from an enemy vessel.

polative way rather than in an undisciplined fashion. This really means they were more intelligent. They also confined their statements to the task in hand and the phenomena they were dealing with; they were much less likely to give their *opinions* about what was or might be happening. When they talked about what might happen it was on the basis of the evidence of their instruments, not on the basis of something inside their head.

The second category was called 'evaluative interchange'. Again this is jargon for the tendency of better crews to provide one another with relevant information. In other words, they would not speak unless spoken to or unless their instruments told them there was something the other crew members should know. They also explained things more fully to one another when the need to communicate arose. They did not indulge in a lot of unrequested information.

The third factor was called 'hypothesis formulation'. Good helicopter crews are not risk-takers but they *do* show a flexibility of behaviour in the light of their past experience on that mission. They weigh alternatives, have a problem-solving attitude to the task in hand.

Finally, the fourth factor was called 'leadership control'. This means that the communications of the good crews showed an awareness of the hierarchy within the crew and that they were more disciplined in their use of communications.

The study also confirmed that those crews in which the communications are more evenly shared were also better at eliminating subs. Figure 24 shows how uneven in general this message initiation was; a direct consequence of this, the naval researchers suggest, is to train co-pilots to take more part in initiating messages. This they feel will lead to better helicopter performance in general.

Figure 24 Average percentage of mission-oriented communications transmitted by crew members.

Two practical consequences of this research are that (a) the communication patterns of poor crews is now automatically monitored and they are instructed in the areas where they are deficient in order to improve their performance; (b) various techniques are being tried out to get the crews to change their behaviour. For example, if poor crews are bad because they take unwarranted risks, then during training they now have to 'bet' when they make a probabilistic inference. Winning or losing the bet is proving a useful and effective way of teaching pilots or navigators what is good evidence to base an inference on and what is not.

Patrol gunboat

The US Navy's other experiment with specialist crews has used as a 'laboratory' the Ashville class of patrol gunboat (24). This boat is designed to patrol, blockade, and to carry out surveillance and perimeter security as well as support missions. The Navy was anxious to know what conditions on board ship brought on fatigue and hampered military proficiency. During any prolonged mission, some men will fall ill, or be injured, and this will place a strain on the rest of the men. What is the best method of coping with this likely eventuality? Two ways suggested themselves: to train men in more than one speciality so that, if someone should fall ill, the other men could take over their jobs as well; or to carry one or two extra hands on a ship to take up any slack once action is seen.

In the experiment the boat's task was a twelve-day mission between several islands near South Vietnam during which some time was spent ashore, while at other times the boat was at sea for twenty-four hours or more. For the purposes of the experiment on different missions the ship was fitted out either with its normal complement of men or else had a few more or less than usual.

During the twelve days on the mission the men had to carry out their normal tasks – take their turn on watch, shoot the stars and navigate, carry out gunnery drill, prepare for landing and so on – but in addition to that they had to repair equipment (rafts, depth sounders, etc.) and, finally, adapt to three types of emergency. These were the arrest of an unauthorized fishing boat, to go to the aid of an ailing ship and to adapt to the sighting of an enemy vessel.

The findings were straightforward but important. By and large, the amount of sickness had no effect on work efficiency until most of the other seamen had to work a longer day as a result. As this approached two hours more than usual then efficiency did drop – by about 10 per cent. Under these conditions fatigue was found to be most likely to show among the top ranks and among the ratings at the bottom. The NCOs, on the other hand, stood up well to this type of stress. It was also found that the officers who could maintain a decent interpersonal distance between themselves and others were the best able to see that the increased work load was being shared. The closer the officer to his men the less well did they adjust to the extra stress. Carrying a slightly larger crew did no harm, provided the crew was efficient; when it was not efficient, the arrangement was unsatisfactory.

The study concluded that an intermediate size crew – as in the experiment – needs to have about 10 per cent slack. Without too much enemy action, performance does not suffer adversely. This is now a useful military planning doctrine.

The Navy Personnel Research and Development Center also looked at 'organization climate' in cruiser-destroyers and amphibious vehicles and its effects on military competence and discipline. Psychologists Kent Crawford and Edmund Thomas found that it was possible to construct, for each of forty-one ships, a human resource management index (HRMI). This index consisted of scores for the way the ship was commanded (how open and frequent communications were, for example, or how hard the commanders tried to motivate their men), how much support was offered to the men by lower level leaders, how much team work was encouraged, how much work in the ship was coordinated, and so on. The important finding was that both 'mission effectiveness' (on simulated combat manoeuvres) and the rate of petty offences were both related to the HRMI – something which now provides an effective check for many administrators (25).

6 Allocating and Inducting Men

CATCH 23: HOW THE INTELLIGENT AVOID COMBAT

The specialist nature of much of modern combat and the central importance of interpersonal relations in small squads mean not only that the right man must be in the right job, but that *replacing* that man is crucial: the replacement could upset the balance in a crew or squad. This is particularly relevant to combat units because, as many on the general staff of western armies have commented, the relatively high standards which have been set for school training in technical courses mean that 'relatively poor personnel have been allocated to the combat arms'.

The main reasons for the imbalance, when it arises, appear to be two-fold. One is rarely talked about – the tendency for any sensible man to avoid allocation to a combat unit if he can. Other things being equal, it is perhaps reasonable to suppose that the more talented individuals will use these talents to avoid such allocation.[1] The other reason is that with the growth of technology in warfare, many jobs in, say, radar or artillery need men with high IQs and aptitudes. As a result the quality of men going to front-line combat units may, and usually does (if it is not checked), suffer.

This neglects, of course, the fact that the job of the combat soldier has also become far more technical. But the US Army at least has from time to time grown worried enough to study how it happens, how much below par its infantry units are as a result and how the imbalance can be rectified.

The Personnel Research Branch of the Adjutant General's Office

1. The evidence on draft-dodgers supports this. They tend to be above average in intelligence.

surveyed 9500 enlisted men from six armoured and infantry divisions in the USA and in Europe (1). Each man, at some time in his military career, had been tested on several aptitude tests and these were now compared. It is generally accepted that the minimum acceptable score on any of these tests (where the average is 100) is 90. The PRB therefore worked out the percentage of men in any job who had scored below 90 on a particular aptitude – a rough measure of how many men in any type of unit could be regarded as ineffective. Table 14 reveals just how badly the combat units were doing on this criterion.

Table 14 Percentage of men with aptitude scores below 90 (1)

Aptitude area	*Total army*	*Combat unit*
Combat A	26	29
Combat B	24	26
Electronic	57	61
General maintenance	28	25
Motor maintenance	23	28
General technical	24	30
Radio code	67	72

Thus, in all aptitude areas except one – general maintenance – the combat units have a higher proportion of ineffective soldiers than the rest of the army. Although the proportions are only of the order 3–6 per cent, in absolute numbers this is a very large difference indeed.

Further analysis of these results revealed that combat soldiers were worse *at combat* than other specialists were at their own specialities. For example, 30 per cent of the combat units scored below 90 on the combat aptitude tests whereas only 22 per cent of electronics maintenance personnel scored below 90 on the electronics aptitude test. Only ten per cent of the radio code operators scored below 90 on the radio code aptitude test.

The research, however, was able to provide a way out of this critical situation. First the PRB decided to test whether the officers and enlisted men did think that combat duties were important. They asked the men in the survey to rank all the military occupational specialities in order of importance and these ranks were then compared with how selective those specialities were; a highly selective speciality was one where the average score of the men doing it was high. The results are shown in Table 15. It is clear from this that soldiers do see combat as most

Table 15 Ranking of occupational areas by importance ratings of field grade officers and by selectivity in assignment of enlisted men to the occupational areas*

Rank order	Occupational area	*Importance as Rated by Field Grade Officers in* Combat branches	Support branches	Selectivity in assignment
1	Combat (CO)	CO	CO	EL
2	Electronics (EL)	EL	EL	EM
3	Motor maintenance (MM)	MM	MM	GT
4	Precision maintenance (PM)	PM	GT	GA
5	Military crafts (MC)	MC	CL	CL
6	General technical (GT)	GT	MC	MM
7	Electrical maintenance (EM)	EM	PM	PM
8	Clerical (CL)	CL	EM	CO
9	Graphic arts (GA)	GA	GA	MC

*This table is based on small samples and is illustrative only. Solid lines show perfect agreement between branches. Broken lines show relation of importance rankings with perfect agreement to ranking by degree of selectivity in assignment.

important but it is equally clear how poor the US Army was in selecting men for combat at that time (see right hand column). To rectify this state of affairs PRB next carried out a survey comparing what the army required in the various specialities with what was

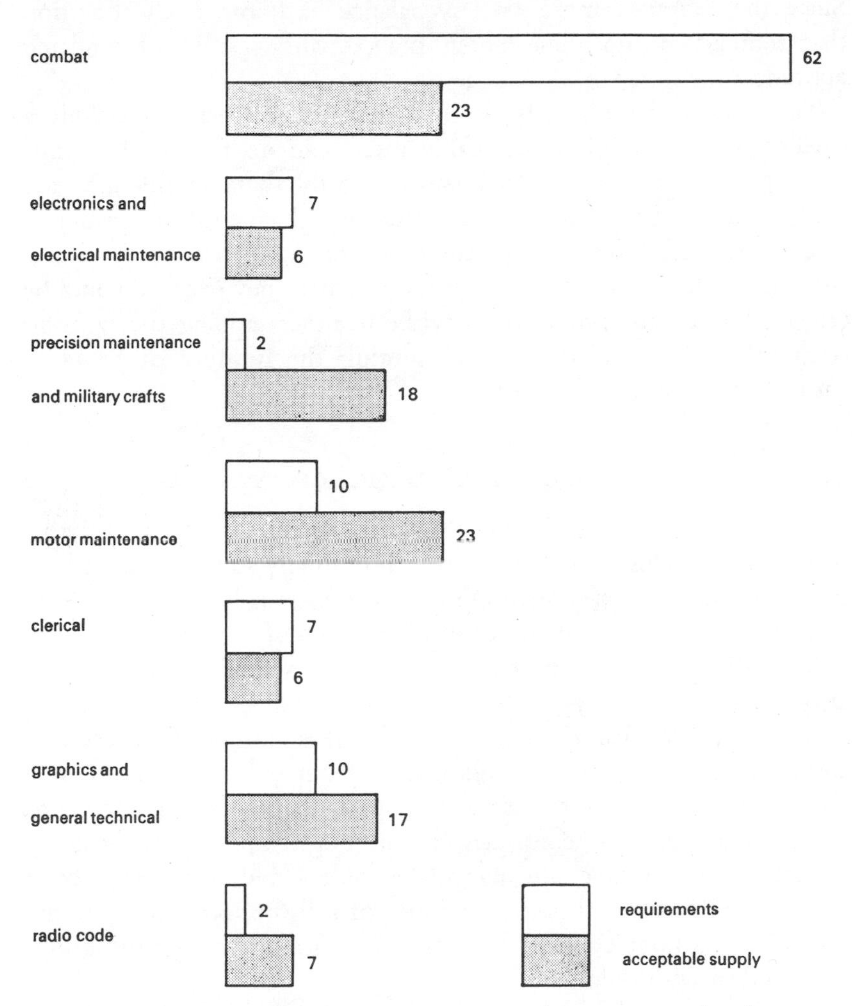

Figure 25 Percentage requirements on occupational areas compared with percentage acceptable supply.

available (in terms of men who reached an acceptable level of expertise in each speciality). Figure 25 immediately suggests what might be done to rectify the army's manpower problems.

It is clear from Figure 25 that there is a shortage of men qualified in combat and a surplus of men qualified in precision maintenance, military crafts, motor maintenance, graphics and general technical and radio code aptitude areas. The shortfall on requirements for combat personnel is 39 per cent and the surplus of others totals 41 per cent. Since these figures more or less balance each other, the question that remains is: do these other men qualify on combat-related aptitudes?

The answer, quite simply, is that these other specialists include a smaller proportion of unacceptable men than do the combat units themselves, in combat terms. In other words, these surplus men are every bit as good at combat, if not better, than the men already assigned to combat units. Rather neatly, therefore, this study was able to show that the surplus men in the specialities in which they excelled could be safely allocated to combat units without either sapping the fighting strength of the combat units or diluting the number of properly qualified non-combat specialists.

THE 'MENTALLY MARGINAL' AND THE DIRTY DOZENS

Attention has now shifted from this problem of combat allocation to other related fields, most importantly how many sub-standard men a unit can tolerate before its fighting effectiveness is hampered.

The concept of limited service developed out of necessity during the Second World War. Men were inducted who would be useful to the army even though they were limited in the kinds of work they could do. (Over five million men had to be rejected as unsuitable for service – almost a third of 18 million applicants.) The term 'limited service' was abandoned in 1943, but the concept of special handling and acceptance of 'marginal' personnel continues, for, although in peacetime the army can afford not to have any marginal people – and selects men on a more or less individual basis – at times of full mobilization marginal men will turn up in considerable numbers and the military must know how to deal with them.

There are various categories of such so-called marginal men. There are the physically marginal; the mentally marginal of one sort or another; recent immigrants who cannot speak the official language; members of allied forces working with the army who cannot speak that language; the emotionally unstable who under conditions of full mobilization, may be unable to adjust to military life; conscientious objectors; and the 'morally marginal' – men with criminal records.

Table 16 shows the percentages of men in various categories in the US Army in the Second World War.

Table 16 Percentage of marginal men in US Army in the Second World War

Illiterate	81·5
Non-English-speaking	12·5
Poor education	18·7
Physically handicapped	6·8
Personality disorders	2·3

Training approaches to the marginal soldier can be grouped into two: development of basic skills – usually educational – on the rationale that this is what a soldier most needs if he is to adapt; and selection and training for a specific set of jobs.

In the Second World War the army coped in three ways (2). At special training units the ratio of instructor to pupils was kept deliberately low; screening tests were developed; and special methods of teaching military skills were invented. For example, one test used was administered in pantomime form for non-English-speakers to see whether they had the ability to learn military duties; a sort of comic book was developed to get across, for example, the system of pay; particular jobs were also explained in this way, as was the general war situation. Language was taught by means of bingo. Drill, guarding and marksmanship were taught almost wholly by demonstration rather than by lectures. Special attention was paid to cooperation with other men in military tasks to give the marginal man a feeling of what it was like to be in the unit before he actually got there.

Since the Second World War things have become more organized. There have been several projects specifically aimed at finding ways of developing marginal manpower. And in addition, there has been a greater tendency to treat the separate groups differently.

During the early post-war period the American Army established a special experimental unit to look into testing and training. Men with an army general classification test (AGCT) score of less than 70 were able to benefit from instruction in the use of hand-tools and simple power tools; further formal education was found to be inappropriate. They also needed a course in elementary psychology to explain the motivating influences on behaviour, why people behaved as they did; it was the men with limited mental ability who showed most intolerance in communal living so necessary in army life. Through this and other training (for example, coordination exercises), many of the trainees managed to raise their AGCT scores to 80 or above and it was concluded that about 50 per cent of this type of personnel might be salvaged as 'partially effective soldiers' (3).

In 1949 the Conservation of Human Resources Department at Columbia University studied the work performance of marginal groups. The research examined military practice in relation to the uneducated and concluded, broadly, that the army policy of not training too many marginal men, because it was expensive, was correct until that time. Projections were made forward to 1952 (thus taking in the Korean War) and it was recommended that research be conducted to determine the minimum intelligence needed to absorb military training. This research concluded that the military over-evaluated formal educational background and argued that there was no positive correlation between education and fighting ability (a conclusion later disproved by better research). The study also concluded that basic jobs were those most suitable for marginal men (4).

The Conservation of Human Resources Project also looked at 'the soldier who was considered ineffective because of emotional disturbance, inaptitude or behavioural disorders'. These, plus the uneducated, were equivalent to fifty-five lost divisions in the Second World War. The study showed that, during that war, many men had been discharged as unfit as a result of bureaucracy. Because men regarded as seriously disturbed were discharged, many others, only mildly upset, were also made to leave the service. At the worst time of the war, in this context that is, the army had to induct 100 men to acquire *five*. The standards for entry were not those for discharge. Eventually this policy was sorted out and in March 1945, just before the war ended, an operational definition of psychoneurosis was developed. It proved too late to test its effectiveness then but examination of the statistical records of personnel by the Columbia team revealed that men discharged for emotional reasons came mainly from two groups. The inept were almost a total loss to the army; they remained in the USA, served for short periods and hardly ever made it above private. The psychoneurotic, on the other hand, were a different kettle of fish. Almost a half were in the army for eighteen months, 60 per cent went overseas, and 22 per cent become NCOs of one sort or another. In fact, 'psychoneurotics' were 50 per cent more likely to be promoted to NCO than 'normal' privates.

The Columbia study was also able to identify seventy-nine men who had broken down yet recovered fully, thus confounding the conventional wisdom of the time. This enabled the team to conclude that the following procedures be adopted: there should be more precise concentration on the combat veteran, closer integration of personnel and medical services, and greater indoctrination of leaders in the individual differences and capacities of soldiers.

In 1953 at Fort Leonard Wood, a systematic study into the special training of mentally handicapped men in *combat skills* was begun. In general, the men in these studies had scores of 70 or less in the standard military entrance classification tests. They were given some basic academic training and then began a specially devised project including not merely military justice, map reading, adaptation to group living, rifle marksmanship, but also more specialized things like spotting mines and boobytraps, chemical, biological, radiological (CBR) warfare, range estimation, and first aid. The men's scores were compared with a control group of average ability soldiers going through similar courses at that time. The results were universally disappointing in that the men improved very little and remained below average on all things. Furthermore, it made no difference whether their training was primarily academically oriented or whether it was more practical. Neither did the special tuition motivate them to take a greater interest (5).

In a similar study at the Air Force Personnel and Training Research Center at San Antonio in Texas, 1000 marginal men were given special training devised to improve their attitudes to air force life, their adjustment to it (to uniforms, to security, to military law), to flight discipline, survival under CBR attack, to marksmanship, camouflage and so on. Here again, however, it was found that although the training made the marginal men happier in themselves, it did not appreciably improve their proficiency in military tasks. Naval studies during this period also showed similar disappointing results (6).

Due to fewer peace-time obligations, the US Army was able to raise its minimum qualifications, which it did, to 90 on the Armed Forces Qualifying Test, in July 1958. This reduced the burden of special training but small-scale studies were continued on the basis that marginal men could be used 'to some advantage by the army during emergencies'. A longitudinal study of certain groups of men was begun in 1959 and followed up eighteen months later, towards the end of 1961 (7). In this study, men who enlisted prior to the raising of the entry standards and who were mentally marginal were compared with average ability soldiers. Their colleagues were asked to rate whether their performance on their particular job was acceptable or not – with the following results:

Table 17 Ratings by soldiers of acceptable performance in colleagues (7)

Speciality	*Percentage acceptable*	
	Marginal	*Average*
Infantry	50	60
Armour, field artillery, air defence	49	65
Military crafts	43	78
Medical care, military police	49	78
Combat	50	63

Whether you think these figures encouraging or otherwise, what we seem to be left with is that (a) marginal soldiers are between a half and two thirds as good as the average soldier at his military duties, and (b) extra tuition is not much good. It seems then that the way to make use of marginal people in the military is not to spend vast sums on training programmes, but instead to plan on the fact that marginals are really only suitable in emergencies; and then you need almost twice as many marginals to do any particular job. A harsh verdict, no doubt, but – in the defence context – probably realistic.

Since these studies we have seen the introduction of computer-aided instruction (CAI) and learning laboratories in all walks of life. These allow individual instruction where the student can pace his own progress. There is some evidence – from outside the military – that CAI is especially beneficial for the mentally handicapped. But the feeling in most services, in the past few years, seems to have been that given the poor research results before, the new methods are not likely to prove much better, especially as many armies have been run down relative to their numbers in the sixties. It is fair to say, then, that at the moment there is not much interest in integrating any more marginal personnel into the world's armies.

Inducting criminals

Before leaving the marginal soldier, however, let us briefly consider one remarkable form of marginality – the man who according to some armies at least is 'morally marginal'. To you and me, he is criminal.

Basic policies on these men tend to be simple: they are kept out of the services except in emergencies. The Second World War provides a good example of what can happen: as the war worsened the authorities relaxed the admission criteria for men with records of misdemeanours until, in August 1944, everyone in the prison population except those convicted of murder, rape, kidnapping and treason was eligible.

Despite this, many were still rejected by the units they were allocated to, especially for neuropsychiatric reasons. But how well did those inducted from prison perform? One study was instituted by the Illinois Division of Corrections in 1950 but never fully completed (8). This showed that of 1300 men paroled to military service (in 1943–4), 5·2 per cent violated this parole compared with 22·6 per cent of a comparable 2000 men paroled to civilian life. Eighty-seven per cent of all those paroled to the military received honourable discharges. Over half the group inducted received battle stars. Of the 30 per cent who were in combat, 98 per cent received honourable discharges, one third received the Purple Heart and the percentage killed in action was one and a half times greater than for the army as a whole. A similar study in New York found much the same but also showed that 40 per cent of those going from jail to barracks were promoted (9).

The question also received attention, in 1951, from the Working Group on Human Behavior under Conditions of Military Service, a project of the US Secretary of Defense. The working group took the view that the prison population should, at all times, be regarded as a source of manpower in emergencies and it went some way towards estimating the strength of that pool (10). Its estimation, on the 1940 and 1950 census, was that 200 000 men were in prisons and similar institutions and that 57 per cent of these were in the eighteen- to thirty-four-year-old age bracket and so eligible. It also noted that the health of the prison population was equal, or superior, to that of other comparable groups and that its intelligence compared favourably too. The working group thus recommended that psychological screening techniques be prepared for emergencies so that criminals could be rapidly but efficiently inducted. It recommended that these be based on civilian parole techniques and that the nature of the offence, behaviour in prison and length of record and imprisonment also be taken into account.

In September 1956 research was begun to identify former delinquents who might prove of value to the military. However, based upon 2209 men enlisted at Fort Leonard Wood in 1953–4, it was found that personnel receiving other than honourable discharges were more likely to have had a criminal record prior to joining up (11). A similar study made at Fort Dix in 1958 reported almost identical results (12). And another study made in 1958 showed that 41 per cent of a sample of military prisoners had been in trouble with the law before they joined up (13). Nowadays, commanding officers are able to waive the embargo against the entry of criminals with minor records – drunks, vagrants, peace disturbers and so forth. Prison longer than a year, though, definitely disbars. A USAPRO study in the early 1960s (14) showed that, in simulated combat, enlisted men with criminal records made poor soldiers.

So, on all sorts of marginal personnel, it is by no means clear whether they are a hidden asset or whether they are, in fact, more trouble than they are worth. One gets the impression that the forces now incline to the latter view though that may have something to do with the fact that, since nuclear or guerrilla war is considered most likely for future conflicts, the role of large armies is a limited one anyway.

RECRUITMENT AND CONSCRIPTION

The ease with which an army can recruit volunteers or introduce conscription depends largely on the public's attitude towards the military in the event of war. Total war tends to unite a nation, but in a limited war such as Korea or Vietnam a proportion of the public may be hostile to government policies and to the military. Conscription, involving coercion of at least some young men who do not want to fight, is an unpopular measure and requires careful public relations.[2] A long-term solution is for the services to attempt in peacetime to keep the volunteer contingent as high as possible. A study by Jerald Backman and Jerome Johnston found that the high-grade equipment that is a feature of air force and navy life means that it is often the army which finds recruitment most difficult (15). For this reason General Yarborough – once in charge of US Special Forces – has argued that the army should get priority in recruitment drives. The survey also showed that the most attractive inducement to recruits is a scholarship to college – a lesson that the British services have learnt well.

2. Several research papers have also considered the feasibility of studying pacifists. In 1965, for example, Clark Abt looked into American attitudes towards the use of nuclear weapons. One of his recommendations was that the number and distribution of pacifists be studied as well as the arguments and techniques most likely to 'mollify' them (16).

Inevitably the advent of conscription will provoke draft-dodging. There seem to be three types of people who attempt to resist conscription: the genuinely psychiatrically ill; those who are afraid of war and try to be certified as mentally unfit; and those who disagree morally with the war – the conscientious objectors.

Ira Frank, of the Neuropsychiatric Institute at the University of California, reported a seven-fold increase in the number of referrals, which coincided with the issuing of draft notices in 1968. He studied eleven out of twenty-three patients and found that all were anxious or depressed and had either threatened suicide or murder. Their chief motivation in coming to see him was to be excused military service. None claimed to be conscientious objectors (17). He diagnosed nine as schizoid or sociopathic. After treatment only two were accepted by the army; the rest were turned away.

Ralph Ollendorf distinguishes two types of draft-dodger (18): the disaffected, whose objections are mainly moral but who come to the psychiatrist for 'draft counselling' – in other words, a clever way out; and the genuinely mentally ill, whom he describes as schizoid or compulsive individuals. Many of the latter had 'lived in a kind of conscription panic' since they were fourteen or fifteen. The psychiatrist's duty, Ollendorf implies, is to make sure the genuinely ill do not get drafted and to see the disaffected are not excused on medical grounds. He points out that the army is often not clear about its own criteria: 'For several months acknowledged potheads, avowed homosexuals, severely neurotic or borderline psychotic and other youths will be inducted – while only weeks later such young men will be rejected.'

Robert Lieberman similarly reported on the psychiatric evaluation of 147 young men facing the draft who wanted to be exempt (19). Of these 50 per cent had strong political feelings; the rest did not think they could adjust to military life. Twenty-one had been 'consistent hippies' and ninety-six had been psychologically dependent on drugs at some time. He divided this group into two: those with no real underlying instability, but who did not treat kindly of authority; and the drop-outs, whose symptoms were unrelated to the draft.

The army should not obscure the difference between the genuinely mentally ill, for whom conscription provokes a kind of panic, those who morally object to war and those who seek to camouflage their fear with psychiatric diagnosis. The truly ill will be useless in battle anyway; but where subterfuge exists it should be exposed to deter others from similar attempts.

Probably no draft-dodgers have received so much publicity as the young American men who, in the late 1960s, left home, either just before or just after their conscription papers arrived, and went to Canada or Scandinavia to avoid being sent to Vietnam, often publicly burning their draft papers. In 1971 it was estimated that there were about 250 000 military deserters. Of these 60 000 were thought to be in Canada and, according to Carl Kline, who went to interview many of the draft-dodgers, forty a day were arriving in Montreal and Toronto, and twelve a day in Vancouver. Kline studied thirty of these men in some detail (20). All were WASPs and well-educated. All, furthermore, had a history singularly free from mental disturbance. Three quarters of the deserters had fathers who had fought in the Second World War, and two had fathers who had themselves been conscientious objectors in that war. Twenty-one were at university and most had researched into Vietnam before they left and this had led them to appreciate the deep differences within the USA over the war. They all knew that they would fall out with their families. Some had allowed themselves to be drafted at first and got cushy home jobs on health grounds. However, like the others, they came to think that it was more courageous to desert than to stay doing what they were doing. A few tried to fake psychiatric examinations or criminal records but most, says Kline, were too serious about their desertion to do this. Most planned quietly and efficiently for their exile. They saved; they acquired information about where to live in Canada. The planning phase ranged from several weeks to two years. Parents, they reported, often denied the crisis, saying, 'You'll go when the time comes.' But the wives all supported their husbands. They seemed to have adjusted well in Canada; only one out of the thirty wanted to go back to the USA and only those with parental ties wanted to visit. They enjoyed Canada and many said they would even fight for it if it became necessary. They were, at the time Kline saw them, mostly inactive politically, but anticipated more activity later. None said he regretted his decision to go to Canada.

This group is the largest of the three types of draft-dodger; and they seem to be highly intelligent, serious, and – in many ways – quite brave young people.

For the military to keep draft-dodging to a minimum, it should never assume that its moral position is understood at home: this should be stated simply, and re-stated time and again. It should be made clear that only army psychiatrists have the power to exempt men on grounds of mental ill-health (there is evidence that many young Americans felt that had the psychiatric service of the military been better equipped and had a higher status, they would have been less assiduous in seeking out 'cooperative' psychiatrists).

7 Military Leadership and the Skills of Command

WINGATE, WESTMORELAND AND BEYOND

Throughout this chapter, I use the word 'command' rather than 'leadership' in describing many activities carried out by officers. The word leadership is seriously deficient as a description of all the psychological tasks which face an officer. He needs personal skills to make the most of his men: this is leadership. But there are also certain military skills which are either the prerogative of the officer (such as making decisions), or are directed by the officer even though they involve all the men (such as the flying of an aircraft). These tasks are more accurately problems of command rather than of leadership: a man can be good at one without necessarily being good at the other. The distinction is possibly more important in the services because officers are not elected leaders, nor do they always 'earn' their leadership by being better at a particular job than their colleagues, as is the case in many civilian walks of life. They are instead designated as leaders by a system that maintains them in that position. Because of this system it is likely that an officer who is a skilful commander will find it easy to be a good leader. But there are now many jobs in modern armies that require men to be skilful commanders without having to possess the traditional qualities of leadership. This distinction is more important now than ever.

I shall not be discussing the leadership behaviour of great generals such as Wingate, Westmoreland, Dayan or Mao Tse Tung. In the first place, there have been many such books. But secondly, as Brigadier Bidwell has pointed out, these men, though they may have great powers of leadership, are public relations figures as much as real people and are not really the leaders whose qualities and capabilities decide an army's behaviour. 'The soil from which the genuine higher leadership springs,' says Bidwell, 'are the NCOs, the leaders of platoons and companies and second-grade staff officers' (1). It is precisely this level of leader which has been the focus of most research.[1]

DEVELOPMENT OF SELECTION PROCEDURES

Military leadership was first studied extensively by psychologists in the Second World War. This was mainly because, in Britain at least, leaders were needed in much greater numbers than ever before. The Germans and the Americans had, by that time, begun to use psychiatry and psychology in an *ad hoc* way to select officers, but Britain lagged behind. Before the Second World War, public schools had been used as the breeding ground of the officer. By the middle of the war, however, this source could no longer cope with the numbers needed. As a result, different sources had to be tapped and in doing so, new methods were shown to be effective, which led to different induction procedures for officers and the ultimate demise of the public schools' influence on Britain's professional army. In making these advances, the British methods became more systematic and efficient than either the Americans' or the Germans'. It thus makes sense to start with them.

By 1942 not only had the supply of officer material fallen off from the traditional routes, but the failure rate of those reaching officer training units was also giving cause for concern, running at somewhere in the region of 20–50 per cent according to Brigadier Bidwell (2). There was also a high rate of psychiatric breakdown among officers and officer candidates under training. In June 1941 the adjutant general, Sir Ronald Adam, decided that a new system was needed, and that this 'might well include' something along the German lines and make use of personality tests.

Perhaps three fundamental differences marked the methods that were

1. It is for these reasons that I do not discuss in this book the extensive research on the psychology of strategy or the innumerable games of conflict and conflict resolution that have been invented to explore the behaviour of entire nations. There have been many books in this field – see, for example, Andrew Wilson's *War Gaming* (Penguin Books, 1968) or Robin Clarke's *The Science of War and Peace* (Hutchinson, 1971) for introductions. Furthermore, they usually relate to the political aspects of international relations rather than to purely military affairs. They are a natural adjunct to the matters discussed in this chapter – but not part of it.

Some of the unpublished military documents on the psychology of strategy appear to be very similar in approach, if rather less imaginative, than the more accessible books. Interested readers are referred, as an example, to *The Psychology of Strategy* by Lt Col Dexter, US Army War College, Carlisle Barracks, Pennsylvania (3).

then introduced from the earlier techniques. First, the selection procedure was taken out of the hands of the commanding officers who had previously been free to choose officers and did so mainly on the basis of social factors like school and family background. Instead, selection was put in the hands of psychologists and psychiatrists. The psychiatrists were given the main responsibility of looking for negative qualities – for trying to screen out the people who would not make good leaders. The psychologists, on the other hand, were tasked with developing a system which would actually show who *would* make a good leader.

Bidwell believes that at this point the contributions of Major W. R. Bion were invaluable. (Bidwell indeed goes so far as to suggest that Bion's contribution was as valuable to the war effort as the Bailey Bridge or the 25-pounder gun, though it went largely unrecognized (4).) Bion recognized that it was important to test men under stress and he thought that the officer interview itself, because it was highly stressful, could be used to see how well candidates could be expected to perform in battle. His method was to put candidates into a group of half a dozen or so and give them a problem which required cooperation but to which there was no immediately obvious solution. He then watched *how* they performed – not simply whether they got it wrong or right. His classic problem – of men finding their way across a shark-infested river using only an oil drum, a plank, a piece of rope and one or two other things – is by now familiar to many in Britain. It has often been described in advertisements for recruits to the British Army. The merit of this approach, in Bion's view, was that the actual solution did not matter: it was the real-life situation of dealing with the group that was watched, not the artificial one. The selectors looked less to the solution and more at the personalities of the candidates: who emerged as a natural leader, who sat outside, who adopted unrealistic procedures and became inflexible.

The new system was found to work. It produced officers who had not necessarily had the classic public school education but who, nevertheless, did well in the various training courses for officers and, more importantly, did well in the field. (Applicants also liked the new methods; the disappearance of questions about where someone came from, what his father did, caused applications to shoot up by a quarter and for the above-average gradings of officers leaving the training units to jump by 12 per cent (5).)

The method proved indelible in the British Army even though it was chopped and changed after the war. At higher levels of command, it had in some senses been in existence even earlier. At the various defence colleges, the basic approach of group work and selection by competitive performance alongside other candidates has operated since well before 1939.

Since the Second World War, however, the psychological approach to officer affairs has had a high credibility and psychologists have been given more scope in this area than in many others.

THE PSYCHOLOGICAL REQUIREMENTS OF AN OFFICER

One post-war reaction was to see in the war-time results the suggestion that an officer needed certain definitive psychological characteristics and that therefore a certain kind of personality was more suited to leadership than others. In 1950, for example, Jay Otis of the Personnel Research Branch of the US Army, published 'The psychological requirements analysis of company grade officers' (6). This was basically a survey of the available literature at that time, including an analysis of the records of many Second World War men in successful and not so successful leadership positions. Combat interviews from no fewer than fifty divisions were obtained; as well as histories of actions by many other small units. Citations for medals among officers were analysed and, in the end, several clusters of personality traits were identified as characteristic of the good officer. A useful distinction was also made between the personality which an officer needs in the barracks during peacetime and the personality he needs at the front line.

Personality *traits* were also distinguished, in this study, from leadership *behaviour*. This is not so academic as later studies will show. The early analyses, however, produced no very surprising results. They mainly confirmed that good leaders should be aggressive, yet calm, clear thinkers, flexible, not too arrogant, yet able to take speedy decisions. The real surprise in the early studies had been in the difference that had emerged between barracks and combat leadership. Good garrison leaders, for example, were found to be aggressive, as were combat leaders, but were also found to do better if they were sticklers for the rule book, athletic, possessed a passion for detail, had a good physical bearing and personal tact. None of these was found to be relevant for an officer to be effective as a leader in wartime.

Analysis and identification of personality traits, however, proved too general for the accurate selection of leaders. What seemed to matter more were the traits or behaviours more directly relevant to the leadership situations. There then followed a series of systematic studies of the types of situation in which military leadership was thought to show itself.

Early studies comprised an analysis of results from rifle platoons and rifle companies and from what enlisted men *said* that they wanted from their leaders. This produced some clusters of traits which good leaders were said to possess (7). Good combat leaders, for example, were described as possessing:

Courage	*Personal integrity*	*Adaptability*
bravery	sincerity	flexibility
fearlessness	flair	rapidity in action
daring	calmness	speedy decision-making
prowess	modesty	disciplinary ability
gallantry		clarity of thought
guts		
intrepidity		
undaunted courage		
fighting spirit		
aggressive action		

In contrast, garrison leadership was seen to involve four clusters: *attitudes* – being tactful, friendly, cheerful, dependable, and with an understanding of human nature; *drive* – being aggressive and enthusiastic; *mental ability* – being experienced, intelligent and original; *physical ability* – showing athleticism and physical courage.[2]

LEADERSHIP BEHAVIOUR

These results, though obtained in a somewhat rough and ready fashion, were useful not because they pointed to different leaders in war and in peace, but because they showed leaders the different things that were expected of them in different situations. It was also worth knowing, from the army's point of view, how far non-military leadership differed from military leadership. The same study went on to explore this and

2. Possibly in addition to these characteristics, a recent study of Communist generals in Vietnam found that the successful ones, even those who belonged to minority ethnic groups, had been in the revolutionary movement for a long while and had often missed the 'romantic' phase of the early days of violence because they had been in prison. In many cases, the implication appears to be that this precaution merely kept them alive to fight another (and more decisive) day (8).

came up with the conclusion that non-military leadership appeared to need the following six clusters of traits: *knowledge and ability* – analytical ability, common sense, foresight, scholarship and judgement; *attitudes* – being fair, moral and patient; *group relations* – being active and lively, possessing a sense of humour, being popular and sociable; *drive* – being aggressive, energetic, thorough; *physical data* – age, appearance, height; *emotional stability* – mood control, emotionally stable.

It was from these early studies, showing that non-military leadership, garrison leadership and combat leadership needed different talents, that the notion that command might differ from leadership first arose.

Following this line of reasoning, the US Army initiated two types of research programme: one that distinguished between the interpersonal aspects of leadership and the organizational and technical aspects of command; and one that explored the differences in leadership according to rank – the pre-commissioned level, company level, field grade and so on.

Interpersonal aspects

A project coded 'Offtrain' began in the 1950s with the development of a leadership course for junior officers using sound films to depict characteristic leadership problems. These were based on the descriptions of crucial situations collected from army officers and NCOs in combat and non-combat areas. Each film terminated at the point where the leader was faced with making decisions and taking action. Group discussion then followed. This led to a major effort to study the actual behaviour of the junior officer as he interacted with his subordinates in a particular mission. The idea was to explore whether the *behaviour* (and not just something less visible like personality) of an effective leader differs from the behaviour of a poor leader (9).

One early experiment involved the study of forty-two platoon leaders drawn from two infantry regiments at an army post in the USA. These leaders and platoon members were asked a series of questions about leaders' behaviour in certain situations (the questions being phrased so that the platoon member did not have to evaluate the behaviour of his own leader, merely describe it). Six to eight platoon members were questioned for each leader on such things as how that leader assigned tasks or planned a mission; what he did if a job was being done well or badly by a platoon member; how he reacted when a new man came into the platoon; what he did when an unexpected event occurred; and how he responded to suggestions from platoon members. The replies of the men were scored on various categories, fed into a computer and related to whether the leader was effective or not. This was judged through ratings of effectiveness, not only by platoon members but also by the platoon leader's superiors.

From this elaborate procedure five important aspects of the leader's behaviour began to emerge. First, the effective leader takes pains to make clear *what* he wants done, and where failure occurs he is equally careful to point out the reasons for this and elaborate his own views on how performance can be improved. Second, he urges high standards of performance at all times when assigning work; this is associated with the fact that he gives explicit rewards for good performance but does not threaten punishments for weak performance. The third factor is related and concerns the actual use of rewards and punishments once tasks have been completed. For example, the effective leader is one who delivers his punishments in private rather than in public. The standard the leader sets is also important: the good leader does not set standards too high nor, just as importantly, too low. This is how he motivates his men. The fourth important aspect of the leader's behaviour is the way in which he handles disruptive influences. This includes talking to men about their personal problems and paying attention to their physical welfare. Though not directly related to group performance, if left untended, these problems fester and eventually impair military efficiency. Finally, the good leader uses his NCOs well. He is good at getting information from them, questions them and is able to accept good suggestions from them without feeling threatened. He is capable of rejecting poor suggestions without giving offence. He gives them support when their subordinates are being difficult.

As a result of this study which, as can be seen, began to identify the more important factors in military leadership behaviour at this level, the HumRRO unit developed a leader activities questionnaire as a way of assessing, in training, which men behaved like successful leaders. By the late 1960s this had been done (10).

Non-commissioned officers

In line with the early studies on commissioned officers was also a series of studies on NCOs. Members of eighty-one rifle squads on the front line in Korea during the winter of 1951 were interviewed to determine some of the factors related to the effectiveness of the squads (11). The men were asked whom they liked to fight near, go on leave with, and so forth, and various sociometric indices were obtained for group cohesiveness, patterns of acceptance among squad mates and friendships with platoon members outside the squad. These were then related to effectiveness as judged by superiors' ratings and combat successes.

For sixty-nine of the squads, for which detailed descriptions were made, five factors were shown to be important for successful NCO behaviour: managing the squad, defining rules and procedures for appropriate behaviour, performing as a model, teaching squad mates, sustaining squad mates with emotional support. The assistant squad leader, it was found, often carried out management functions, while defining, modelling, teaching and sustaining were done by leaders.

In emergencies or extreme conditions it has been found that NCOs provide emotional support for the squad. Israeli psychologist, Meyer Teichman, was part of a small unit near the Sinai front during the 1973 Yom Kippur War (12). He noted that to begin with what the unit most wanted was information and that at this time the formal leader was very much in command, as he was in charge of communications. Later on, however, other more informal leaders took over – they provided better emotional support for the men. He studied these men for two hours solid on each of seven consecutive days in battle. In a later analysis of these men Teichman showed that whereas information about things like money, food and equipment comprised about 70 per cent of all conversations in the beginning, this sank to about 30 per cent after a week. In contrast, concern with and talk about status, feelings and the giving of mutual support rose in this time from less than 10 per cent of conversations to as much as 64 per cent (13). Some American studies now tend to see this function particularly belonging to NCOs. Indeed, it may be said that this is one of the chief functions of the NCO – the commissioned officer normally being the man most concerned with tactics, communications and so forth. But there are no direct research results on this.

A number of studies of naval chief petty officers have shown that in CPOs a high regard for the navy is a major factor in their effectiveness (14). On the other hand in the navy showing a lot of consideration for the men does not appear to have much effect, certainly on whether or not the men say that they wish to stay in the force. Apparently, the group dynamics aboard ship are somewhat different to those in an infantry platoon. There is also some evidence that junior CPOs tend to pass the buck on many interpersonal issues up to senior petty officers;

this may reflect navy practice or may indicate that petty officers are not well trained in handling men (15). The technical competence of the men in the navy also appears to be a factor relating to their acceptance, more so than in the army.

Later studies with army NCOs suggest that, by and large, men who see their NCOs as similar to themselves make more effective soldiers (16). Richard Snyder studied how orders should be given for maximum effect. He found that the NCOs who know their men rarely give orders they would not or could not obey. He thus recommends that an NCO be chosen for his understanding of others. This can be done, he says, by having the other men in a platoon or squad choose him. Other experiments show that, at this level, authority and responsibility are linked, so that an NCO who cannot hold down a technical or specialist job is likely to have his interpersonal authority in the squad undermined as well (17).

Particular attention has also been paid to the conflicting situation in which the NCO often finds himself – having to take orders from above and pass them on, yet being aware of the way those orders are received by the men.[3] If the NCO does not habitually mediate and 'interpret' the orders, and if the officer does not allow him to do this, tension arises and effectiveness is impaired (see also chapter 9, pp. 208ff.).

On the other hand, recent evidence also suggests that the lower ranks distinguish, when orders are given, between the person who actually gives them the order and where the men see the order originating. Paul Bleda, of the US Army's Research Institute in Arlington, Virginia, found that a unit's general leadership policy – its fairness, its suggestiveness, its willingness to explain – all affected how the men reacted to orders. In other words, the higher levels of command can make things easy or difficult for lower level leaders like NCOs (18).

NCOs are in a difficult position in another way. They are neither an elected leader nor an appointed leader. To be sure, their rank is given from above, after recognition from above. Yet there must be an unwritten understanding from superior officers that only men who are natural leaders can survive in the rank of NCO.

The Israeli Defence Forces have explored the possibility of peer

3. The problem of obedience has been raised in a series of experiments by Professor Stanley Milgram (*Obedience to Authority*, Harper & Row; Tavistock, 1974). He found that, in conditions of relative safety, ordinary people will commit an atrocious act. The situation in battle, however is, as Bidwell says, persuading a man to take the more dangerous of two *already* dangerous courses of action. So Milgram's work may say little about obedience in this context. For its relevance to atrocity research see below, pp. 244ff.

nominations (nominations for promotion by a man's colleagues) as a predictor of advancement in rank in 125 platoons, involving 3897 soldiers (19). These studies showed that soldiers could predict with remarkable accuracy who would get promoted from among themselves. This applied not merely to men being promoted to NCO, but also to NCOs being promoted to officer. The predictions were quite good even when the men had been together in a group for only a short while. Even more remarkable, the peers who judged which men would become NCOs and then officers were actually *better* at it than the officers who also made such predictions. This suggests, with these ranks at least, that the elective method of leader selection may actually be more efficient than the more traditional way.

A textbook for leaders: training developments

Having moved from the study of leaders' personalities to that of leader behaviour, and having isolated that behaviour which in both commissioned officers and NCOs seemed to be related to military effectiveness, the next step was to convert these findings into a form of training (and/or selection) to improve the quality of leadership. One of the most popular methods has been the development of training films either posing leadership problems or showing models of required leadership behaviour. On the basis of the petty officer experiments described earlier, for example, fifteen role-playing situations were developed for training.

In the army, and bearing in mind the basic results of leadership for NCOs (managing the squad, defining the rules, performing as a model, teaching, sustaining emotionally), other tests were developed. These included the command presence test (in which the examinee has to lead a group of men in close order drill through a clearly defined course) and the field leadership test (where the examinee has to lead his group of men through three field problems on a patrol mission). In addition, peer nomination was taken more into account and a military speciality, the LNCO – standing for Leadership NCO – was created for those who devoted all their energies to training NCOs. These changes were all introduced into the army fairly quickly and with apparent success (20).

With the commissioned officer, the aim of the psychologists has been to develop techniques based on research for teaching the trainee officer the important areas where his behaviour will affect his authority as a leader. The idea is to teach him the essentials of appointive leadership. This is done through discussions, films, and tape-recorded skits from actual situations in units. In addition, a textbook is at times now used which presents the basic problems uncovered in the research. The course emphasizes 'the study of the leader's interactions with his men in the accomplishment of assigned tasks, and the effect of his actions both on the motivation and morale of his men and on the unit's ability to perform their missions'. The methods have now been used to train helicopter pilots, warrant officer candidates, the officers of the security agency training centre, quartermasters, the armour school and the WAC school (21).

COMMAND SKILLS AND SIMULATED COMBAT

The more technical aspects of command, however, are equally, if not more, important than the interpersonal aspects of military leadership. When he is not wet-nursing his platoon, the officer commanding a unit has to attend to many things, from actually firing his own weapon, to assessing tactics, communicating with headquarters and so on. The army has been interested to see how the various skills of command break down and how they relate to each other in much the same way as the interpersonal skills of leadership. The US Army Behavior and Systems Laboratory therefore embarked on a study of this which was released in 1971 (22). Four thousand lieutenants entering active service between late 1961 and early 1964 formed the pool from which the guinea pigs for this study, 735 in all, were drawn. They were followed for five years from recruitment through training and into active service either in combat (Vietnam) or as combat-ready troops in Europe, Korea, Alaska or in the continental USA. Hundreds of them were brought back to the US Army's Officer Evaluation Center at Fort McClellan for a simulation exercise. There the officers went through an intensive three-day exercise in simulated combat. Each officer was presented with an array of leadership problems – five administrative, five technical, and five combat. In order to make these plausible, and still maintain the simulation as a stressful emergency, the tasks were integrated into a sequence in which the officer was 'assigned' to a Military Assistance Advisory Group (MAAG) in a 'friendly foreign country'. A highly trained corp of seventeen officers and forty-one enlisted men played the roles of US, allied and enemy personnel in the exercise; and at the same time these men made precise notes and checks on many selected aspects of the officers' behaviour and actions (23).

During this three-day exercise, the experimenters recorded more than 2000 observations and judgements on *each* officer which were rated by the experts. In addition, there were the written products of the officers – organization charts, military orders, requisitions, plans and so on, which could be judged objectively in many cases. The officer was also rated according to whether, for example, he kept his head, the clarity of his orders and so on. All these evaluations were then fed into a computer and statistically analysed.

This analysis showed four important points. First, there are, in the military, two types of command. One is combat command and the other is technical/managerial. But in turn, each of these two types had two components. One of these is a personality-motivational component. In combat, forcefulness in the command of men, personal resourcefulness and persistence in the accomplishment of a mission proved to be related to one another and formed the most important factor; for the technical/managerial side, what mattered was the ability to direct well and again persistence in mission accomplishment. These results broadly support the research on interpersonal skills discussed earlier. The successful combat leader was the one who gave clear directions, showed consideration for his men and was a good example. In technical/managerial leadership, it was found that face-to-face situations were not so common but that where they occurred, the good leader was the one who persevered, communicated clearly and in general made a good impression on the men by his technical mastery (24).

The computer analysis also uncovered a 'cognitive element' (as opposed to the interpersonal element) in the two types of command. In the technical/managerial command this element consisted mainly of technical skills (of understanding vehicles, for example, or enemy weapons, or radiation effects) and in the combat command this related mainly to tactical skills – decisions relating to the deployment of troops, combat zone communications and so on. We shall return to this tactical side of affairs in just a moment as it marks the other major research interest of the military psychologists over the last ten to fifteen years. For the moment, however, the general picture of the skills involved in military command – the factors required – are shown in Figure 26.

The next move was to see to what extent army selection methods took account of these distinctions between men. The Differential

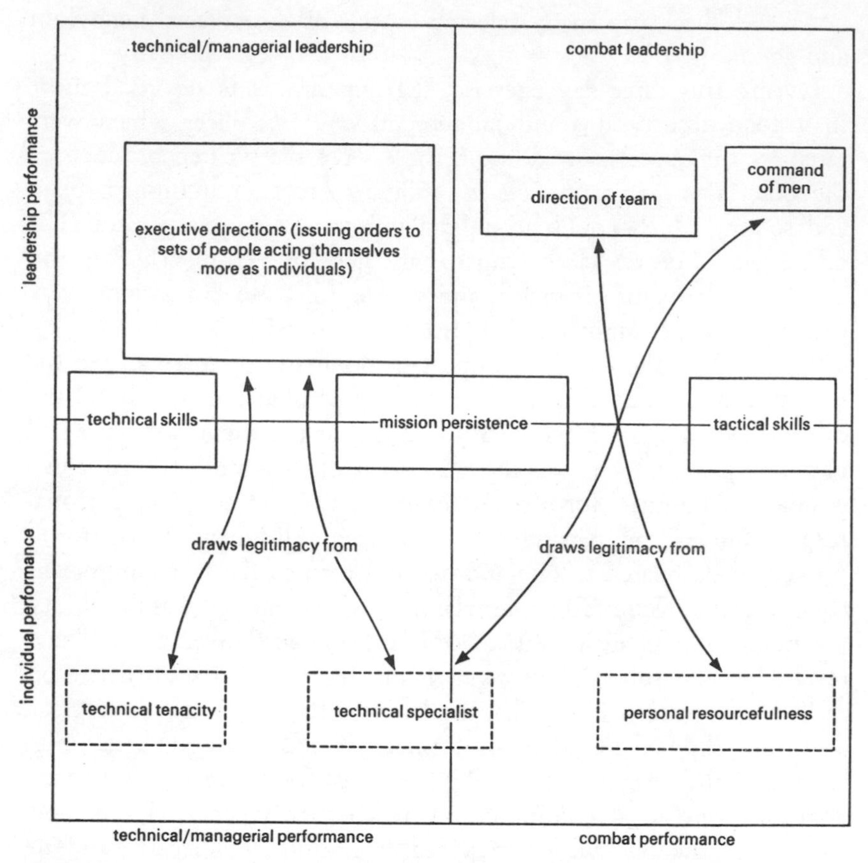

Figure 26 The factors in leadership and command in simulated combat. The figure shows most importantly how the leadership/command role of the office depends in part at least on personal qualities as well as the situation.

Officer Battery (DOB) is a standard selection test for US Army officers which tests a number of background variables, interests, attitudes and knowledge, among them outdoor activities, 'manual' versus 'white-collar' interests, economic-sociological knowledge, political interest, interest in details and so forth. The same team which carried out the simulation exercise related performance of the men who did the exercise to their scores on the DOB (remember that they had all been chased through from their commissioning). This showed, as can be seen in Figure 27, that when used properly the DOB can predict, at a reasonable level, the various characteristics measured in the simulated combat exercise, and that therefore a better placement of

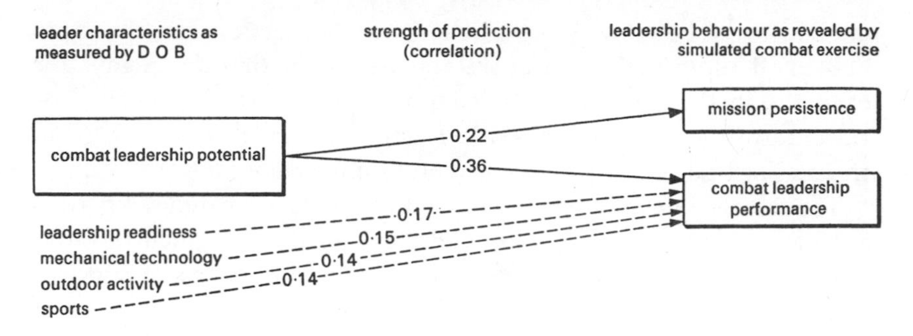

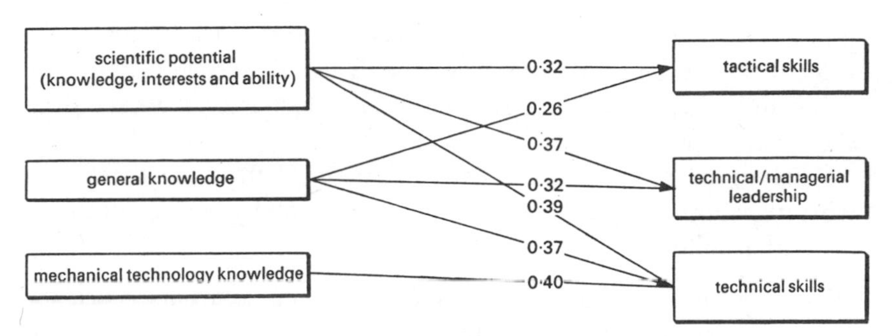

Figure 27 Prediction of performance as assessed on the Differential Officer Battery (DOB) compared with actual performance on three-day field test.

officers in, for example, technical, administrative or combat leadership positions is now possible without too much of a change in the selection procedures (25).

THE PSYCHOLOGY OF TACTICS

Arguably the most important function of a leader, in any military group, of whatever size, is to take decisions to achieve advantage or at least avoid disadvantage. This is the question of tactics and much time and effort has been expended on attempts to improve the tactical ability of officers.

Partly, this has been done through simulation training; for example, tank commanders or platoon leaders play miniature war games, or even full-scale war games in which blank ammunition is used. In general, these have been regarded as successful – in fact the games, because they allow the tank commander a wider range of 'experience', are said to be better ways of teaching combat tactics than the more limited, if more realistic, full-scale activities. In the US Army, for example, in the mid-1960s, tank commanders were in general considered to be low on tactical skills. HumRRO was therefore called in to assess tactical performance and then to improve it (26).

The test they devised was developed to measure six duties regarded as basic: preparation of the tank and crew for their assigned mission; navigation of the course and use of the terrain for cover and concealment; accuracy and speed in detecting targets; accuracy and speed in initial aiming and ranging onto these targets; accuracy and speed in reporting the events and situations encountered; and use of radio procedure. The test HumPRO devised incorporated all these features plus those of surprise, threat, and urgency, and the need at times for more than one thing to be done at once. It was then tried out on tank commanders at Fort Knox, Kentucky. The layout of the exercise is shown in Figure 28.

Analysis of the tank commanders' performances on these tests (enemy personnel and tank crews were all carefully trained umpires) showed as follows:

Preparation for the mission: commanders checked their tanks well, but briefed their crews inadequately, failed by and large to get to the

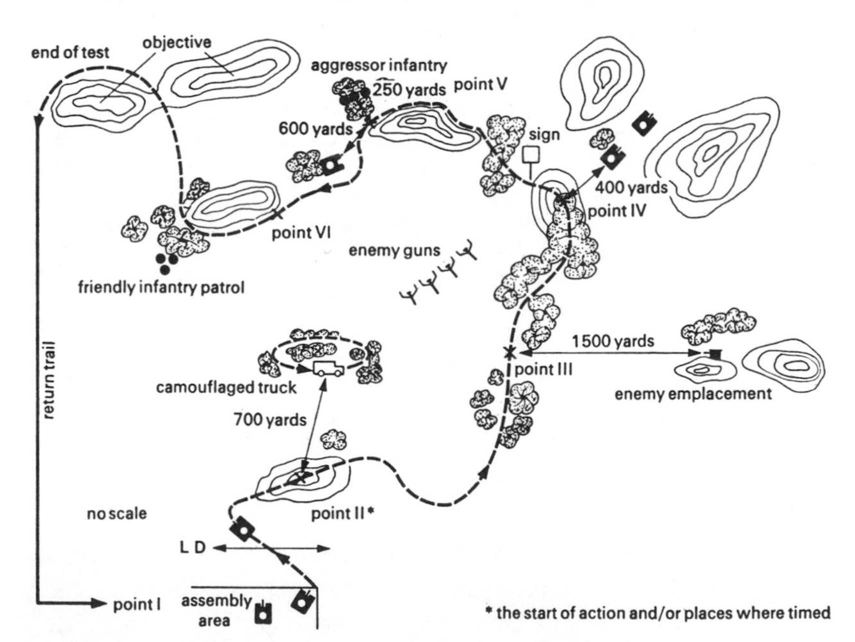

Figure 28 Map of the tank commander's tactical proficiency test.

departure line (LD) on time and did not report their departure to their commanding officer as they should have done.

Navigation: performance was generally excellent on this, except in one aspect – emerging from a wooded area. This was usually done so rapidly that commanders over-ran their field of fire.

Target detection: even on relatively easy problems, 30 to 40 per cent of tank commanders failed to spot targets within ten seconds.

Fire commands: about 60 per cent of fire commands were correctly given; but only 20 per cent of commanders, when confronted with two tank targets at once, correctly specified the priority of fire.

Gunnery: measured by the ability of the commander, having spotted a target, to give adequate instructions for the gunner to 'identify' the target. This was done in less than five seconds less than 50 per cent of the time, and was considered to be a poor state of affairs.

Accuracy of reporting: only 50 per cent of reports contained half the information about an event sufficient to qualify as a success. Considering the liberal standard, this performance is also regarded as poor.

The tactical exercise was thus regarded as a success in isolating the areas of tactical ability where tank commanders fall down. When these were pointed out to the men, and the implications spelled out, they were then given a second exercise. This time their areas of deficiency were much reduced. But these deficiencies are the ones which are likely to recur and must be stressed in training. The lag in reporting events and the low level of reporting details must affect spot reports which, as we shall see in a moment, are now more important than ever since the advent of computer-aided tactical warfare.

In one small way, however, these exercises ignore the real aspect of tactics – for all tactical situations are by their very nature emergent situations in which commanders have to produce new solutions and behave in new ways. Ultimately, it follows that the psychological analysis of the way in which individual commanders actually take tactical decisions should be most helpful in improving the training and tactical ability of future officers.

As a field of study it is very much in its infancy. So far two basic aspects of tactical decision-making have been explored. At its most rudimentary, the psychologists have explored the way new tactical information should be presented to a commander – normally this

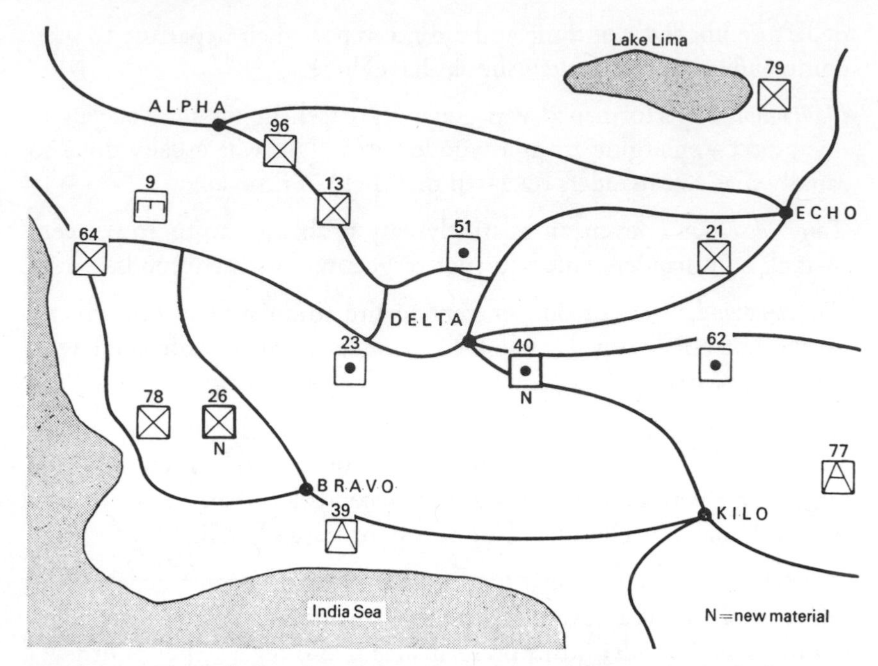

Figure 29 Tactical map of operations showing new information. To facilitate identification new information should have two distinguishing features. In this example the new units each have only one new feature.

would be added to his tactical map of operations. Quite simply this showed that each new symbol denoting a tank as a unit, for example, should have *two* features different from the existing symbols for ease of spotting (see Figure 29) (27).

The psychologists have also explored whether maps are, in fact, the best way to present new tactical information to a commander. Another method tried was to present information in alphanumeric form (28). The commander has before him columns of figures and symbols (see Table 18). Again a gradual military build-up was presented to the men and they had to render a final decision on action as soon as they felt confident enough to do so. Surprisingly, perhaps, there was no difference between the two methods. In other words, the maps, on which much care and devotion are often lavished, and which cost so much are, when it comes to it, no better than a simple set of figures.

Other tactical research has been initiated because, with technological advancements, increases in speed, mobility and destructive power, commanders have to take decisions much faster. It is not surprising, therefore, that the use of computers in tactical warfare has been

Table 18 Military build-up depicted in alphanumeric form

Enemy Unit Status					
Section A	*Unit*	*Section B*	*Unit*	*Section C*	*Unit*
I	TRANS CO. A D BN MTR RIF DIV TANK DIV ART RGT **MTR RIF DIV** (ADDED)	I	MTR RIF RGT ENG RGT TANK DIV MTR RIF DIV ART RGT **TANK RGT** (ADDED)	I	INF BN ART RGT MIR RIF RGT ENG RGT TANK DIV **MTR RIF DIV** (ADDED) **TANK RGT** (ADDED)
II	ART RGT ART RGT MTR RIF DIV TANK DIV TANK DIV **MTR RIF RGT** (MOVED FROM I)	II	TANK RGT TANK RGT CHEM CO MTR RIF DIV TRANS CO	II	MTR RIG RGT ART RGT TRANS CO MTR RIF DIV TANK DIV **MTR RIF RGT** (MOVED FROM I) **TANK DIV** (ADDED)
III	INF BN INF BN CHEM CO ART RGT ENG RGT TANK RGT **MTR RIF RGT** (ADDED)	III	INF BN ENG RGT ART RGT **MTR RIF RGT** (MOVED FROM II)	III	INF BN ART RGT INF BN

exhaustively explored. One area in which psychologists have a prime interest is whether computers are able to help in soldiers' certitude judgements. What is the role of confidence in the decision process?

In one experiment at BESRL, a sequence of nine slides depicted to the subjects the gradual enemy build-up of forces (29). Within the sequence each slide represented an independent sighting of equipment and contained varying combinations of three types, totalling six per slide. The relative *proportions* of the three types of equipment provided the only clues as to which of three kinds of regiment were being built up. The trainee commander's task was to indicate after each slide what type of unit was being built up, to estimate the probability that it was that type of unit and to commit himself to action when he thought he had enough information for a decision. The sequences were arranged so that the true possibilities attached to each unit were increased from trial to trial. The trials were also arranged so that action became more difficult each time (so there was pressure to act as soon as possbile). Success on the test was, of course, credited only if the correct action was taken.

The results showed two things quite clearly. First, using a computer to carry out the clerical calculations relating to the build-up relieved the officers of a heavy chore and left them free to concentrate on the heart of the matter; this undoubtedly improved their tactical ability. Secondly, if the men waited until the picture had changed somewhat, and then made a calculation based, at least in part, on these changes, they were more accurate than if they calculated, as is normal, the various probabilities after each change.

A similar study by Michael Strub, at BESRL, looked at officers' abilities to predict, at any given time, whether they thought the enemy would attack or not (30). The experiment lasted a hundred 'days' and on each day the enemy either attacked or rested. Over a longer time scale, however, there was a hidden pattern of attack and rest. If the men spotted these patterns, they would obviously stand a better chance of being effective commanders. The patterns were quite simple 'second order' ones – that is, from the performance on two consecutive days the action on the third day could be predicted. For example, 'attack', 'attack' was followed by 'rest'; or 'attack', 'rest' was followed by 'rest'. Some patterns occurred more frequently than others to see if they were easier to recognize. The results showed that a pattern was recognized only if it occurred often enough. Two days of attack, for instance, could be followed by either attack on the third day or rest. If the enemy attacked two out of three times, the pattern was never learned. The pattern had to occur roughly four times out of every possible five before the men cottoned on. The importance of this finding is two-fold. First, it shows that for relatively rare events computers are much better at spotting patterns than people and gives a guide as to when computers should be used tactically. Second, it means that, where you know that your enemy is not equipped with a computer, then by planning your operations on a statistical basis in which the probability of one action leading to another never rises above two in three, you can disguise your intentions effectively.

A computerized system for information-processing in tactical decision-making in the field has been developed by the Automatic Data Field Systems Command, US Army HQ Europe and the Seventh Army, in Germany. A mobile central computer handles the data, and each field station has a data-accessing terminal. Additional data-storage capacity is provided by remote data terminals. The information handled relates to the enemy situation, deployment of friendly units, nuclear fire support and the effects of enemy nuclear strikes.

BESRL has been studying the human factors involved in using this system (31): for example, they analysed over 2000 messages in one field exercise and found that intelligence reports comprised 70 per cent of the traffic. Half of these reports lacked the required reliability and accuracy ratings; hence the confidence placed in on-the-spot reports was not justified. Reliability and accuracy tended to be highly correlated. This suggests that reliability and accuracy are quite possibly the same thing psychologically. BESRL have therefore tried to develop a method to improve reliability by attaching numerical values to words or phrases expressing probability – to make their meaning more explicit (32). In the experiment men were asked to attach numerical weightings to such sentences as: 'The CIA reports that from satellite photographs it is very probable that anti-missile sites are being constructed around Moscow.' (Non-military situations were also used to see if life-and-death matters change things; they appear not to.)

It was found that individuals tend to use probability phrases consistently but that there is a wide difference *between* people in the values they ascribe to such phrases. This can give rise to misunderstanding.[4]

4. The *Pueblo* incident highlights this danger. Prior to its mission, a routine 'risk assessment' had been done. This concluded that the mission had 'minimal risk', the lowest of four categories of risk assessment then used in the US Navy. A Rand Corporation study concluded that this label caused several factors, which might have prevented the mission, to be overlooked (33).

BESRL therefore suggest that in such computer systems values should be attached to phrases expressing uncertainty, and have devised a standard lexicon using a restricted number of alternatives (see Figure 30).

The Lab is now using tactical games to study how commanders use the military field computer system, how they classify information and what criteria they use as a basis for action (34). So far the results of this study do not appear to have been released.

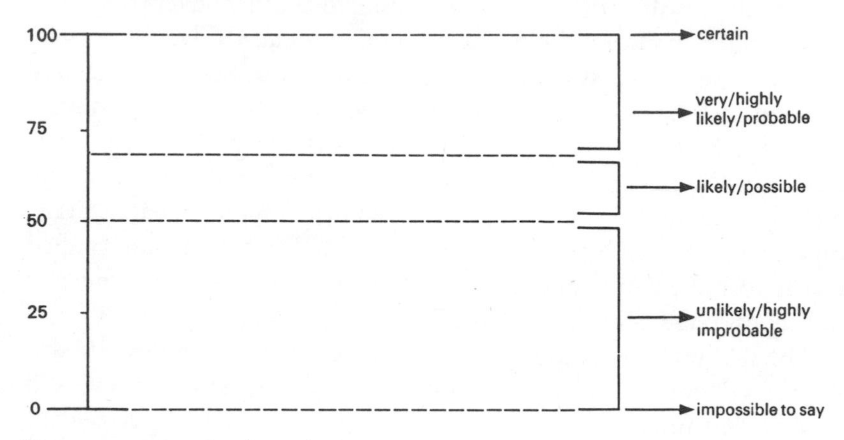

Figure 30 Numerical weightings for probability phrases for computer information processing in the field.

Let us just end this section on tactics by getting away from computers. There have been interesting attempts to do this on the basis that if computers can handle information, then so can people – much more cheaply. In one study, a batch of second lieutenants were given training in information handling. In the exercise, they had to imagine that they were in charge of a tank platoon. They received all their information over a tank radio (even though they were in fact in a barrack room) and they also had the use of maps and other aids. However, one half of the lieutenants had been given instructions on the principles of information-processing – storing, patterning and so on – using a card which is shown in Figure 31. The officers had these cards with them on their manoeuvres and in a subsequent real-life exercise with proper tanks, they outperformed the control group of officers who had not been trained to use the card (35).

Techniques to bring about the improvement of officers' tactical abilities have, therefore, been effective in a few areas. Now let us

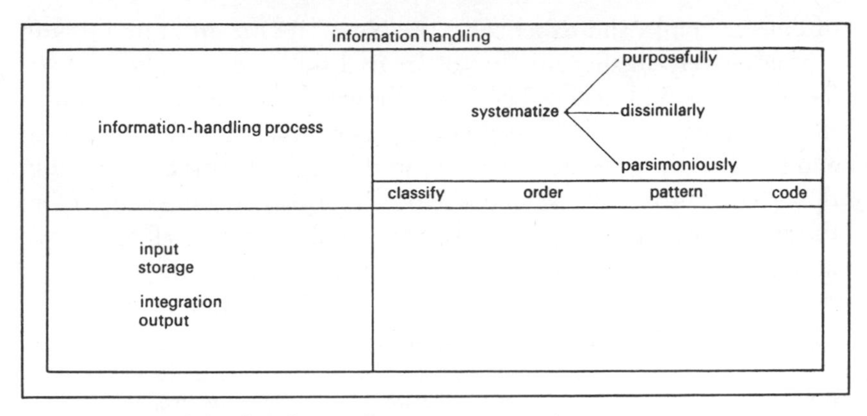

Figure 31 Card for classifying information in tactical manoeuvres.

consider higher levels of leadership than the captain and second lieutenant we have just been discussing.

LEADERSHIP AT SENIOR LEVELS OF COMMAND

In some people's terminology, senior levels of command usually applies to ranks equivalent to or higher than a division commander – usually a major general. For our purposes, however, senior level of command is from battalion commander and his staff upwards. At these levels, in general terms, leadership means three things. There is the personal contact between the commander and his immediate staff, a smallish, close-knit group of highly professional men. Here the qualities of leadership will be those interpersonal qualities which we have already discussed and which we know to affect the way small, independent groups operate. Second, there is the factor of executive action – as part of the leader's technical ability. This is likely to be somewhat different since the battalion commander and above has his own staff who will perform some of these functions. And third, there is the special nature of military organizational leadership in which the commander has to give orders to men who are themselves commanders of smaller units and therefore need to be able to interpret them in order to maintain their own authority. We shall concentrate here only on those types and problems of leadership that we have not yet met.

One major thrust of HumRRO research has been in the field of military job analysis. By a process of gradual selection, using officers of different ranks and experience, the HumRRO team came up with a progressively smaller number of 'critical skills' for battalion leaders (commanders and staff). The research was concerned with combat manoeuvres only and followed both types of task the leaders had to do and the thought processes they used in their jobs. It was carried out in the USA, Europe, Alaska and Vietnam (36). Briefly, and simplifying the results, the jobs of the battalion commander and his staff are shown in Table 19.

Table 19 Job analysis of senior level leaders (36)

Battalion commander	*S1 (first staff officer)*	*S4 (fourth staff officer)*
Setting and maintaining standards	Officer-efficiency report	Maintenance (vehicle and aircraft)
Enlisted men's safety and welfare	Sympathy letter	Daily battle-loss report
Organizing	Military justice	Unit-readiness report
Inspecting	Morning report	Logistical estimate
Accessibility to men and officers	State of discipline	Feeding plan
Eliminating uncertainty among men and officers	Awards and ceremonies	Water
Delegating authority	Medical service	Traffic control
Warning/reprimanding subordinates	Admin reports to HQ	Captured enemy material
Great work output	Replacements	Installation property book
Briefing	Morale indicators	Salvage material
Knowing enemy's capabilities	Duty officers	Statement of charges
Reducing conflicts between members	Educational development	Decontamination squads
	Civilian employees	Area damage-control plan
		Maintenance of medical equipment
		Cash-collection voucher
		Maps

Though far from exhaustive, these lists (the jobs higher up the lists are considered more important by the men) give a good idea of how the jobs at the top of a military unit like a battalion differ. There may be nothing particularly surprising about this: but that was not the army's intent. The idea is that by making these things explicit in a systematic way, world-wide experience is pooled. This way of looking at the job used not to be taught in army schools before this study but as a result many of the specific tasks in the list above are now tackled explicitly so that the young officers get more experience more quickly. (HumRRO has since produced five booklets outlining the job characteristics of the battalion commander and the S1–S4. At one time military men would have resisted the notion that such jobs could be categorized in this way, but HumRRO researchers were themselves categorical that, however much senior command may seem a unique job to one man, in practice the roles are very similar the world over – at least in the US Army (37).)

In the second part of the experiment, the thought processes of the five roles were examined. 'Thought process' was a deliberately chosen term, as opposed to 'thinking'. This time, the officers took part in a simulated airmobile assault in Vietnam which involved twelve other people, each of them an experienced game player and simulator. The exercise was divided into three phases: in the first, the battalion was engaged in routine company level patrolling; in the second, a warning order was issued for an airmobile assault; and in the third, the airmobile assault was made – the battalion then being engaged in extensive tactical operations.

All officers' responses, orders, requests for information, and actions were observed and recorded in the way normal in these simulation exercises. But in this case, the officers' thought processes were also studied as revealed in their behaviour – in handling information. Three patterns of behaviour appeared to matter. First was the 'storing' response. In this context, storing means that the communication input was merely accumulated in some form – sometimes in the officer's head but also in a file perhaps, or marked on a map. No real decision-making was usually needed at this point. The second pattern was the 'automatic' response. This is an activity in which the stimulus – the new piece of information coming in – automatically activates the standard military response, pre-patterned due to training and the experience of the officer. For example, when the S4 receives a request for certain supplies, he immediately sets in motion the orders to get them off. The third pattern was labelled 'analytic' by the researchers. The officer has to consider various alternatives in order to select the one that solves the requirement or he may even have to come up with something new. The thought processes of each of the top five men in the battalion were then examined, using this type of breakdown, in each of the three phases of the exercise. Some of the analyses are shown in Figure 32 (38).

The commander is predominantly the man who does the analytic thinking, the S4 handles most of the automatic replies. But the picture for S1 was a surprise. In general, the army has always assumed that the

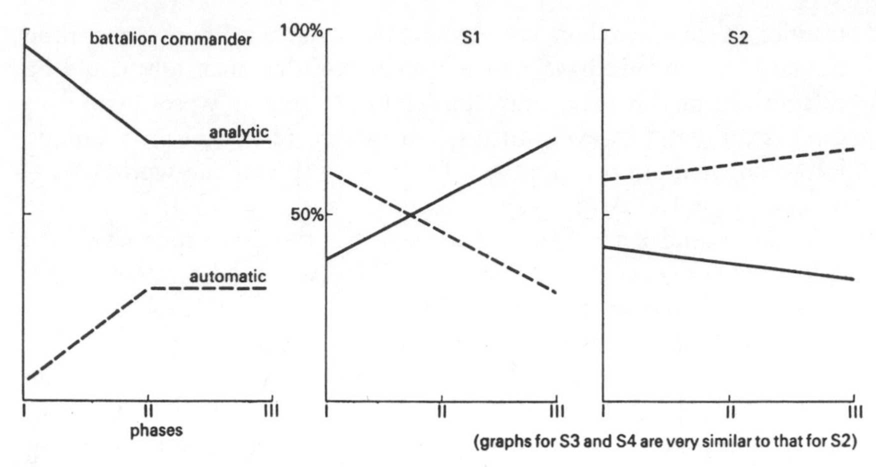

Figure 32 Analysis of thought processes of battalion commander and staff officers in field exercise.

S1, like the other staff officers, was mainly an automatic role; but the research clearly showed that more creativity is involved in this role and that one half of all responses by the S1 are analytic. As Figure 32 shows, the analytic thought process actually became the main one for the S1 in the third phase of the exercise. In future, therefore, the research implies that the training for S1s should include more preparation for this type of thinking than is currently the case (39).

The final aspect of higher command that we shall consider is one which arises out of the hierarchical nature of the military and the difficulties this can pose for commanders of intermediate level who have both to take and give orders. HumRRO tackled this by conducting a study with officers of differing rank but all of whom had seen quite a bit of service in Vietnam (40). Half of them were colonels or lieutenant-colonels and had commanded battalions or their equivalent; the other half were majors or captains who had commanded companies. Both of these groups were asked whether or not they thought a number of selected leadership actions were suitable for battalion commanders and company commanders. The actions could all be divided into one of the four classical ways in which psychologists group leader activities:

Task-centralized actions: mainly concerned with the mission, which serve to increase personal control of the leader or otherwise to centre authority or responsibility in the command level being evaluated (e.g. the leader keeps sole control of the communication facilities).

Task-decentralized actions: mainly concerned with the mission, which serve to decentralize authority or otherwise to increase the contributions of subordinates (i.e. delegation).

Social-emotional positive actions: which principally affect the interpersonal, emotional and motivational relations of the leader with the other men and are usually interpreted as rewarding.

Social-emotional negative actions: are as above, but are usually interpreted as punishing.

The battalion and company commanders showed quite remarkable agreement about the nature of leadership except in one marked way. Company leaders were far more in favour of decentralized actions than were the battalion commanders. Battalion commanders are more likely to want to centralize their own actions and to want the company commanders to do the same at their level. The company commanders, on the other hand, felt that battalion leaders should decentralize things much more whereas, for their own company level, they wanted to keep things much more centralized than even the battalion commanders did. In other words, the army is a loose network of fairly autonomous groups according to the battalion commanders but is a very loose collection of highly autonomous groups according to the company commander.

Two points may arise from this. First, however much people may differ individually, the role someone plays and his level in an organization very much determines his view of what leadership is. And second, following on from this, part of the training in leadership should take into account how rank affects one's concept of leadership. This is likely to reduce a good deal of misunderstanding and even antagonism in future battles that might otherwise lead to chaos, as they have done in the past, when the chief conflict has often seemed to be more between the different levels of command on one side than between the enemies.

ONE IN SEVEN AGAINST WAR: PUBLIC CONSIDERATIONS

Finally we look at an area of concern to the highest levels of military leadership – and to national governments also – the situation in which a nation is about to engage or actually is engaged in war. What are the public's attitudes to the military, to mobilization, to war itself?

How can the leadership assess and, perhaps, manipulate public opinion? At the outbreak of war public opinion varies widely. Joel Campbell and Lelia Cain, who looked at the two world wars and the limited conflict in Korea, found that at the most only 86 per cent of the American people supported the war (41). This means that there is always a rump of one in seven who never want to go to war whatever the circumstances. Initially, opposition comes from those who are constitutionally against war rather than politically opposed to military policy. It is as a war proceeds that political opposition grows. According to Campbell and Cain, in the first days of a war, an enemy attack on the home troops is the most effective rallier of public opinion. The breaking of a treaty has nothing like the same effect. Their results also showed that it is important that the public see an end to the war. War-weariness sets in surprisingly quickly, according to Campbell and Cain, and once this happens, recruitment drops and conscription becomes all the more necessary, causing further opposition to the military to grow.

The Israelis – understandably – have been interested in the effects of living under repeated threats of war. Given their small geographical size and population, their economy can be ruined by the huge costs of military mobilization and their morale seriously reduced by having to live under perpetual threat. Dr Shlomo Breznitz, of the Universities of Haifa and Maryland, has for some years now been looking into the psychology of false alarms. The research was paid for by the US Department of Defense and, though fairly fundamental and laboratory-based, was clearly designed to help the Israelis work out how to keep up motivation among soldiers and civilians following, say, a series of war alerts that did not in fact lead to armed conflict.

Breznitz's research is in its early stages so far but one of his experiments has shown, for example, that the higher the chances that a threat will go ahead, the greater the false alarm effect if it is cancelled. Breznitz concludes – and this could well be the Israelis' attitude to approaching wars – that 'in order for a warning system to protect its credibility it . . . might even be useful to *deliberately understate the announced probability of a danger*', (Breznitz's italics) (42).

Possibly more immediately useful to a military leader at a time of crisis is some prior understanding of public reaction to the tense build-up of events which may well lead to war. The main lessons appear to be two-fold. First, the leader should not underestimate the amount of confusion and lack of knowledge about even the most important events. One study carried out in America in 1962, immediately after the

Russians had just exploded several H-bombs and when there was widespread anxiety expressed in the press about nuclear fall-out, found that only 39 per cent of those polled had strong attitudes towards any kind of policy; the great majority – 61 per cent – either could not be bothered or could not make up their minds which course of action to take (43). The second lesson for the military commander (or civilian leader for that matter) is that he should beware what he reads in the newspapers. What stands out from the polling studies is that vocal or militant expression of views on a war – or an approaching war – does not usually reflect the view of the majority. However much the media may not like it, there *is* a silent majority. This does not mean however, that this majority is inevitably a right wing one. To be sure, Jerome Laulicht's study of the Canadian public's views on disarmament and defence showed that on balance it did think that military power is the only deterrent (44). But studies in America, at the height of the Vietnam War in 1969, showed that contrary to what anyone thought, the youth of the country was actually *more* conservative about ending the war than were the older people (45). In spite of the fact that the older generation had a higher proportion of people with no opinion about the war, there were still more of them who opposed the American involvement. The unreliability of the media in this respect is further underlined by the studies of Hazel Erskine, an American pollster, who showed that opposition to the Vietnam War at that time (1969), though it dominated the newspapers and broadcasting, was actually *less* than the opposition to the Korean War at its height and to the First World War (46).

Differences *do* exist between the generations, but not in the way they are normally stereotyped. Older people prefer more extreme, short-term solutions *but this preference can go in either direction*. That is, a higher proportion of older people than younger people advocate fighting but a higher proportion advocate giving in as well. Younger people, on the other hand, prefer longer term, more moderate measures say foreign aid, rather than bombing or surrender as a solution to the world's problems (47).

The most sophisticated military public opinion machinery of any kind now exists in Israel. It was set up after the Yom Kippur War in 1973 and is now primed and ready for any repeat performance (48). Within forty-eight hours of that war breaking out a group of social scientists at the Hebrew University in Jerusalem, led by Professor Louis Guttman, the famous attitude-researcher, presented the government with

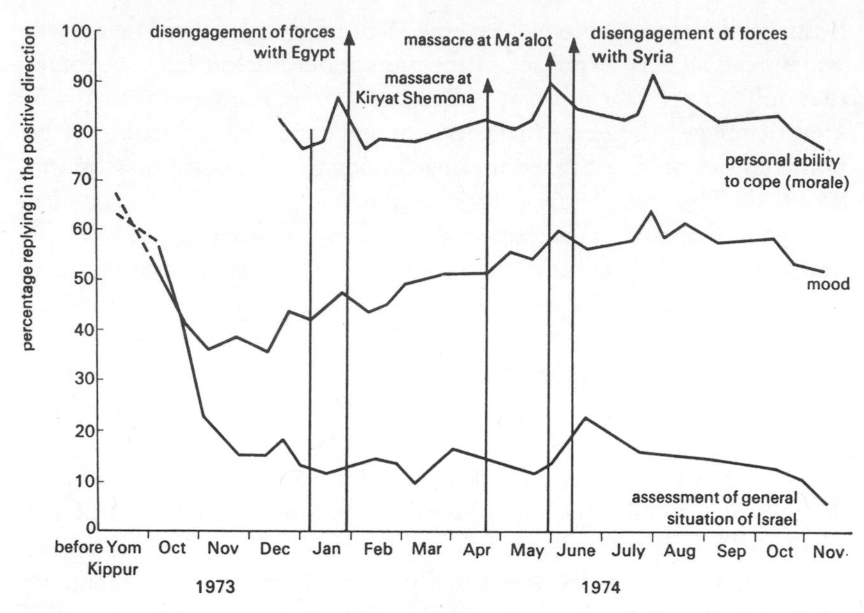

Figure 33 Time series for morale and mood.

a scheme for monitoring Israeli public opinion as the war went on. The scheme seems to have been approved for a couple of surveys were carried out during the war; but afterwards, the idea was followed more enthusiastically and now, even in peacetime, a continuing survey is mounted every week. Moreover, the system is ready to be adapted at a moment's notice to a daily service should war ever break out again in the area.

Most important from the point of view of governments and military commanders, the Israeli study found that public opinion does not often react dramatically to particular events; instead, opinion changes on a slightly longer term. In general, the public did not seem to have a great deal of faith in its government even though it was prepared to accept whatever hardships the government had to impose. The lesson for governments is that mood and morale are not necessarily the same thing, that an unpopular government can, in a threatening situation, still get away with unpopular measures. And that the general trend of opinion is more difficult to control than the odd changes due to specific events. In a chronically tense situation such as Israel finds herself, or as is likely to precede any major war anywhere, the Israeli results suggest that a government should not attempt to change the overall mood of public opinion but always make sure that its own measures are fully explained so that opposition to them is minimized. And it is also clear, from the Israeli results, that reassurance on military capabilities is also needed regularly. This particular worry seems to preoccupy everyone.

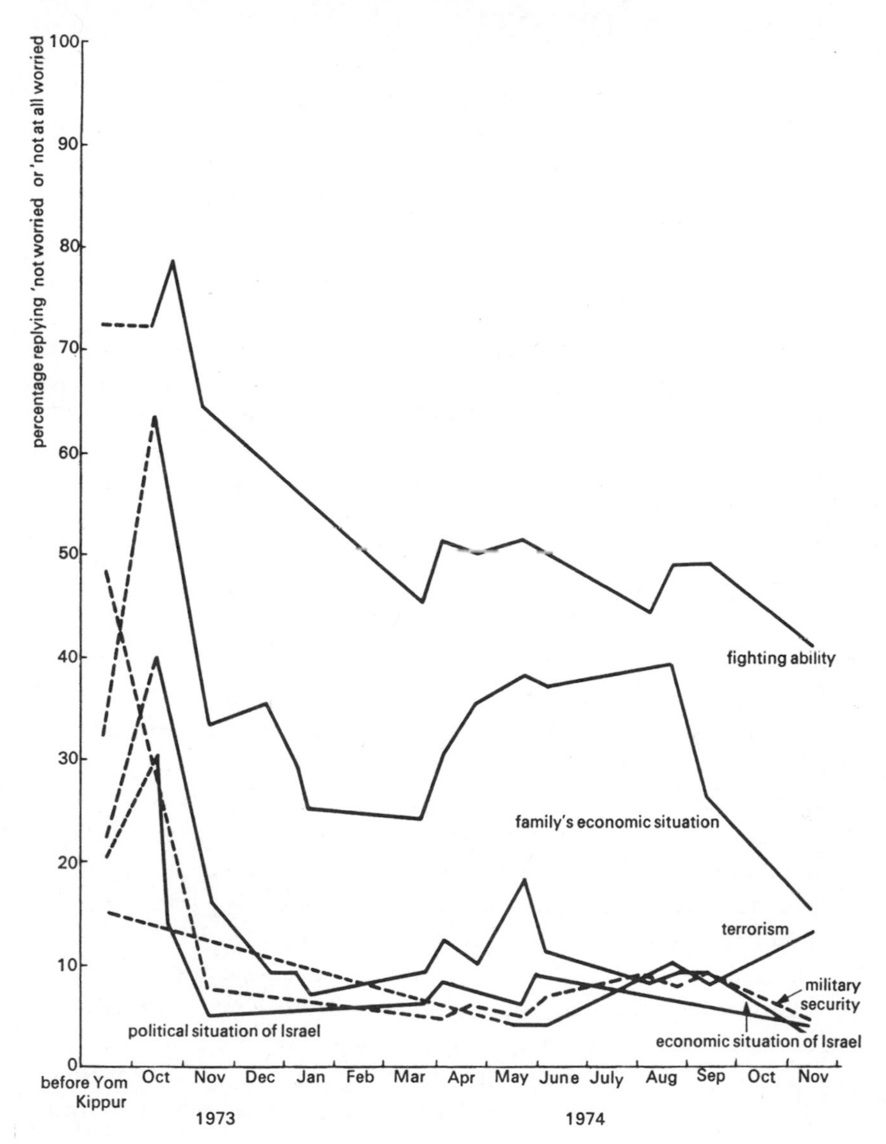

Figure 34 Time series for 'worries'.

8 Whales as Weapons? Animals in a Military Context

WHALES AS WEAPONS?

Some ideas catch the imagination more easily than others and in the field of military psychology perhaps the one idea that, over the past few years, has persisted most is that the American Navy has been using marine animals – whales, porpoises or dolphins – as weapons (1). That is, the animals are said to be trained as '*kamikaze* fish': they are taught to spot enemy personnel or ships, fitted out with explosives and so are able to swim up against the side of some enormous battleship, scuttling it; or to guard harbours from enemy frogmen. It is fair to say that, although these reports persist, no hard evidence has ever been produced in support of such allegations. It *is* true that the US Navy has used dolphins in classified research. They do not deny this. For example, they admit that in Can Ranh Bay, a US base in Vietnam, dolphins were part of an experiment carried out in the early 1970s (2). But the research was classified only because *all* research carried out in an operational area is classified; there was nothing, according to Harris Stone, director of the Navy's Research, Development, Test and Evaluation Program, to do with *kamikaze* dolphins or porpoises. One would not expect him to admit it even if it was going on but it is one of those cases where the official explanation is plausible and convincing. For Stone does not deny that dolphins and other sea mammals have exceptional powers: he simply makes the telling point that it is very expensive to train them and that therefore it is 'inexcusably wasteful' to use animals for tasks such as detection and aiming which are much better done by mechanical devices.

This does seem to be one of those areas where attention has focused on one area, in the belief that something nasty was stirring, when in fact the true situation is rather different, and as a result other contentious research has gone on unnoticed.

Three types of animal appear to have been studied in a military context: whales, dolphins and porpoises have been used as aids to men in maintaining vessels or in research; dogs, whose olfactory abilities can be especially useful in combat, have been used to detect explosives and guerrilla personnel; and birds have been trained to locate enemy personnel or structures. This is by far the most bizarre and aggressive use of animals that I have come across in researching this book.

TUFFY AND THE TORPEDO

The US Navy has quite successfully trained whales and dolphins to retrieve experimental torpedoes and other inert ordnance from a great depth, beyond the reach of human divers (3). Human divers are normally limited to depths of 300 feet, whereas one whale, Morgan, achieved a depth of 1600 feet in 1972. Morgan, a pilot whale, and Ahab, a killer whale (who was less reliable), made routine dives of 1000 feet. Each whale was fitted with a series of devices which guided him towards a metal object on the ocean floor. Then, when he pressed against a torpedo, for example, a pair of claws attached to the animal fastened around the missile and a gas balloon inflated, lifting the torpedo to the surface. The Navy now anticipates that eventually the larger whales may be able to dive to 6000 feet to retrieve objects, like experimental weapons, which are too expensive to waste (4).

Other research has involved the use of a porpoise, Tuffy, in an experiment with divers (5). Tuffy was trained to ferry equipment from the surface station to the sea-laboratory on the floor of the ocean 200 feet down. Tuffy had no problem in completing the dives in a matter of seconds whereas the humans had to come to the surface very slowly. In another experiment, divers who strayed too far from the laboratory and got lost on the seabed were told to click a 'cricket'. Tuffy could hear this, pick up a line and deliver it to the diver who was lost. The diver would then simply follow the line back to the laboratory.

The other main naval use of the sea mammals appears to have been an even more straightforward study of their zoological characteristics for the information which this may provide about how to design torpedoes that would be better at swishing through the water (6). The animals' hydrodynamic capabilities are also the subject of much interest with a view to understanding how it is that they can change depth so rapidly without any of the side-effects which a human diver would suffer. Besides diving capabilities, the sonar systems of the sea mammals are of particular importance to the navy, as are the lifestyles of organisms which can adversely effect naval operations. Included in this are not only the dangerous sharks and sea snakes but even snapping shrimps whose noise underwater can disrupt delicate naval listening devices (7). Nothing particularly sinister in all this.

DOGS AS DETECTIVES

The abilities of the dog to distinguish one smell from another are well known and military attempts to use this highly developed sense go back a long way. As with other research reported in this book, however, the use of smell to detect people took on a new urgency in the early 1960s with the realization by many military agencies that it was yet another technique that might be especially suitable for use in a guerrilla war. The Institute for Defense Analysis (IDA) called a special one-day seminar in Washington in December 1964 to consider 'Olfaction and its potential applications in personnel detection' (8).

The aim of the seminar was to discover among other things what substances in or on people provide the greatest smell or taste sensation; whether the use of mixtures accentuates the detection of specific odours; whether insects or mammals can be instrumented to facilitate the detection of odours; whether drugs can intensify the ability of animals to detect odours. A series of lines for further research and military adaptation were suggested. For example, one idea was that dogs might be taught the characteristic smell of the inhabitants of a particular village: then when newcomers or intruders arrived the animals would be able to pick them out. It was noted, also, that dogs could readily tell Negroes and Caucasians apart; the fact that the Japanese had a fatty acid spectrum different from Caucasians also probably meant that dogs would be able to distinguish those two races as well, according to scientists at the seminar.

Several Russian papers were also considered. The Russians have found, for example, that twenty minutes after the administration of amphetamines or caffeine the ability of police dogs to detect smells is markedly improved (9). Various races – Ecuadorian Indians, Egyptians – who use their own sense of smell in tracking were also noted as being of possible military significance, as were a whole range of animals (from rats and pigs to fleas and cockroaches) who live in the vicinity of man and whose presence, signified by their smell, could also indicate the presence of humans (10). Workers in other countries where animal research could have military significance were mentioned. These include Professor Cheesman in Tasmania, Sir Solly Zuckerman and Dr William McCartney in Britain (11).

The question of sensory deprivation and sensitivity to smell was also alluded to in the official account of the IDA seminar, it being noted that humans and animals who are blind or deaf from birth sometimes have an even more developed sense of smell than sighted organisms. 'Conceivably,' ran the summary, 'dogs blind and deaf at birth could be trained to rely on olfactory clues. Their performance might then be abnormally high and they would not be distracted by auditory or visual stimuli.'

The scope of the seminar gives some indication of what was considered important by defence experts in the mid-sixties. We know now that mechanical 'people sniffers' were brought into use in the Vietnam War in the late 1960s, but these did not really use the olfactory organs so much as the ability of infra-red rays to respond to the heat generated by human bodies and to distinguish them from the cooler surrounding jungle. Research into the actual uses of smell appears to have proceeded in two directions since the IDA seminar.

First, at its Natick laboratories in Massachusetts, the US Army has developed a substance, 'squaline', which, when squirted onto a man even in small amounts, persists for a long while (12). Even if he washes or wades through a stream it can still be distinguished by a dog. Even 0·001 milligrams on a person is sufficient for it to be detected by alsatians. It will even withstand one laundering. It cannot be detected by humans by odour, colour or its effect on vegetation even when applied at several times the amount required for successful contamination. It can therefore easily be sprayed from the air or from conventional ground-spraying equipment. In certain circumstances it becomes a useful substance with which to encircle an area which an army wishes to keep secure. Anyone who has passed through the contaminated ring of vegetation from outside can be detected by the dogs.

The other use for smell also occurs in guerrilla war. I was shown this at the US Army's Aberdeen Proving Ground in Maryland. There dogs have been trained to respond not merely to the smell of people but also to the smell of explosives *and* to respond differently according to the state of the person they come into contact with (13). The idea is that, in the jungle, a dog is better than a man at spotting boobytraps – and more expendable. For example, the dog will be sent out on its own and trained to sit when it comes up against a boobytrap, to lie down if it meets a wounded person and stand still if it sees a person standing or moving about. The Aberdeen Proving Ground Land Warfare Laboratory also fits the dog with a radio transmitter which indicates to base not only where the dog is but also in what position it is holding itself. To a certain extent therefore the dog's handler in comparative safety can get quite a lot of simple intelligence about the area around the base by the judicious use of the dog.

I am grateful, at this point, to Dr Robert Lubow who was kind enough to let me see a proof copy of his book, *The War Animals* (Doubleday, 1977), in which he disclosed two further uses for dogs. In work for the US and Israeli military authorities Dr Lubow has trained dogs as 'bomb detectives'. Dogs can be used in harbour protection to detect the presence of divers. They can be trained to discriminate the smell of the gas contained in the bubbles given off by divers. The gas is released when the bubbles break on reaching the surface. In Israel, dogs have also been used to detect, from among a sack of 500–600 letters, just one that is a letter bomb. The dogs are maintained at 80 per cent of their body weight and are fed as soon as they make a correct response.

Dr Lubow also confirms that more than one military authority has considered far more imaginative uses for animals in war – birds as spies.

'PROJECT PIGEON': BIRD OF PREY

The pigeon is capable of distinguishing man-made objects, as seen from the air, from the natural shapes of lakes, rivers, mountains, trees and so on. Dr Robert Lubow's work in this field has been at least partly financed by the US Air Force (14) and is not his first. He appears to have carried out for them an early experiment, which he shared with Dr E. Carr-Harris at the US Air Force Avionics Laboratory – the report of which is still classified (15). Dr Carr-Harris is known among military scientists for his work with dogs at the Land Warfare Laboratory described in the preceding section. Lubow has trained his pigeons, also fitted with radio transmitters, to fly behind enemy lines and land on man-made objects. In a remote place like Sinai or the Golan Heights this invariably means military objects – runways, missile silos, even mobile lines of artillery and tanks. The enemy soldiers never guess what is going on.

One of Lubow's reports is entitled 'The perception of high order variables by the pigeon'. In it Lubow says that it takes about a month to train the birds to distinguish man-made from natural objects with an acceptable and usable accuracy. The particular breed of pigeon used in the reported experiments is the white carnaux and the teaching is done, quite simply, by showing the bird aerial reconnaissance photographs, some featuring roads, railways, bridges and so on, and others which contain only mountains, rivers, trees and other naturally occurring features. The birds are first put on a restricted diet until they are down to 80 per cent of their normal weight and as a consequence are very hungry. They are then placed in a small chamber with, in front of them, a food hopper and a screen. One by one the aerial photographs are flashed onto the screen for a short while with the man-made objects and natural features shown in random order. If the pigeon pecks at the photographs which contain man-made features or if it refrains from pecking at a picture without man-made features, it is fed through the hopper. If it makes the wrong response it gets no food. Eighty photographs are used in training, forty of each type, every day for a month. By the time a month is up, it has been found, most birds can learn to respond to the pictures in the appropriate way (and those which have not learned are dropped). As a final check, after the training period is over, the birds are shown a further set of reconnaissance photographs to see whether the concepts they have learned of 'man-made objects' can be transferred to other pictures they have never before seen. (Most of the birds, it seems, retain the concept without further training, for up to two months.)

Dr Lubow's paper, which was apparently completed sometime in 1974, also analyses the features of man-made objects to which the pigeons seem most ready to respond. His conclusions are that pigeons see most clearly pronounced straight lines and symmetrical curves and he appears to take the view that, with longer, more systematic training, pigeons could be taught to respond not just to man-made objects in general but also to specific types of installation – probably runways, lines of trucks or tanks and circular fuel tanks would be most readily perceived on the basis of his research so far.

The use by psychologists of birds in a military context is by no means that new. Professor B. F. Skinner was involved in a project in the Second World War which was, he says, 'a crackpot idea that eventually became intellectually respectable' (16). 'Project Pigeon began as a search for a homing device to be used in a surface-to-air guided missile as a defence against aircraft in 1940. The idea was to put a pigeon in the nose (or 'beak' as it was called) of the Pelican missile. The pigeons were to be trained to keep the missile on target by pecking the image of the target which they could see through a lens at the front of the missile. The US Navy was more interested at the time than the other services but the idea was not officially taken up.

However, one company, General Mills Inc., were willing to provide money for research, motivated, say Skinner, by the 'national interest' and in 1943 they were ready to go back before the Washington boffins in charge of the scientific war effort. This time they got a modest grant to continue their research and the project was classified. Although they made progress (with at times more than one pigeon in the nose cone) the project was some time in coming to fruition. Only after the war, when it became known that there had been other Allied plans for such things as 'incendiary bats' (to be released over enemy cities, to nest under eaves, and then explode), Swedish plans for *kamikaze* seals to blow up submarines, and Russian plans for dogs to blow up tanks and for sea-lions to cut mine cables, did the Navy turn once again to Skinner's still classified project.

Project Orcon, standing for Organic Control, was then developed at the Naval Research Laboratory in the late forties and early fifties. It was an up-dated Project Pigeon and this time the birds were trained with a film simulating the approach of a 600 mph missile onto a ship at sea. Again Skinner considered the idea workable but in time mechanical aids to guiding missiles became more sophisticated and the pigeon idea was never made operational.

Given this background, though, the Israeli use of the bird as a scavenger of military intelligence should come as less of a surprise than it might otherwise do. During the Second World War the US War Office actually went so far as to ask Skinner to estimate the pigeon population of the entire United States.

There is, of course, no evidence to show just how, when or whether the Israelis have yet used pigeons as spies. One of the problems of this kind of work is that though the animals themselves may be very cheap, their training is usually either long, costly or both, and even after training their performance may be unreliable. So though it is a bizarre field, and one which earns the armies and navies concerned a lot of public ill-will (from people who appear to be not so much against war as against cruelty to animals), and regularly strikes the imagination of the press, it should not be considered a major field of research, nor as one which, with the exception perhaps of tracker dogs, is likely to be very important in the future.

Dr Lubow has denied that he was training – or ever had trained – pigeons to act as spies for the Israelis.[1] He claims that his research was mainly theoretical to see what exactly distinguishes man-made from naturally occurring features. However, in his book he does include details of a project directed by Richard Herrnstein of Harvard to explore the use of pigeons in ambushes. The idea was to train pigeons to fly on ahead of patrols in the jungle looking for men or man-made objects. The birds were trained to land on or near these objects/people and each was fitted with a radio direction finder triggered by the rapid deceleration of the bird on landing. This project was abandoned after all the trainee birds caught an infection and died. But in his book, Lubow devotes a whole – allegedly speculative – chapter as to how a 'pigeon-spy' system would work – and it is exactly as outlined above. He also admitted that the Israeli military authorities did ask to estimate the pigeon population of the country. The still-classified documents are listed in the references for this chapter (17).

1. In an interview in 1977. Dr Lubow was then a professor of psychology at Tel Aviv University. He has previously worked for the US Department of Defense and is at the Department of Psychology, Yale University, New Haven, Connecticut.

PART TWO *Stress*

9 In the Danger Zone: Testing Men Under Stress

In earlier chapters we have examined research designed to test who will make a good fighter in combat. Armies are also concerned with the related question of who will break down under stress of battle. This breakdown rate has varied greatly over time, from war to war, battle to battle, nation to nation. It also varies with age, rank and the time that men have been in the field. In this chapter we look first at the army's general screening procedures, and then move on to the various techniques to find ways of measuring stress and of battleproofing men – to ensure, as far as possible, that they can stand up to all the frightening events that come their way in wartime. Finally we look at studies of specialist groups performing under stress in the field to see how they are selected and what techniques have been devised to help them cope.

GENERAL SCREENING PROCEDURES

In general the armed forces have attempted to screen out, at an early stage, all those likely to be unsuitable for the dangers of battle. The psychiatric effects of the stress of war is a continuing field of study and so screening processes are continually being up-dated, but it is usual at present for men to be screened on entry to military life. Three things are the main focus of attention. Is there a history of psychiatric illness? a history of crime? low intelligence? Psychiatric rates of breakdown during the Korean War were found to be far below those encountered during the Second World War, and this is now believed to be due to effective screening in the 1950s, not just for those psychologically unsuited to battle but for those with low intelligence. When rejections of men both for psychiatric reasons and low intelligence are added together the total rates are found to be similar in both the Second World War and the Korean War – about 12 per cent (1).

After their initial screening the men are usually watched closely in the first few months of service life. They are given regular medical check-ups which invariably include a psychological component. Military psychiatrists will be particularly concerned with the types of problems which service life gives rise to, even in peacetime, such as communal living, rank, separation from family. In some cases psychiatric films are shown to new recruits on subjects such as psychosomatic disorders, combat fatigue, the role of the officer or the division psychiatrist. All men who go absent without leave are carefully screened the first time to see whether they are going to make suitable combat personnel. In general, those who fall foul of military law too often have been found to make poor fighters. Soldiers whose jobs are in some way special – for example, handling top-secret information – will have extra attention.

Armies are well aware that, even in peacetime, emotional reasons should never be regarded as acceptable criteria for being excused duty; although this makes for hardship on a few individuals who genuinely cannot cope, the psychiatric referral rate of a unit as a whole is kept down (2). It was also found in Japan in 1953 that the hospital admissions among American Marine units were affected by whether or not a psychiatrist was present with the unit. Those units *with* psychiatrists had one fourth the admissions that units without psychiatrists had. This is a catch that many officers do not understand: if you have a psychiatrist, he will not be needed; if you do not have one, he will (3).

In general then, the soldier faces two problems – to cope with the stresses built into military life itself, but also to be prepared for the ordeal of battle. This means that, even in peacetime, discharges for disability due to psychiatric disorders run at roughly 6 per cent (half that in wartime) (4). However, in spite of all the studies, it has repeatedly been shown that out of every three individuals predicted by tests as likely to break down in battle, only one actually does so. Conversely, only one half of those who do break down are recognizably predisposed on the tests (5).

CONFIDENCE VERSUS DESPAIR

The central issue is the nature of stress itself. There are two distinct views on this. One is that all types of stress are the same, and that those who cope well with one kind will be able to cope with other kinds. The other view is that combat stress, with its real fear of death, is quite different from other kinds of stress – say, where someone has to complete an exam in a certain amount of time. The results of research are by no means clear as to which view is the correct one.

The general ethics of stress studies must also be considered. Should men be unnecessarily subjected to stressful events? How do we judge what is unnecessary? It is necessary to give a soldier an idea of the experience of battle so that he does not turn and flee at the first explosion; but his training should probably not involve exposing him to dangers equal to those of battle itself. Many psychologists and others have felt that military stress training has been far too realistic to be safe.

According to Dr Richard Kern, of HumRRO, performance under hazardous conditions is a combination of technical skills and the relative strengths of two opposing attitudes – confidence and despair (6). If training contributes unnecessarily to despair it does more harm than good and will undermine a man's ability to cope in combat. What matters is a man's attitude to danger. Probably little can be done to change this – and all that armies can do is hope to screen out the worst people from their battle units. Everyone, however, needs practice in stress situations; the more specific training a man has, the less he will worry about specific actions. He must be sure that he has the necessary skills. This means giving him feedback about his actions in the simulated battlefield (see chapters 2 and 3). So training should build confidence – but not entirely at the expense of less positive things. Kern cites the case of a downed airman who was eventually found half-dead by a rescue party. When asked why he had not used his flares to signal his position, he said he was nervous because he had never fired one in training. This kind of immobility can be avoided. Battleproofing is not perfect yet and probably never will be, but the breakdown rates show that it is better now than in the past.

DEVELOPING TESTS OF STRESS

In October 1953 Psychological Research Associates produced a report entitled: 'The development of experimental stress-sensitive tests for predicting performance in military tasks' (7). This is the first published account in a long-term search by the US military for a test that would simulate, properly, the stresses and dangers of the real battlefield. PRA's aim was to find a test which could be given to soldiers under non-stress and again under stress conditions and which would show how his performance had or had not been affected – the hope being that this would actually predict how well the soldier would perform under fire. In the first instance they had to find a test and a form of stress.

They found nineteen tests which they thought might be 'stress-sensitive' but concentrated on only three which need concern us here. These were:

The critical flicker-fusion test at dim intensity: in this test the soldier is placed in a dark room faced with a light about the size of a 5p piece. He simply has to move a lever one way or the other to indicate whether he thinks the light is steady or flickering.

The trembleometer: this was developed by the British Army Operations Research Group and consists of a ring into which the soldier places his finger. Every time the finger touches the ring the contact is recorded electrically.

A word fluency test.

A test which involved crossing out the 'Cs' in a field of 'Os', as follows: OOOOCOOCOOOOOCO.

PRA then attempted to isolate several types of stress situation (other than battle itself) which would be covered by their tests. Ten types of 'stress situation' were identified:

Muscle fatigue
Sleep deprivation
Noise
Pressing an individual beyond his limits
Real physical danger
Conflict situations
Social disruption
Perceptual distortion
Habit disruptions
Monotony

But where the PRA work really initiated a line of research was in its way of producing realistic danger. Eight ways were originally devised:

Gory battle movies showing what can happen to a man in battle.

Simulated dangerous (though actually harmless) situations achieved by locking a soldier in a darkened room that is apparently (a) refrigerated or (b) suffused with a noxious gas.

Abusive and derogatory remarks concerning the inadequacy of the man's performance on a test by the experimenter or by a master sergeant, the abuse being given while the man is actually taking the test.

Use of sodium amytal (the 'truth' drug) on combat-experience troops with the suggestion that they relive stressful combat experiences.

Diving in the training tower for submariners.

The first experience of an infiltration course or a simulated village battle.

The first jump from the jump tower used in training paratroopers.

The first live hand grenade throw.

In fact PRA concentrate on using the jump tower, and later on a method not on the initial list – painful noise. They found that the critical flicker-fusion tests and the trembleometer were the most stress-sensitive tests, though they were never able to confirm that the tests had combat validity.

What they did find was that most individuals were affected by stress and that the rank order of abilities only changed for a very few people. This was an under-rated finding at the time, played down by PRA. But it implies that only a small proportion of people do either unexpectedly well or unexpectedly badly on stress tests and so might give an unexpected performance in battle. This is an important minority, however, and the accurate prediction of such people is a crucial aspect of battle screening. PRA appears not to have carried on with its research, but its legacy was clear when the field suddenly blossomed in the last half of the 1950s; the research is discussed in the following section.

A recent example of research to find a test for stress is a paper by Wiley Boyles (8). He has been interested in spotting in advance those men who will make good combat pilots – helicopter aviation, particularly, being a dangerous activity. The technique he has explored is how these men react to physical harm – or at least to the threat of it. This is because it is felt by many psychologists that performance in combat may be more a function of emotional and motivational characteristics than of aptitude.

The first of Boyles's tests is known as the background activities inventory (BAI), which attempts to assess the degree of exposure to adventurous activities that young men might encounter during childhood, adolescence and their later teens. Three hundred trainees at Fort Ord were asked to rate the degree of physical harm that could result from each of fifty different activities and to estimate how likely such harm was, the ten activities judged most harmful then being chosen.

The second part of this test reports the degree of confidence which a person feels about carrying out these activities. The second of Boyles's measures is derived from a test used to assess the rifleman's ability to use the M14 rifle; known as the situational confidence test, it was adapted to measure what is generally regarded as the two most stressful aspects of learning to fly – the first solo flight and the first autorotation descent (essentially this is a fast form of descent – like falling in a lift out of control – which requires a rapid series of movements near to the ground, to bring the helicopter back under power and control if a disastrous landing is to be avoided).

Boyles first found that the background activities score appeared to have some validity in that men who volunteered for aviation training had a higher mean value on this score than men who did not (and a useful cut-off point on this test was thus obtained). He also found that men who failed their flight training for some deficiency of flight ability also had a lower background activities score than those who were successful. Finally, he also found that the situational confidence score worked well. In other words, those men who failed their flight training (and about 10 or 11 per cent do) had a lower score on the situational confidence scores – i.e. were worried about solo and autorotation flights. It is very expensive to train a military pilot and therefore any improvement in prediction of those who will fail is very valuable. The failures referred to in Boyles's work are those who fail their flight tests – that is fail *after* they have learned to cope with the instruments. Many, of course, never learn to cope with the flight instruments but this shows up easily in their training and is less expensive. Boyles's research is therefore most promising.

After the Korean War, the US Army introduced 'Task Fighter', a study of the effective versus the ineffective fighter, part of which we have already examined (see chapter 2). The aspect of the study which we are concerned with here was the attempt to develop a substitute criterion for combat and to explore the relationship between performance under stress and combat effectiveness.

In this study ten measures of performance under stress were used (the first subjects being 110 soldiers at Ford Ord in California):

Jump tower: the men's latency time was measured between being told to jump and actually doing so. The tower was 30 foot high and the men were fastened to a parachute harness suspended from an overhead carriage. They also had to traverse a steel cable to a landing 11 feet away and 13 feet high.

Fire fighting: in two tests the men had to put out an oil fire in an open tank 15 feet in diameter and in a replica of a ship's boiler room using standard fire suppression methods.

Hand tremor with and without electric shock.

Time estimation with and without electric shock.

Hand co-ordination with and without electric shock.

Pencil mazes without and with electric shock.

Dark room: two tests in which the men had to go through an almost darkened barracks individually, following a luminous path and under instructions to bayonet all dummies they came across. Reaction times were measured from the moment the dummy appeared, moved and screamed to the time it was bayoneted (9).

The performance of the men was rated as high or low on a 'stress index' depending on whether or not they came in the top half for performance on both stressed and unstressed scores. The high scorers were then compared with low scorers and with fighters and non-fighters as already distinguished in phase one of the project (see pp. 45–52). It was found that there were differences and that these agreed well in some cases with the results of the fighter study. For instance, so far as background characteristics are concerned: fighters and high scorers (stress resisters) had more financial experience in life, made and spent more money than low and non-fighters. High scorers and fighters had more interests, of a more masculine nature than low scorers and non-fighters – they liked playing poker, sports and cars more. Fighters and high scorers preferred body-contact sports; non-fighters and low scorers tended not to.

But the study also found that intelligence was *not* related to performance under stress, nor was high school education. The fighter studies in Korea had found such a difference: but the matter was complicated by the fact that in Korea the average intelligence of the combat troops was 85 whereas in this last study the average was 102. Next Tor Meeland, the scientist in charge of these experiments, factor-analysed the performance of the men on the various stress-sensitive tasks. He found that tests of stress in the laboratory do *not* correlate with field tests but that it is the field tests which correlate best with actual combat. He also identified six factors which appear to alleviate the experience of stress.

Intelligence: fighters are more intelligent than non-fighters.

Deliberate accuracy under stress: in rifle firing, fire fighting and under extreme fatigue there is a group of people who still maintain accuracy although reaction times may slow down.

Stress performance index: there is likewise a group of people who also are not affected in their reaction times under stress – or are affected much less than most. This applies in paper and pencil tests but also in fire fighting and in rifle firing surrounded by explosions.

Eosinophils stress index: measured by a simple blood count.

Pulse rate change in field stress situations: basically due to exertion.

Autonomic efficiency: pulse and other responses react quickly not just due to exertion but to shock and to the sudden appearance of targets. It appears as if in some men their bodies are more prepared to react swiftly to dicey situations. There are also some men who tolerate more shock in the presence of another in the same situation – they rise to the occasion.

There was also evidence of two specific factors: that is, some men did very well on the fire fighting, and others in the dark – but performance on one was to a large extent independent of the other. So it may be that there is opportunity here for selection for certain special dangerous missions.

BATTLEPROOFING: A SERIES OF BLOODTHIRSTY EXPERIMENTS

As a technique, battleproofing can be regarded as a way of exposing men to a direct experience of stress using simulation of actual danger. The rationale is to examine how men stand up to stressful situations, rather than to screen out those who are likely to breakdown, and to present the men with situations which will accustom them to stress.

In 1958 began a series of controversial experiments. A study carried out by Kan Yagi, Robert Knox and Patrick Capretta, unremarkably entitled 'Development of a verbal measure for use in stress study' (10), revealed what must, at that time, have been the most extraordinary experimental methodology ever carried out in the social sciences.

Army trainees were flown in a DC3 aircraft on the pretext that they were taking part in a study of altitude effects on psychological processes. The experimenters were themselves disguised as flight crew. The men first had to complete one irrelevant psychological test. The plane then changed altitude, suddenly lurched violently, one engine stopped completely and other malfunctions occurred. The men were told over the intercom that there was an emergency and were given special 'official emergency data forms' to complete as part of the ditching procedure. This form was actually a stress measure, consisting of twenty-three categories of items concerning personal description and disbursement of private possessions in case of death. The form was deliberately put together badly, referring the men back from one part of it to another at different times.

Debriefing of the men showed that the deception worked.

As control, two groups were used – one set who stayed on the ground, and filled out the form, and another who flew in a plane where no emergency occurred. The men filling out the form under the emergency conditions performed significantly worse than those in the other two conditions. The scientists finally seemed to have achieved their aim of a realistically dangerous test – though it must be said that it involved an extremely complex and expensive procedure. However, there was no assessment of how valid this test was in reproducing the experience of battle. Nor would it ever have been easy to maintain the element of surprise in other groups.

But this experiment paved the way for other equally bizarre attempts to test for stress. It was left to Mitchell Berkun, of the leadership human research unit of HumRRO, to present in all its fine detail the wondrous experiments that the organization had got up to in the years that followed. He did this at a symposium on 'Performance Reserve' at the Human Factors Society of Los Angeles, on 18 June 1963. Again his paper had an unremarkable title which underplayed its true content: 'Psychological and physiological criteria for stress simulation research' (11).

Berkun's paper is almost disarming in its frankness about what he and several others had been doing. He first made his case for the stress experiment to be genuinely stressful: 'I am particularly concerned with the role played by fear, that is by a concern about possible death or injury in the response to adverse environments.' He went on to confess that early experiments along these lines had been frustrated by sophisticated soldiers who realized that they were in fact in an experiment and therefore the danger was only apparent. An early experiment had the soldiers traversing a high and apparently unstable bridge, which threatened to collapse throwing the men into a ravine below. But the men spotted the deception, though the experimenters for a while did not know this and were more deceived than deceiving. This was because the men acted *as if* they were frightened in some sense, although of course their performance did not drop off as it otherwise would have done.

Berkun and his colleagues were moved to design even more machiavellian tests where the danger was more realistic. They did this using several special situations (though there are published reports on only three of these, HumRRO documents refer to six situations in all). Prior to actually carrying out the experiments, however, Berkun and his colleagues did try to find out whether what they were about to do would have any adverse consequences on the men they used as guinea pigs. They put men through the fear-provoking situations, with a clinical psychologist in attendance, and gave the men long psychiatric interviews before and after and again weeks later to see whether there were any hidden effects. None were found.

Of the three reported tests, one consisted of an aircraft ditching – a method similar to that used by Kan Yagi and his co-workers. Subjects had to fill out two forms while the aircraft was being prepared for an emergency landing. They showed a 10 per cent reduction in performance in correctly filling out the forms compared with the controls.

In the second test a group of soldiers were stationed singly at isolated outposts as part of an allegedly large-scale military exercise. Each man had a two-way radio and had to report to base any aircraft in the vicinity. Each soldier was briefed on using the radio and then left alone. He could hear the command post's messages to himself and to other outposts. After twenty-five minutes of normal transmissions, the command post reported in alarm that artillery rounds were bursting outside the designated impact area. At about this time a charge some 150 yards from the subject was detonated. After forty-five minutes the subject was informed that the command post had received no messages from him and he was to repair his apparently broken transmitter so that he could be located and rescued from the impact area. Six more charges were exploded increasingly closer to him, giving the appearance of artillery rounds landing just by him. The repair of the transmitter was in fact a stress-sensitive performance test in which competence could be measured objectively.

A control group of seventeen soldiers underwent a similar exercise without the emergency situation. There were no differences in intelligence, aptitude, education or previous military performance between the two groups. The control soldiers earned an average score of 6·05 points out of a maximum of 10 for speed and accuracy on the repair task. But the soldiers in the life-and-death situation earned only

4·05 points – a decrement of 33 per cent. Statistically this was very significant. It is also worth pointing out that eight of the twenty-four men in the dangerous situation actually *defected* – that is they abandoned their posts in specific defiance of orders from their command post. There were no control defections. (Part of the later objections to these experiments were that in defecting the men could have walked into the TNT explosions that were simulating the shells.)

In the third test each member of the experimental group, again in an isolated position, was led to believe that he had accidentally injured someone by mis-wiring demolition equipment. This time as a test of performance each soldier had to repair his apparently broken field telephone. The group was compared with controls again matched for intelligence, aptitude and educational background.

Out of 10 points maximum, the control group earned on average 6·1, very close to the 6·05 of the control group in the artillery experiment. The experimental group earned a total of 5·01 points, significantly below the controls. The decrement due to stress this time was 18 per cent.

In the three experiments not reported in detail by Berkun, soldiers were presented with a simulated atomic disaster, a forest fire, and ambulance failure in a life-and-death dash to save a person injured in a car crash. Berkun says that these tests did not completely meet the requirements of simulation necessary for proper stress studies. One hesitates to think what the effects on soldiers would have been had they really thought that they were in the middle of a nuclear holocaust.

Beside the obvious drop in performance, there were several subsidiary findings. One was that older men, with more military experience though with less education and in some cases less aptitude, behaved rather differently on these tests. Figure 35 shows how this operates. The experienced troops actually did better under the stress conditions, poorer in the control situation. Berkun attributes this to the fact that in the control conditions the experienced men quickly became bored – they could not see the relevance of the experiment and had no motivation to do well; it was not what they came into the army for or to do. But, says Berkun, 'when the situation becomes personally important, then their maturity and experience are brought into play, and they cope with their environment somewhat better than their untrained counterparts'.

He thought that these results also showed that many experienced soldiers became so used to playing war-type games that even stress

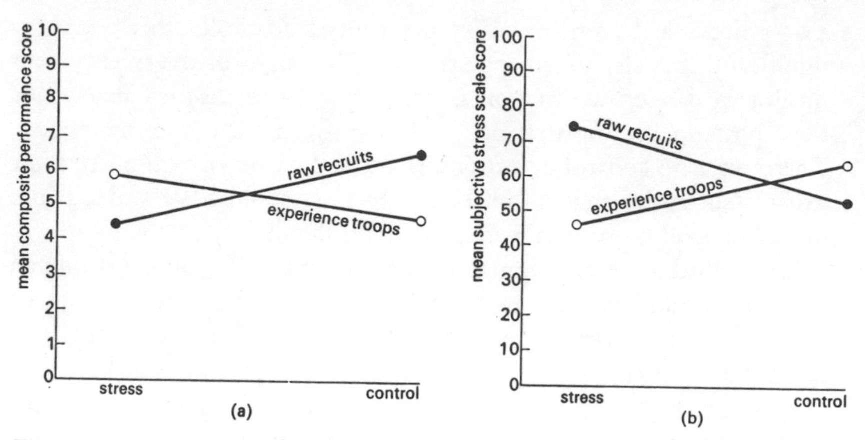

Figure 35 Comparison of performance of experienced and less experienced troops: (a) on ditching experiment; (b) on artillery experiment.

situations were not as stressful as they were for raw recruits. He therefore recommended that stress experiments take this into account.

Berkun was also interested, of course, in contrasting the effective people with the ineffective people in his samples. One of the original paper and pencil tests he gave his men was the classification inventory, a combat potential sub-test of the US Army Classification Battery (ACB). The test is an autobiographical personality, interest, attitude and activities questionnaire. The more effective personnel, Berkun found, achieved a higher score on this test. In other words, poor performers are more anxious, achieved less in their early lives, have less general aptitude, less self-confidence and lower intelligence than their better performing counterparts.

FEELINGS AND PHYSIOLOGY

Did the situations devised by Berkun actually produce stress? In addition to the changes in performance that we have already recorded, Berkun recorded changes in attitudes and physiology. The following four figures show the stress which the men were under. Figure 36 shows that the men in the stress situations actually *felt* as though they were under stress. The words on the left-hand side of the scale represent a gradient of feelings, from 'fine and wonderful' at one extreme to 'frightened and scared stiff' at the other. In each of the three situations the men in the stressed situations felt worse than in the controls.

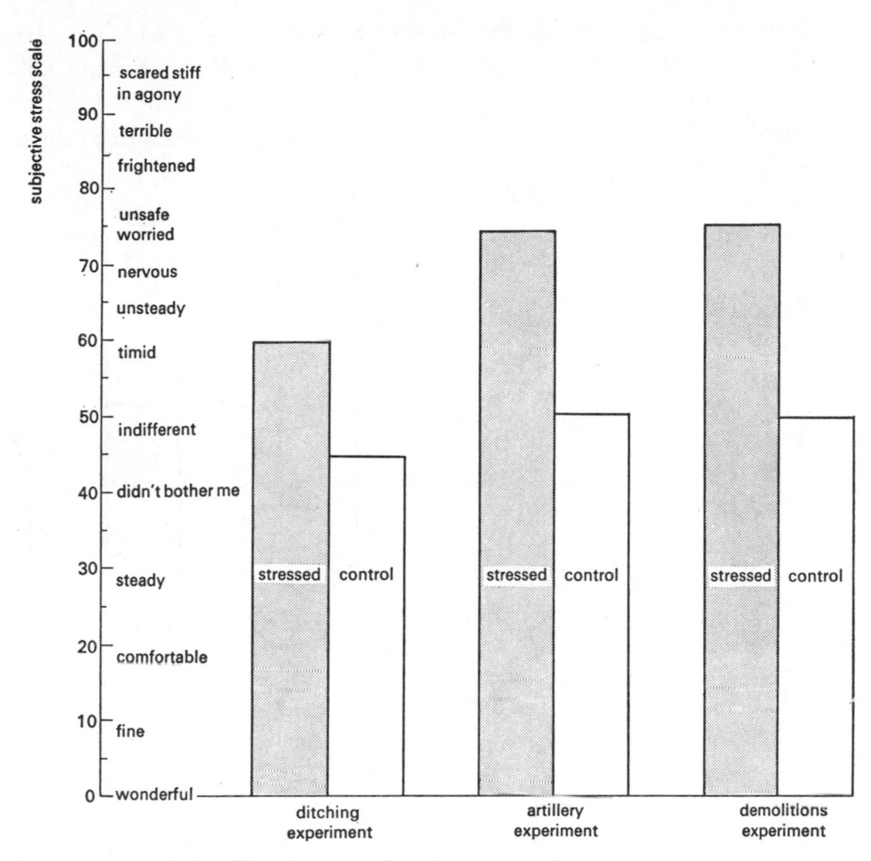

Figure 36 Men's self-ratings on subjective stress scale.

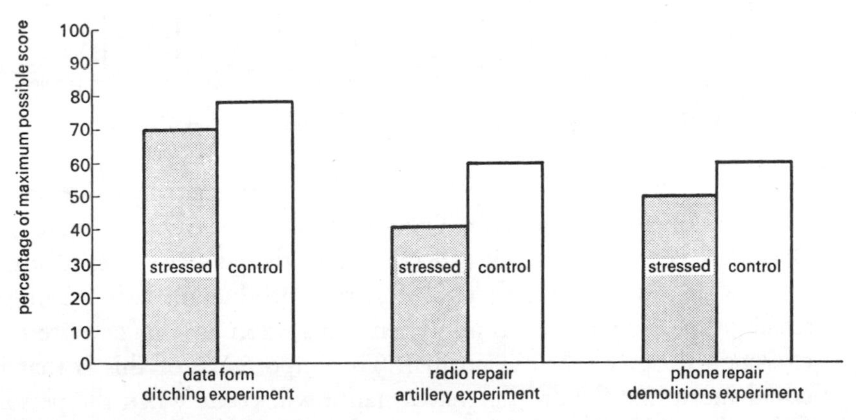

Figure 37 Performance under stress.

Figure 37 shows graphically the performance changes which we have already discussed. Figures 38 and 39 show that in all experiments either the eosinophil count was lowered or the urinary steroid level was raised, or both. In other words the stress situations did indeed disrupt body physiology.

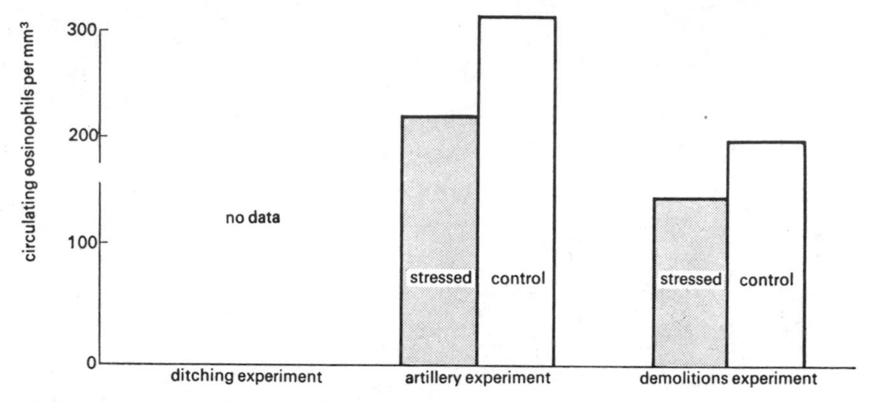

Figure 38 Eosinophil count for stressed and unstressed groups.

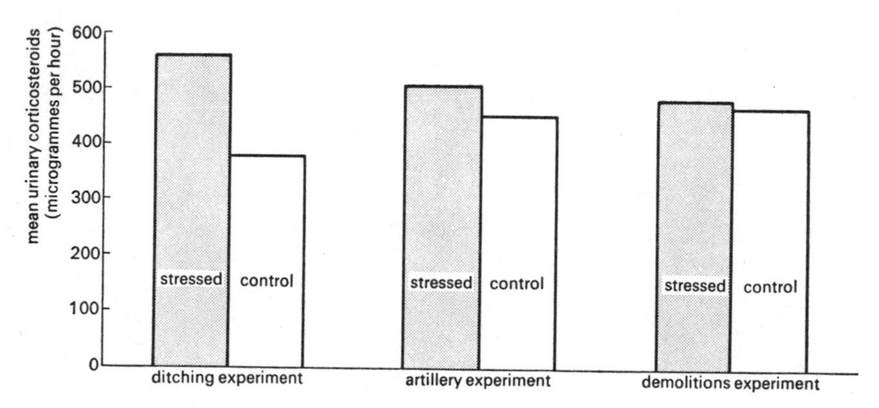

Figure 39 Urinary steroid levels for stressed and unstressed groups.

Berkun compared the physiological data with the subjective stress scale data on an average basis. The relationship is shown in Figure 40. Berkun's theory is that this is a curve and that under extreme stress there is an immediate high rate of steriod production lasting only a small proportion of the time. Then exhaustion sets in and steroid excretion drops far below normal. The importance of this is that it could mean that the drop in performance will come when the person feels only mildly stressed (between timid and unsteady on the subjective

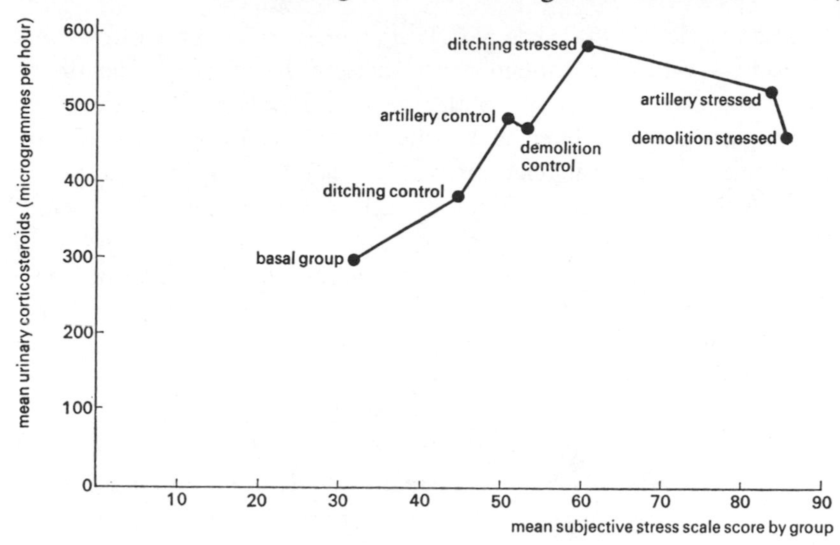

Figure 40 Steroid excretion as a function of subjective stress scale.

stress scale) and not when he is 'scared stiff' (a contradiction in terms, in fact).

A series of studies by Peter Bourne has looked directly at the relationship between psychological and physiological reactions to stress of men actually operating in the field (12). Bourne was attached for a time to a Small Independent Action Force (SIAF) in Vietnam. These units worked with local forces to establish small, isolated, highly fortified camps from which they attempted to widen the American and South Vietnamese sphere of influence. The unit which Bourne studied was located in territory controlled by the Viet Cong and situated so as to obstruct the flow of arms and men from the Ho Chi Minh Trail into the central highlands of South Vietnam.

The basic structure of the unit – a twelve-man 'A' team – consisted of two officers and ten men, including two medical specialists, two communications specialists, an intelligence sergeant who recruited and trained agents from the local population, a man in charge of civic action projects, an expert on demolitions, and two weapons experts.

Because of the highly dangerous nature of the work, Bourne was not able to undertake a full laboratory investigation, but he did take daily urine samples to measure cortico-steroid excretion, interviewed the men every day and also gave the men a mood questionnaire for a month.

Bourne specifically looked at the differences between the officers and men in the unit. The captain was definitely the formal leader of the team: but he was usually young, with little combat experience – a crucial factor for acceptance by the group. So long as the captain maintained the formal relations of his assigned rank his status went unchallenged. However, this forced him to keep some social distance from the other men. His greatest asset was his control over information coming into the camp and apart from the intelligence officer he knew more than anyone about the group's tactical position at any one time. The way he released this information afforded him some control in the group before he had earned it through combat. The captain, however, could not escape the value system of the group, which placed a premium on the ability to prove oneself in combat. Led by the team commander, the senior NCO, the team was always urging the captain to attempt highly dangerous missions in order to prove himself. This social pressure on the young captain frequently remained the chief social interaction of the group for a very long time. When the captain did make an exploit it was not unknown for some other member to equal or better this with some other daring act. Escalations often followed and, according to Bourne, this might well have been one reason why the mortality rate of these special forces troops was higher than any other group in any war involving US troops. Bourne adds that the subconscious awareness which many team members had of this process was reflected in the guilt they felt if the team leader was killed. This showed itself in excessive praise – in sharp contrast to the criticisms of him during life – but this only made things more difficult for his successor and the whole process more likely to be repeated.

Bourne also adds that although a team captain could function on his assigned rank alone, it was almost impossible for the team sergeant to function without having gained the full acceptance of the team. He had to maintain his position by force of character alone, as he could not easily fall back on the formal command structure. On the other hand, he was also under strong pressure from the enlisted men to get control of the whole group and so was in perpetual conflict with the formal number one, the captain. Any officer who tried to overcome this by forming an alliance with him could well have destroyed him because of this in the eyes of the other men. The best relationship, Bourne concluded, was when the sergeant could symbolically challenge the captain without the captain feeling that it was a real threat.

Bourne found, not unnaturally, that the possibility of enemy attack overshadowed all other matters in these units and coloured even the most routine activities. The tension fluctuated according to intelligence reports but a characteristic pattern developed: an air of expectancy developed during the late afternoon and early evening, reaching a peak between sunset and midnight and then dissipating as the threat of attack receded again. By morning the tension was at its lowest.

Every man, Bourne reports, prided himself on his individuality and independence. By selection and training the men had an intense faith in their own capabilities which went with an active aggressive behaviour to deal with any threat to their well-being. They would constantly reaffirm their invulnerability in the hazardous missions they kept undertaking. The camp was used as a base and the units preferred to keep the initiative by actively seeking out the enemy. When they had to stay inside the camp perimeter they would fret; competition to be allowed out on patrols was intense. As a result, except at times of actual attack, conflict between the men themselves was sharp. Labelled 'cabin fever', rows over trivial issues were the order of the day, transcending rank, but quickly forgotten. The men would rarely express anger towards the enemy, but instead would treat him with respect. They got on poorly with the Vietnamese special forces, the less so the more they were under attack. They were also constantly railing against central command whom they regarded as not really fighting the war. Within the camp the men were individuals, rather than a team, and they would fiercely mark out their own territory rather than share what little they had. The communications bunker, for instance, was closed to everyone but the communications officers and the captain. The dispensary on occasions was even closed to those who were ill.

This was the way the men coped, somewhat successfully it would appear, in terms of military results, but no one has studied the long-term effects of these constant battles.

Now to the physiological measures. Collections of the urine of the men in 'A' unit were obtained on a twenty-four-hour basis and they were all tested for a steroid called 17-OHCS. It is known that secretion of this hormone from the adrenal gland varies under stress. Figure 41, based on Bourne, shows that the officers had a significantly higher mean level of 17-OHCS in their urine, indicating that they were under greater stress. Next, urine collections were made during a few days in which an attack was made on the unit. Figure 42 shows how the officer and the radio operator (whose job at this time was to work closely with the officer, sending and receiving messages) were totally different in

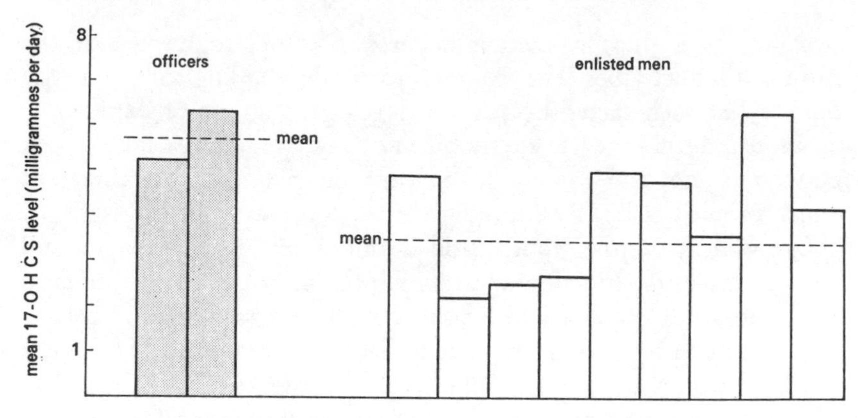

Figure 41 Level of 17-OHCS secretions in individuals in 'A' unit under stress.

their physiological responses when compared with the other men. Either one has to assume that the officers were adapting less well to their role or that the role of officer is more stressful than that of enlisted men. Bourne inclines to this latter view. As an attack approaches, the men can busy themselves with technical matters and in so doing cope with their anxiety – hence their lower level of 17-OHCS secretions. The captain, on the other hand, is at these time in constant communication with the base camp; he is in full possession of all the available facts so knows how bleak their position might be; he also knows that in the forthcoming action he, more than anyone else, will be on trial. He has no way of dissipating his tension at this time so his 17-OHCS secretions build up. The secretion of the radio operator also builds up due

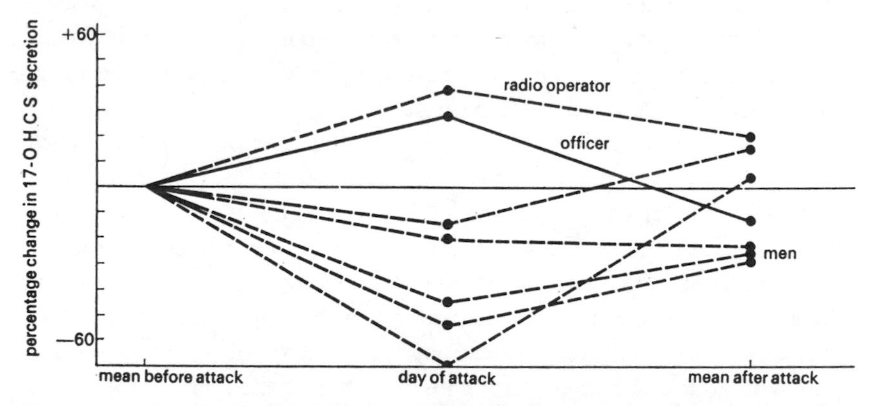

Figure 42 Secretions of 17-OHCS in individuals in 'A' unit during an attack lasting several days.

no doubt to the fact that at this time he works closely with the officer, sharing the decision-making and also being unable to get rid of his tension in more activity. V. H. Marchbanks, who studied 17-OCHS levels among the crews of B-52 bombers, also found that the leaders had much higher levels of secretion. It thus appears that it is the assigned role of leader in a combat crew that makes for added stress, rather than personality differences between officers and men (13).[1]

Bourne also looked at helicopter ambulance crews because he felt they characterized better than anyone the nature of stress in battle – long periods of inactivity interspersed by short patches exposed to intense danger. These crews, operating from Tan Son Nhut Airport, manned the medevac helicopters and their job was to fly into combat areas, land in clearings and evacuate the wounded. Since these clearings were invariably in areas where fighting was either still going on or had just finished, the missions were exceptionally dangerous and the helicopters, flying low and slow, highly vulnerable targets.

Bourne flew several of these missions in acts of striking bravery in order to obtain his material for data analysis. He was able to take daily urine samples for a month from seven of the medical personnel flying on the missions. The men also filled out the Minnesota Multiphasic Personality Inventory and a mood adjective checklist every day. Every third day they reported to Bourne anything unusual which they felt had happened to them since the previous listings were made. At the end of the study period the eleven pilots with whom they flew were asked to rate their effectiveness.

Bourne found, first, that the 17-OHCS secretions could not be related to whether the men were flying or grounded, or to whether they were under fire or not. On the other hand, when compared with what was normal for the men's body weight, in all cases the men were secreting less than they would have been in a non-stress situation. Most of the men denied that their jobs involved any real danger and were reluctant to talk about it. (However, these crews are reckoned by others to have had one of the most dangerous jobs in the business.) The men also emphasized the prestige they had among other men, as if to minimize the danger they had to face. But, says Bourne, three of the seven expressed strong (and on the surface perfectly rational) desires to leave their units. To what extent was this due to the danger, even

1. Different military specialities, not just rank, affect breakdown. One HumRRO study has calculated the psychiatric attrition of various sorts of personnel in a conventional war (see p. 328).

though it was not acknowledged? Bourne is hesitant on this. All of the men, however, had sophisticated defences which enabled them to minimize the danger and keep going. These ranged from belief in God to what Bourne calls a 'statistical defence', whereby the man had calculated the figure for being killed on any one day. This was for him reassuringly small – though of course he had not made the calculation for being killed during the course of the months he was actually there – a much higher figure. What was also interesting, however, was that those medics whose secretions came nearest to their body-weight averages (that is, were depressed least because they were defended least) were rated by the pilots as the best at their job. As Bourne concludes, 'The man who structures his perception of the environment to make it more acceptable may achieve greater comfort in his denial but may not perform as well.' This is an important comment. It means, in effect, that those who seem to have adjusted well, and do not appear afraid, may well have done so at the expense of being less effective at their job. If true, it poses a difficult problem for command. It suggests that units difficult to handle – like the 'A' squad – may be more effective than more placid units (14).

Given the rather unsatisfactory results of the psychological studies mentioned at the beginning of this chapter, and the rather controversial methods used to carry them out, this latter line of research by Peter Bourne seems to hold out most hope for a greater understanding of stress in the future. It also seems more likely, on the face of it, that a physiological way of controlling fearful reactions may be easier to attain than the more psychological approach – through selection or training. The ethical problems that such a line of research could raise are discussed at the end of the book.

10 The Psychological Effects of Weapons

LIFE AS A TARGET: THE SOLDIER AND HIS WEAPONS

The modern battlefield is usually a pretty deserted place. Only an occasional tank or aircraft may be seen, and not too many people since all are taught to take cover. The immediate impact of war is noise of many kinds; usually incredibly loud, often painfully so – small arms, shells, bombs and aircraft during an attack. This can be prepared for but the first time in real battle is always different. Danger and noise are not always the same thing. Soldiers also become acclimatized to the sight of dead and wounded (though there are two exceptions to this: death by fire or napalm, and the death of a close friend).

The best review of the military research on the psychological effects of weapons which I have come across is that by John Naylor of Ohio State University and was his contribution to the first symposium to be held on 'The psychological effects of non-nuclear weapons' at the Eglin Air Force base in Florida in 1964 (1), but organized by the University of Oklahoma. Much of this symposium's proceedings were classified but Dr Naylor's review was not. From his paper it emerges that rudimentary studies into the psychological effects of weapons were carried out during the First World War but that the first detailed survey was made by John Dollard who interviewed 300 combat veterans from the Spanish Civil War. One of the things he asked the men was: 'Did you experience fear when going into your first action?' Seventy-four per cent of the men said yes and 26 per cent said no. Only a few men were unafraid and many continued to experience fear when going into subsequent actions. In reply to the question as to whether they were most afraid before, during or after the action, 71 per cent said before, 15 per cent said during and 14 per cent after. Dollard also found that in this war at least (and we shall see that fears do change from war to war) the major fear was not of being killed but of being wounded. Furthermore, the probability of a wound was not related to how much it was feared – the rare wounds to eyes and genitals, for example, were the most feared. In that war, the most feared weapon was bomb shrapnel, with trench mortars, artillery and expanding bullets less so, and grenades, strafing, machine guns and tanks being least feared.

Dollard's view was that techniques of fear control should be explored: fear should be discussed with the men before battle, with the normal reactions explained; the result of surveys showing that most men experienced fear before but not during battle should also be pointed out; and fear should then be suppressed once the fighting had started since, in Dollard's view, it can be contagious.

In the Second World War, Finan, studying US soldiers in Tunisia, found that by and large weapons perceived as the most dangerous were the most frightening (2). Certain weapons, notably the dive bomber and the horizontal bomber, were judged frightening even though they were not that dangerous; the chief fear of the dive bomber was the noise that it created.

Perhaps a better measure of fear is the extent to which weapons not physically maiming are judged to contribute towards psychiatric breakdown. A study of 120 US psychiatric casualties in the North African theatre in the Second World War showed that 97 per cent feared shell fire most; of these dive bombing accounted for 42 per cent, artillery for 35 per cent and high level bombing 11 per cent (3).

But just as the picture changes from war to war, so it also changes as a battle progresses. Figure 43, for example, shows how soldiers get

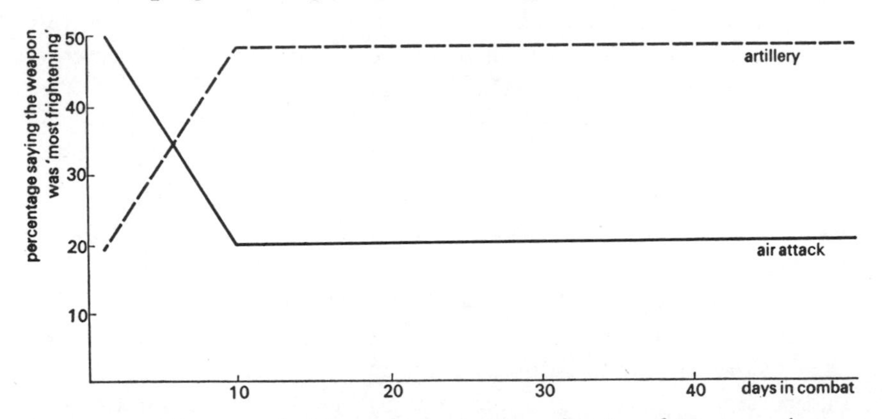

Figure 43 Weapons judged 'most frightening' as a function of time in combat.

It appears from case reports and surveys that the battle incident most likely to 'break' a soldier is the explosion of a shell in his immediate vicinity. One study of 115 consecutive patients diagnosed as blast-concussed showed that 105 were suffering mainly from a form of acute anxiety resulting from exposure to a nearby explosion.

The reasons why men fear weapons also produces some interesting insights (7). Consider Table 22:

Table 22 Percentage of men's reasons for fear of weapons

	Mortar	*88s*	*Machine gun*	*Dive bomber*	*High bomber*
Accuracy	37	31	16	11	23
Lack of warning	19	11	15	5	11
Rapidity of fire	8	7	42	0	1
Noise	11	19	6	48	21
No defence	1	2	0	4	14
Other	24	3	21	32	30

Fear equals accuracy, noise, rapidity of fire – noise being the only truly non-lethal characteristic that is feared. In general then it is the lethal inaccurate weapon that is feared more than the non-lethal but more accurate weapon. (Such elementary knowledge is useful, apart from other things, to make the most of in psychological warfare.) The fact that accuracy does not always count in fear is also shown by the fact that most soldiers fear night attacks far more than day ones.

Being wounded alters a man's perception of a weapon. Men tend to overestimate shell wounds whereas they will, for example, risk a bullet as a way out of battle (8). It has also been said that a man wounded by a particular weapon does not have his priorities changed so far as fear is concerned. On the other hand there do appear to be psychological factors bound up in whether or not a man dies from a weapon given that he has been hit. A study by the HQ Army Ground Forces (USA) showed that the machine gun on this score appeared to be most lethal since 50 per cent of hits were fatal, whereas for artillery the figure is only 20 per cent and for mortar even less (9). There are, of course, more total deaths from fragmentation weapons because there are more hits: but, says this report, 'Possibly the probability of being killed if hit by a weapon can be related to the psychological effect of that weapon.'

By the time the Korean War came round, the Americans in particular were ready to study the psychological effects of weapons in a rather more systematic way. Studies were also done with prisoners of war to see what weapons they feared most.

North Korean and Chinese POWs were interrogated by personnel from the Operations Research Office of Johns Hopkins University as to their fears of enemy weapons (10). They found that artillery was the most feared American weapon but that aircraft, especially strafing, was the second most feared. They also found that the Chinese feared air attacks when marching and were very wary about the use of flares which could betray their whereabouts.

From Korea on, research became even more systematically organized. Donald Mills and Wesley Yale, for example, undertook a study of human reactions to fragmentation weapons (11). This was really a study of neutralization fire which seeks to negate the enemy ability to fire, though at the same time seeking to inflict maximum casualties – in other words keeping the enemy's head down for purposes of manoeuvre. Mills and Yale were particularly interested in the distances over which various weapons would cause men to keep their head down – distances greater than the lethal area. They found that this was roughly speaking 40 per cent more than the lethal range – a useful statistic for a defending company.

W. S. Vaughan and P. G. Walker looked at the psychological effects of small arms. They interviewed in depth fifty combat veterans after Korea about the relative dangerousness of six weapons under conditions of assaulting or defending. They looked at the M1 rifle, the BAR, the light machine gun, 60 mm mortar, hand-grenade and 57 mm recoilless rifle. The mortar was considered the most dangerous in defending whilst in attacks the light machine gun and BAR were considered more dangerous than the mortar – the M1 in both cases being considered the least dangerous (12). Finally, pattern of fire. In general the studies suggest that short bursts of fire are thought more frightening than continuous fire and especially so when these bursts are random. Using targets it was found that more hits result from this random pattern also (13).

LIFE ON A TARGET: THE CIVILIAN'S FEAR OF WEAPONS

It should be remembered that there are significant differences with regard to weapons for soldiers and civilians. Both can get bombed, gassed, shot or captured. But the soldier usually has the opportunity to fight back more than the civilian – he has his own weapon, is trained in the appropriate skills, and is part of a unit which will support him. And finally, the fighting for the soldiers is often much less intermittent. The civilian, on the other hand, can normally not fight back; moreover, he or she has to live with the uncertainties that stem from relatives, who are soldiers, being out of touch in dangerous circumstances for weeks on end. And not least, a large proportion of the civilian population will be children, many too young to understand fully what is happening but perhaps only frightened the more so because of that.

Early in the Second World War there was some anxiety among the relevant authorities as to how the civilian population would react to the anticipated new tactic, the air raid. Some predicted mass emotional disturbance; others expected widespread panic. In fact, these fears proved largely unfounded. It would be wrong, however, to think that, because of this, bombings, or other weapons, produce no psychological side-effects other than short-lived fear. Many surveys were done during the Second World War into the social and psychological effects of bombing and many of the results were released a few years later. A full picture did not emerge, however, until the conference organized by the University of Oklahoma (14).

Part of the conference was held in secret, but the unclassified part had two main aims. The first of these was to develop 'a model' of the way weapons affect whole communities. The second aim was specifically to build a 'psychological index of weapons effectiveness' – in other words, to arrive at a rank ordering of weapons to be used under certain circumstances because of their ability to frighten a particular enemy when, because of the nature of the warfare, these weapons were perhaps not very efficient killers. The conference dealt with the effects on both civilians and soldiers.

First then, the effects of bombing. Most of the unclassified work in this area was done in the Second World War, though some is available from Korea and Vietnam. By and large the results suggest that bombing has less of a psychological impact than might be expected. Atkin, writing in the *Lancet* in 1941, for example, analysed 300 admissions to a mental hospital in Great Britain between September 1940 and April 1941 to determine in how many cases air raids were a cause of the admission (15). He concluded that only in some 13 per cent were air raids a major cause and in a further 2 per cent a contributory one – 15 per cent in all (16). Aside from these particular admissions, most other studies of bombings have concentrated on the general effects on morale, work-rate and other changes in behaviour. Fred Ikle, for instance, observed that there is a change in behaviour after bombing (17) – but others have noticed how quickly people adapt to this. The main point is that bombing leaves many more homeless than it actually kills, and this is what produces the change in behaviour.

The Rand Corporation, in its study of the psychological impact of air raids, published in 1949, found that there is only a very slight increase in the 'more or less long-term' psychological disorders such as psychosis and chronic neurosis, and that these occur primarily among already predisposed persons. With increased severity of attack there is increased incidence of 'emotional shock', but recovery usually takes only a few hours or days at the most. There is also a slight increase in psychosomatic disorders, such as ulcers. Psychological changes – as opposed to psychiatric – were: an increased communicativeness, an increase in fatalistic attitudes, an increase in religious interest and the growth of superstitious rituals (18). Other studies have confirmed another Rand finding that people rapidly begin to discriminate real danger and that though psychological morale does drop in a fairly straight relationship with bomb volume, behavioural morale – work-rate, for instance – is unrelated to this. The Rand people, and others, also noted that anti-aircraft fire, however bad it may be in terms of the number of enemy planes it actually brings down, performs a very useful psychological function of enabling people to feel that they are at least attempting to fight back.

Ignacio Matte, a psychiatrist working in Britain during the Second World War, noted also the value of anti-aircraft firing and commented, in a paper in the *Journal of Nervous and Mental Disease*, that the gradual build-up of the Germans' air raids in fact enabled the British to adjust to them – implying that a faster build-up might have produced more psychological effects (19).

One of the most ambitious programmes in the Second World War was the US Strategic Bombing Survey group in Germany which followed closely behind the advance of the Allied armies. The group attempted to assess all aspects of the effects of strategic bombing from physical damage to psychological effects – this being done by a Morale Division whose objective was 'To determine the direct and indirect effects of bombing upon the attitudes, behavior and health of the civilian population.' The division interviewed some 3700 German civilians and also examined selected documents and made in-depth interrogations. The documents included papers on industrial absenteeism, Gestapo arrests and captured German mail. The results are perhaps best presented in comparison with the same division's work in Japan, using similar methods. In general, it would seem that bombing had a much greater effect on the Japanese than it did on either the Germans or the British. The Morale Division's conclusions on Germany were not very clear-cut but on Japan they were able to state that air attacks were the 'most important single factor' in causing the Japanese people to have doubts of victory and also the single most important thing to make them feel certain of defeat. They were also the greatest worry during the war. 'A small but consistent difference' in morale was also found between bombed and unbombed populations. There was no evidence that the experience of bombing 'ever raised the fighting morale of the Japanese' (20).

P. E. Vernon, who later became a very well-known British educational psychologist, made his own survey of some fifty psychologists and doctors during the Second World War. He found, as did others, that adaptation to the night raids was very rapid but he also pointed out that this was much more trying for those living alone than those who had others to be with. People whose houses were demolished were very nervous of raids for a few days but recovered quickly after that. He also noted that there were other psychological reactions to air raids such as exaggerated accounts of raids and the damage they do. It is obviously well worth the authorities being aware that this may happen and quickly broadcasting the truth to stop any rumours. Vernon noted that children and young people adapted to air raids better than did adults (with women being the most fearful), but also made the observation, confirmed by others, that it is people living in areas on the edge of or near to disaster spots who are most prone to break down (21).

In Korea the US Air Force was more systematic in its approach to bombing and its research people developed the concept of 'elasticity' regarding its effects. This concept concerns the link between physical destruction and social change. Thus the USAF found that 50 per cent of housing has to be destroyed before 20 per cent of the population will leave a city, say, or a township (22). Even so the work capacity of the town will not fall off very much because most of the early leavers are the non-workers – children and the old. A rule of thumb, therefore, became that you have to destroy at least half the housing before you can hope to have any effect at all on the social structure of the enemy and begin to affect his morale and work capacity in any permanent fashion. On the other hand, it was found that disruption of the food supply affects the society much more quickly – it seems to be far less 'elastic' than the housing. This type of social-psychological study is more advanced than that carried out in the Second World War and clearly leads to priorities in bombing targets. But it is worth noting that surveys of civilian attitudes to weapons were, so far as I know, not a feature of research in Vietnam and other conflicts such as Ulster.

THE 'WEAPONS EFFECT'

Turning from the effect of weapons in wartime for a moment, there is some evidence that weapons are themselves aggression-eliciting cues. Leonard Berkowitz showed this most clearly when he had subjects give an electric shock to people who were apparently getting a research task wrong (23). In some cases a gun 'happened' to be present on the table of the experimental room. Berkowitz found that when the gun was there, the subjects administered a stronger, longer shock than when no gun was present. This suggests that a soldier or policeman is likely to be more aggressive when carrying a gun than when he is not, as events at the Chicago Democratic Convention in 1968 proved. Police with truncheons and guns were highly likely to use their truncheons, whereas police without guns were unlikely to use any other weapon.

In laboratory studies it has been found that if a person inflicting electric shocks has a shallow gradient – that is, a large number of discrete shock strengths between the least and the most powerful – then he gives strong shocks more easily than when the gradient is steep, that is, when there are only a few points between the two extremes (24).

We can perhaps cautiously extrapolate from these findings and suggest that by possessing a nuclear capability, a nation is more likely to use less devastating weapons, while, at the same time, by possessing a whole range of weapons, from the so-called 'less-lethal' to nuclear missiles, a nation is more likely to move towards a nuclear engagement.

PSYCHOLOGICAL ASPECTS OF CHEMICAL AND BIOLOGICAL WARFARE

By and large chemical and biological weapons produce relatively fewer physical than psychiatric casualties. They provide a good example of a weapon with a psychological impact greater than its real lethality (25).

The chemical weapons that are of psychological significance are the nerve gases and the psychochemicals. Nerve gases are organophosphorous compounds which inhibit tissue cholinesterase in man in small dosages – essentially a disruption of nerve impulse transmission. The gases produce a variety of symptoms including vomiting, blurred vision, fatigue, convulsions, and coma. Psychological effects are tension, anxiety and jitteriness, emotional lability, excessive dreaming, nightmares and insomnia, apathy, depression, abnormal EEG patterns, difficulty in concentrating and remembering (26).

The two main nerve gases, Sarin and agent VX, can both be used to create short-term respiratory hazards; VX is more effective over a long period. Explosive shells have been developed to carry nerve gases. VX can apparently contaminate the vegetation of an area and even military equipment, so that gradual systematic exposure long after the shell has burst is possible.

There are also two main types of psychochemicals: LSD and agent BZ. LSD creates severe problems in a military context. It has been found for example that drugged soldiers may behave in an apparently normal way if there are undrugged men in their unit – but not always. Motivation appears to be a crucial factor: under some circumstances military performance could actually be improved after ingestion of the drug. It cannot, therefore, be regarded as a reliable 'weapon'.

Some specific military implications of LSD use have been spelled out in a document which has not yet been the subject of much public attention. This refers to experiments carried out at the US Army's Edgwood Arsenal in Maryland and was part of a general series which looked at 'The human factors effects of chemical agents' (27).[1] The research is unique in being the only one which actually gives figures about the incapacitating effects of LSD in a military context. The men had to fire their rifles at eight fixed-silhouette targets, at 200 metres, firing from left to right. Figure 44 refers to this experiment; similar results were found for grenade throwing and rifle loading.

All these factors make LSD unsuitable for use on the battlefield, though it is generally thought that it could be useful to knock out, say, a whole city, by contaminating the water.

More promising than LSD is agent BZ. This is a psychochemical that was developed specifically as an incapacitating agent for chemical warfare. In small doses it causes sleepiness and decreased alertness;

1. The letters LSD are not actually mentioned in the document, but I have reason to believe that this was the substance under study.

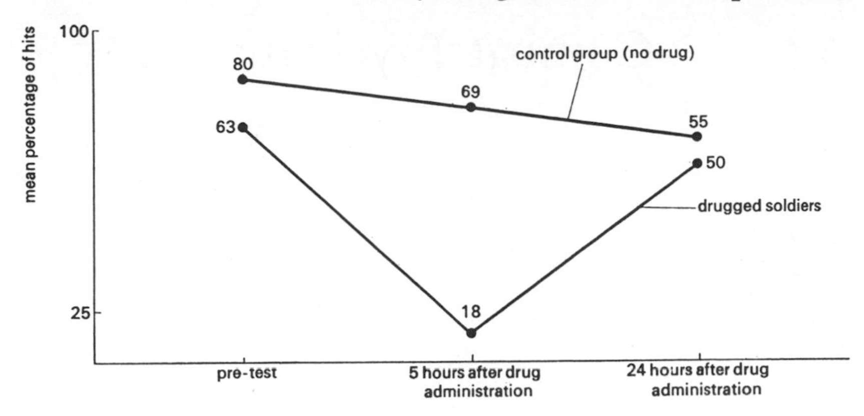

Figure 44 Effects of LSD on combat efficiency.

within four to twelve hours there is an inability to respond effectively to the environment or to move about. In some ways it is the perfect military weapon.

The World Health Organization has published a booklet which examines the physical effects and psychosocial consequences of biological and chemical weapons (28). The WHO team clearly foresees that the use of chemical weapons is much more likely than a hasty nuclear episode, for with the threat of chemical warfare, it will not be at all clear whether hostilities have actually broken out because the symptoms resulting from the use of chemical weapons are not always immediately apparent. The WHO also considers the horrifying possibility that the use of chemical weapons might lead directly to nuclear warfare. Troops exposed to psychochemicals may begin to behave in an unusual way, may retaliate without provocation, for example (this can happen with LSD), and press the nuclear button. Chilling.

11 Combat Psychiatry

GETTING TO THE FIGHT

Right from the declaration of war, soldiers' worries may be expected to increase, even though it could be some time before they are actually in 'hostilities'. Psychiatrist Franklin Del Jones is the only man I know who has studied this phase of operations and that was when he was with the US Army 25th Division in the summer of 1965 (1). The division was then in Hawaii and it was not known for certain whether it would stay there as part of the strategic reserve or go to Vietnam. This uncertainty was naturally the chief topic of conversation while it remained. For security reasons, confirmation that the unit was in fact going to Vietnam was only given on the day before departure.

However, it had become obvious to most that the unit was going and reactions accordingly set in. Jones says: 'Some wives were inconsolable. Some individuals discovered physical defects in themselves which they thought would make them vulnerable, such as decreased hearing, a childhood heart murmur, mild hypertension, a "trick" knee or simple obesity.' Only rarely did soldiers go so far as to inflict wounds on themselves, but Jones does recall one case of a medic who anaesthetized a friend's foot with xylocaine and then shot him with an M14 rifle. More popular was the committal of military crimes (usually AWOL or insubordination) in an attempt to achieve medical or administrative separation from the service. 'They just wanted out,' says Jones, 'any kind of discharge would do.' But the unit did not stand for it: it even reprieved the stockade prisoners and took them with it.

Jones emphasizes though that the number who attempted to manipulate their way out of danger was quite small and the more common response was for a soldier to express relief finally to know for sure that he was going to Vietnam. Most were not happy about it but took it in their stride and sensibly began preparing themselves. Some took courses in Vietnamese, or read books on tropical diseases, insects and reptiles. Others bought hunting knives, special water-repellant clothing and enormous amounts of soap – rumours had already got to them that there was a shortage. Exercising became fashionable. These activities served a useful purpose, Jones believes, for while you are 'busy learning a language or practising with a knife or running to increase your lung and leg power, you don't have as much time to consider the thought of dying or being crippled or of being separated from your loved ones'. It is therefore helpful, he suggests, for units to make available relevant courses, books and equipment that keep the men usefully preoccupied. Even if, for security reasons, it has to be on an 'unofficial' basis.

Before it can actually start fighting, however, the unit will normally have to travel to the battle zone. Two types of difficulty are posed by this. One is the travel itself – rapid time-zone changes have pronounced psychological effects on some people, for example – and the other is the psychological significance of travel itself. There is little an army can do about some of the effects of travel. Seventeen per cent of people, for example, suffer from motion sickness and, despite much research, we still do not know how to cope with this. But changes in time-zone and climate can be predicted and their debilitating effects can, to some extent, be tempered with the right procedures.

At the end of 1972, a contingent of the Special Air Services, the British Army's crack counter-insurgency unit, left its headquarters in Hereford bound for Malaya on a continuous twenty-two hour flight. Half a dozen psychologists had gone out before, armed with basic information about the men, collected in Hereford in the weeks prior to their journey. On arrival at Malaya (the exact location is still an official secret) the men were continuously monitored doing various combat and support tasks. As a result of these tests the psychologists were able to come to two important conclusions. Again the details are classified but in general the psychologists found that the men who performed best on the first couple of days after the flight were by no means the men who performed best later on (the experiment went on for several weeks). Second, the psychologists found they were able to predict how well men would perform in a tropical climate on the basis of several psychological tests given in rural, and normally chilly, Hereford. The tests themselves are secret but the results have now been put into effect in the selection of British SAS men who may have to serve abroad at short notice (2).

Robert Bernstein, a colonel in the US Medical Corps, reviewed the emotional and physical problems encountered in moving troops into combat in a report prepared for the US Army War College (3). Certain procedures are followed on staging posts deliberately to minimize psychiatric casualties (the staging post is, after all, the last really safe location the men will know for an unspecified period). For example, the stockade is made very much into a psychiatric prison so that soldiers know that a simple crime will not excuse them from battle. But the officers are listened to more than psychiatrists in the hospitals at staging posts: in this way habitual malingerers are not allowed to parade their illness to unsuspecting medical officers (4).

Even so there will be situations where these tactics will not be enough because other features of the destination make further adjustment necessary. The most important from the military standpoint is high altitude, because guerrillas often hide out in the mountains. At the 12th Annual US Army Human Factors Research and Development Conference, held in October 1966, W. O. Evans from the US Army Medical Research and Nutrition Laboratory in Denver, Colorado, reported the effects on troops of being rushed from sea level to Pikes Peak – altitude 14 110 feet. Most interesting, perhaps, was the finding he reported which showed that there was a 35 per cent decrement in the handgun-firing ability of the men during the first few days. Normal functioning did not return for two weeks – and the decrement applied both to speed of firing and to accuracy (see Figure 6, p. 63) (5).

More basic factors were also examined in the study. Different types of strength, as shown in Figure 45, were differentially affected.

In such a situation it would appear that the capacity of the troops is

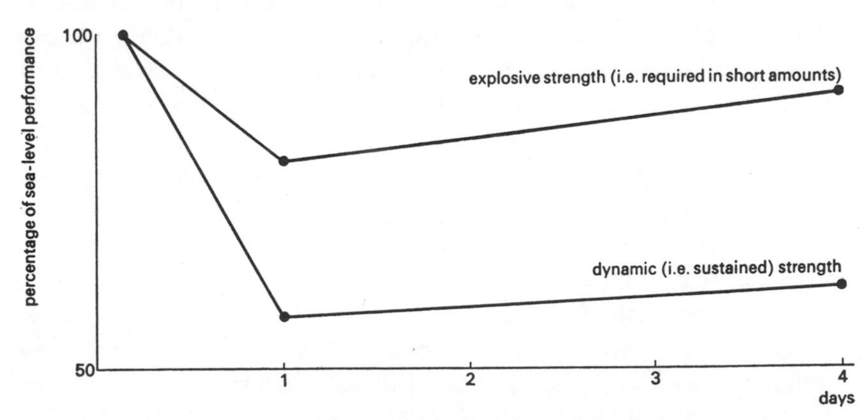

Figure 45 Effect of altitude change on different types of strength.

significantly depressed for a few days. The symptom scale (Figure 46) also leads one to expect changes in mood of the men and a corresponding change in group attitudes. This may well mean that the tasks required of a leader will change in these circumstances. The men become noticeably more irritable (the officers as well), lazier, and sleep suffers and makes matters worse. It should help if the men are told what to expect; the working day should be organized differently as well, to begin with. For example, watches should be shorter than usual, with more men on watch together; and stints of work involving physical strength should be shortened – in the first few days at least.

SAIGON PSYCHOTICS: PSYCHOLOGICAL REACTION PATTERNS IN BATTLE

However much they have been trained, battleproofed or indoctrinated, most men are scared stiff when they first arrive in the war zone. Let us rejoin Franklin Del Jones's account as the US Army 25th Division arrived in Vietnam (6).

The initial period, he says, lasts from three weeks to three months. He refers to it as the 'arrival period' and it is characterized, he says, by *apprehensive enthusiasm.* (We must remember throughout that the US soldier was in Vietnam for exactly one year and he knew this from the moment he arrived.) 'The apprehension', says Jones, 'relates to fear of death, separation from family and stateside surroundings and expectations of a year of discomfort and deprivation. The enthusiasm is a reaction formation against these feelings.' The men intended to 'get this little war over with'. Jones says this reaction was reminiscent of the first few months of the Korean conflict – before the Chinese entered the war.

However, the immediate arrival period, says Jones, can be quite stressful for some men. Fatigue after a long flight is partly to blame but group identity also takes time to settle. In 1966 at the 3rd Field Hospital in Saigon, a centre for incoming GIs, Jones would see a supposedly psychotic patient about once a week. 'Whether he was agitated, fighting and panic stricken or mute and unresponsive the story was generally the same.' He had been normal until the aircraft bringing him to Vietnam landed at Tan Son Nhut Air Base, but on the way to the truck or helicopter which was to take him to the 90th replacement centre or to his assignment station, 'he would suddenly faint or fall and thrash about or just sit down and not respond'. Jones developed a

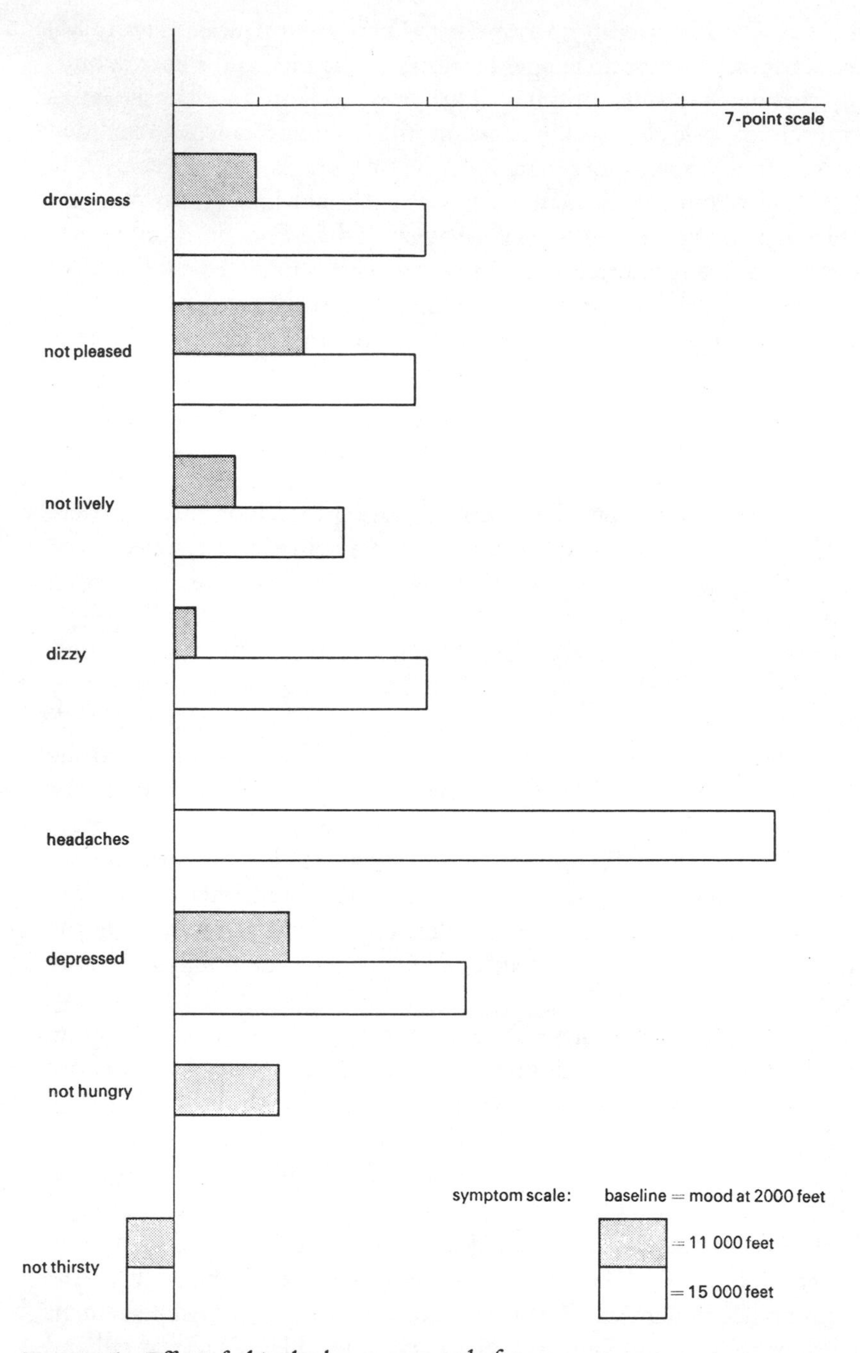

Figure 46 Effect of altitude change on mood of troops.

standard operational procedure which he found worked but he freely admits was designed not to drag him from the bar on his free nights. The medical personnel would carry out as much of a physical examination as possible (it always turned out normal), use a tranquillizer injection if the patient was agitated, say a few soothing things and let him know he would soon be going to his duty station (this was important). If this was not successful they kept him overnight and invariably he was ready to go on the next day.

During this initial period Jones also stresses the importance for the soldier of culture shock. In Vietnam this ranged from the distinctive smells of villages, to the sight of men walking down the street hand-in-hand. Coming on top of other stresses, and with the anticipation of even more danger in the offing, these cultural stresses are real enough and were apt to generate psychiatric patients with a benign sort of illness: the first few weeks saw a relatively high incidence of sleep-walking, bed-wetting, nightmares and other behaviour patterns aimed at environmental manipulation. Treatment at these times was a firm stand by the medical officer plus the possible order of night duty or a rubber mattress.

Next comes the middle period, says Jones. This lasts until about the last month of the tour and has been called the period of *resignation*, in which the soldier is in a chronically depressed state. Jones, however, also describes this period as that of maximum productivity. Somehow, he says, the soldier has managed some form of adjustment. The artillery barrages at night no longer awaken him, his mail is arriving regularly and he watches a film every night. He has established a productive but hopefully not over-tiring work routine. He takes his turn guarding the perimeter with some 'serenity', firing no more rounds at shadows and water buffalo than anyone else. His enemy now is boredom. Weekly or monthly card games become nightly; problems of alcoholism arise. The dispensary physician begins to see what Jones calls *diseases of loneliness*. Loss of appetite and weight is almost universal. Psychosomatic illnesses make their appearance – abdominal complaints mostly. Usually these are put down to the weekly anti-malaria pill yet Jones found that they respond usually to a good chat or tranquillizers and not, for example, to withdrawal of the pill. Even VD, Jones says, could be seen as a disease of loneliness when one considers how it is acquired. These types of illness are not peculiar to the Vietnam conflict. In the Second World War the illnesses known as 'Greenland stare' and 'combat ship exhaustion' were much the same thing. It is people in this state, particularly, Jones says, who should be given the chance to see entertainment from back home – humour especially.

The final month, Jones describes as the terminal period. This is marked, emotionally, by *anxious apprehension*. In part this is due to the lessening of group identification and the withdrawal of investment in friends. There is sometimes a feeling of guilt, says Jones, as men leave behind their friends – and rows can blow up over nothing at all during this period. The other men are of course jealous. Another group of symptoms seen in Vietnam also started in Korea and is called the 'short timers' syndrome'. This is found in almost everyone and consists of mild anxiety and some phobic feelings. A seasoned and battleproofed soldier will suddenly refuse to go out on a mission for no apparent reason – 'If I go out there, I'll get it for sure' sums up the feeling. Some units recognized this and kept men off active ambush patrols during their last thirty days. But since this was not universal it was bad for morale. Also if this terminal policy is in any way widespread all that happens is that the short timers' syndrome appears a month earlier, i.e. two months before the journey home.

INTERPERSONAL FACTORS WHICH AFFECT MORALE

Once in battle, the modern trend is for wars to be fought by men in small groups: decisions come frequently to each man – decisions in which he has to balance survival, duty and loyalty to his group. As Brigadier Bidwell points out, the choice for the soldier is not between danger and safety; it is easy to desert and leave the field of battle. The main problem is in persuading a man to take the more dangerous of two already dangerous courses of action (7). These days the basic fighting unit – the infantry battalion – goes into action not as a closely controlled body, but as a loose conglomeration of combat sub-groups. A spin-off from this is that nowadays humble corporals and others have to take decisions that only commissioned officers would have taken years ago. When directed towards a group goal as they invariably are, they always involve some risk or sacrifice.

In a guerrilla war such as Vietnam, combat for the average GI is, according to Franklin Del Jones, a fleeting and frustrating experience, though it always has to be at the back of his mind. Engagement may be seeing a buddy fall to a sniper bullet or stepping on a mine oneself. The usual mission lasts only a few days, with the intervening time spent at a fortified base camp. There is, as a consequence, a certain amount of behaviour aimed at getting into the base and staying there. Sleep walkers, talkers and bed-wetters crop up from time to time. There is a 'normal battle reaction' for soldiers in combat for any length of time: 50 per cent experience a pounding heart, 45 per cent a sinking stomach, 30 per cent a cold sweat, 25 per cent nausea, 25 per cent shakiness, 25 per cent stiff muscles, 20 per cent vomiting, 20 per cent general weakness, 10 per cent involuntary bowel movement, and 6 per cent involuntary urination (8).

In addition there is a tendency for some men to report certain hypochondriacal complaints on the eve of battle, such as headaches, toothache, indigestion and worry over healed or nearly healed wounds. Helmet headaches also come into this category and, more consciously, deliberately broken spectacles and dentures. Soldiers usually believe their symptoms are significant. These reactions are found in all sorts of wars; but in guerrilla wars, with men in tight groups and intermittent action, minor 'epidemics' of these ailments become more likely (9).

Given these normal reactions, it is crucial that morale should be sustained. In conventional warfare the first encounter will have a great effect: if it is successful, morale will rise even if the overall picture is gloomy, because a personal battle has been fought and won. Everyone fears that his nerve will fail and he will let his comrades down. An early victory helps get over this (10).

In combat also the emergence of informal leaders who provide emotional support is an important factor in sustaining morale. Both American and Israeli studies have found this (11). Morale fades as the number of available leaders suffers natural attrition.[1]

Morale also depends on how a unit is used and its fate in combat. One thing that matters is keeping physical casualties down. According to Bidwell, a unit which suffers casualties to a third of its fighting force is wrecked psychologically if this happens more than once (only about half a battalion of 500–600 men will actually fight these days). The best remedy is to populate the unit with new men – make it a new unit under an old name (12).

In Vietnam Jones sometimes found that the breakdown of a whole squad was related to loss of a leader as a battle casualty or to rotation.

1. The rapid promotion in wartime can produce paranoia in others and produces 'a pronounced disregard for the abilities of . . . subordinates'. The allegedly 'poor' records of subordinates under a rapidly promoted superior officer should always be looked at sympathetically.

Also men serving a second term often chose to return to their old unit rather than go to a safer post among different men (13).

In Korea David Chester of the US Army's Personnel Research Branch had found that morale was much higher if men were trained in groups of four and if these four were kept together at all times, particularly when replacing other men in other units. Compared with the individual replacement system the four-man teams had a better *esprit de corps*, and were much less likely to face new difficulties with a resigned attitude (14).

Those who do survive well change their battle behaviour. They become canny – skilled in battle crafts. They also devote more time to staying alive rather than to achieving unit objectives. Units begin to rely more and more on massive fire power than manouevrability. Tank units tend to stop at the first casualty and to fire from longer ranges than hitherto.

Before a unit finally disintegrates, the actual fighting is carried on by a small fraction of men. The rest become inert – they dig a hole in the ground and are reluctant to use their personal weapons for fear of retaliation. On patrols they merely lie up and return saying they were unable to contact the enemy or were stopped by fire (15).

In the final stages of disintegration, men will desert if at all possible. Sometimes desertion will be collective; if humane treatment is certain, men will surrender freely. Old hands discourage reinforcements from taking any action that will disturb a quiet life. Overzealous leaders may suffer mysterious 'accidents'. In Bidwell's understatement: 'The unit is then no longer an entity.' It is good only for non-combatant duty.

Many strategists believe that wearing an opponent down without necessarily killing him is one of the aims of modern warfare. One example of complete breakdown was the 14th Indian Division (British and Indian troops) who suffered nothing but reverses against the Japanese in the Arakan in 1943. Morale became so low that, even when the division had been withdrawn and was ten miles from the nearest Japanese, its units were still liable to panic at night and fire indiscriminately in all directions. The command psychiatrist described the entire division as 'a psychiatric casualty' (16).

A distinction must also be made between those troops who are actually in combat and those who act behind the lines as support. In Vietnam, non-combat troops tended to suffer from diseases of loneliness: alcoholism, loss of appetite, weight loss, VD, psychosomatic complaints (17). Jones suggests that alcoholism particularly can be regarded as a form of 'behavioural licence' seen as acceptable in war.

COMBAT FATIGUE

Combat psychiatry grew out of the need to care for individuals who, under the stresses of battle, suffered psychological disintegration (18). The most frequently used term for this condition is combat fatigue.

In the American Civil War, William Hammond, surgeon general of the Union Army, described a condition he called 'nostalgia', the incidence of such cases being calculated as 2·34 per 1000 troops in the first year of the war, rising to 3·3 per 1000 in the second year. Hammond treated such cases by keeping them busy with non-stressful work (19). Anderson, in *Neuropsychiatry in World War Two*, says that during the Franco-Prussian War and the Spanish-American and Boer Wars, rates of combat fatigue were in the region of two or three per 1000 (20).

Fatigue seems to be a natural effect of being under fire. Even in a slit trench a man tends to brace himself continuously when under fire and this is exhausting. Perhaps the best example is the taking of Hill 440 in Korea by Love Company of the 27th Infantry. This was in general one of the most successful operations of the campaign. Men were in good physical shape and the point of engagement was only 1100 yards from the point of assembly. Fire engagement began at 8.30 a.m. and was continuous until 5·30 p.m. From 11.30 onwards, however, Love Company leaders found that their greatest problem was simply keeping their men awake – in broad daylight under intense bullet and mortar fire (21).

A similar reaction has been found in assault troops under any conditions. Parachutists and amphibious personnel particularly, it has been found, experience a subtle desire for relaxation once the effort of the assault is over. It is important to understand exactly what is meant here. The parachutist feels tired – exhausted – a few moments *after* he has made a safe landing. He has been keyed up in the plane, and during the drop. Once on the ground, he relaxes – he cannot help it. Often this sense of relaxation is so pervasive that he may drop off to sleep without realizing (22).

A number of methods have been used to overcome this type of fatigue. Some troops have tried carrying dried fruit or glucose. The more adventurous armies have tried to train their men to give each more canny as time goes by and what the effects of experience are. British evidence from the Second World War in general supports the American surveys. For example, 264 British soldiers who were wounded in North Africa were interviewed, 88 per cent saying that a shell-throwing or dive-bombing weapon was most disliked. Only 17 per cent of the men reported that small arms fire was the most disliked. In another study, 461 British troops were interviewed and this time 92 per cent said that shell-throwing weapons were most disliked, 4·1 per cent said machine guns, and 3·5 per cent said tanks (4).

Some weapons have more effect than others on the readiness of wounded soldiers to return to battle (5); these results confirm what we have seen already – that shells are the most effective in eliciting fear (see Table 20).

Table 20 Readiness of wounded soldiers to return to battle in relation to their fear of weapons

Attitude to weapons	*Percentage who want to return*	*Percentage who do not want to return*
Mortar		
Less afraid over time	31	12
More afraid over time	40	56
Machine gun		
Less afraid over time	43	32
More afraid over time	24	42

The results suggest little more than commonsense at first sight. You would not expect men who get more frightened as battle goes on to want to return after a bout in the hospital. But this is clearly what courage is. It does not appear that brave soldiers are those who have nerves of steel and do not know what fear is. In fact, according to one study the exact opposite is the case (6). In a comparison of ninety Silver Star winners with ninety matched non-award winners the figures were quite surprising (see Table 21).

Table 21 Comparison of fear in award and non-award winners

	Percentage fears increased	*Percentage fears decreased*	*No change*
Silver Star winners	66	20	14
Non-award winners	55	32	12

other hypodermic injections of stimulant drugs at the crucial time. Sometimes this is done in the plane just before the drop; sometimes the men do it to each other once they are on the ground. But judging by the fragmentary reports following such studies, none of these procedures provides the complete answer. One solution would seem to be to move the men in a day or two before they are needed but there is, of course, often no way of avoiding combat in the first day or two of a unit's arrival in an area. The best system would seem to be one where the men who *can* travel and then fight straightaway are moved into the fight zones in a specialist capacity for a few days, then withdrawn when the follow-up troops are ready to take over, so the specialists can be held available for later assaults. This may be the way British Special Air Service operate (see pp. 225–6).

In some senses modern guerrilla combat is less stressful than earlier wars: there is usually a fixed term of service as in Vietnam or Ulster; there tends to be an absence of heavy artillery; and in Vietnam there were excellent evacuation facilities by helicopter. The nature of guerrilla warfare itself, also, allows frequent respite. Bourne's work (see pp. 207–12) has shown that following periods of stress corticosteroid secretion rises and hence men need a period of about three days after exorbitant demand to recover fully. This was quite possible in Vietnam; in earlier wars, however, there was often no such chance.

GROWTH OF MILITARY PSYCHIATRIC SERVICES

Effective treatment of breakdown in battle will depend on recognizing certain states of combat illness and providing adequate services to cope.

Obviously the attitude of commanding officers is crucial. General Westmoreland took this so seriously that he wrote articles about it: he advocated that commanders should familiarize themselves with the rudiments of battle psychiatry, and be concerned especially with the way psychiatric services could be organized and centralized to minimize the number of casualties.

Mental health specialists first found their way into combat during the Russo-Japanese War in 1904–5 when, according to Anderson (23), Russian psychiatric casualties ran at such a high rate that the Russians' own services were inundated and the help of the Red Cross had to be sought.

The scientific breakthrough came during the First World War – mental health specialists became part of the military medical corps and systematic study was begun of mental illness in battle. Some battle casualties were seen as organic in origin: shell-shock was thought to be a state of chronic concussion resulting from continuous artillery bombardment. Early on, the French distinguished between the shell-shocked, who were sent home, and the emotionally shocked, who were treated near the front line, often by electric shock or threats of death. The British sent everyone home and did not get as good a rate of return as the French (24). The Americans drew on the experience of both French and British when they entered the war. They devised a system that had three new features:

(a) Examination of officers and men suffering from exhaustion, concussion by shell explosion, and war neurosis, at advanced sanitary posts. This minimized the numbers evacuated from the front lines.

(b) Treatment of light cases of exhaustion, etc., by divisional sanitary formations to keep as many as possible available for duty.

(c) Mental examination of prisoners and men suspected of having self-inflicted wounds.[2]

So successful were these innovations that 65 per cent of men were returned to their units.

The lessons of the First World War were not fully appreciated, however, and in the US Army it was not until March 1944 that psychiatrists were reconstituted at divisional level (25). The result was that the casualty rate of the US First Army in Europe was 101 per 1000 troops. In comparison, the British, who were more prepared to learn from past experience, had a much lower casualty rate. In the early months of the Second World War psychiatric casualties were 23 per cent of all soldiers evacuated, but with the later establishment of front-line treatment centres, this dropped to 6 per cent (26).

In that war it was the general medical officer who made the diagnosis, and some may have confused neurosis with character and behaviour disorders. This may be the reason for the greater number of psychiatric diagnoses made in the Second World War compared with the number made in the Korean conflict. As shown in Table 23, the percentages of

2. It seems that men go AWOL predominantly for neuropsychiatric reasons and a more permissive attitude to discussion of fears in a unit may lessen neuropsychiatric desertion (27). A study by Harold Clark in 1944 of twelve men who had shot themselves suggested that they were neurotics motivated by fear. Clark diagnosed eleven out of the twelve as hysterical (28).

neuropsychiatric evacuations from Korea and Vietnam are similar, and in both instances evacuated cases were screened by psychiatrists (29).

Table 23 Neuropsychiatric evacuations during combat as a percentage of all medical evacuations

Second World War	23
Korea	6
Vietnam (June 65–January 66)	5

TREATMENT TECHNIQUES

As we have seen, war casualties begin long before the battle zone is reached. In the initial stages of a war it is probable that units will move onto the offensive pretty quickly, and various procedures can be adopted to minimize psychiatric casualties.

During the Normandy offensive 'exhaustion centres' were established on two beach-heads. Only casualties who could not carry on were eventually taken to a 'holding unit' for transport away to safety. As soon as soldiers were ready for combat they were returned to the front. It was found that different types of patients needed to be segregated to prevent behavioural contagion; they were divided into low-morale psychopaths, psychoneurotics and those suffering from sheer tiredness (30).

The exhaustion centres produced some worthwhile wisdoms for treatment in early offensives: the greater the delay in bringing to treatment, the poorer the chances of a return to duty; the more doctors or specialists seen, the poorer was recovery; early casualties recover less well than later ones; psychiatric rates vary with overall casualties *but* the greater the casualty rate, the better the psychiatric cure rate.

Once in battle proper, however, the treatment, the effective treatment, of combat fatigue or combat reactions depends on four principles. These are: immediacy, expectancy, simplicity and centrality. *Immediacy* simply refers to the need to implement treatment as early and as far forward as possible. If prompt treatment can be given in the individual's combat unit or its immediate vicinity, this tends to catch the patient while the reaction is still in its initial reversible stage and while he is still in conflict between the interest of his group and his self-preserving interest. The patient is quite suggestible at this stage and tends to decide in favour of the group. *Expectancy* means that treatment is best carried out in a milieu where the soldier is expected to return to battle. Not only the doctors but all the treatment personnel must take the attitude with the patient that he will be returned to duty. This approach tends to minimize the importance of the patient's symptoms in his own eyes – patients otherwise always respond with behaviour and/or symptoms in accordance with what they think is expected of them. There should also be an attempt as much as possible to avoid the atmosphere of a hospital for similar reasons. *Simplicity* is not only helpful in the battlefield where long and complicated treatments are not possible but it too serves to suggest that the illness is not really an illness if it does not require elaborate treatment facilities. The treatment, such as it is, should also aim only to explore the immediate reasons for the psychiatric breakdown and not go into more long-term, possibly life-long problems. The final aspect – *centrality* – refers to the advantages which are now known to accrue from the passage of all psychiatric evacuees through one centre where they are seen by a psychiatrist. Not only does this ensure that the more serious cases are seen early on by a trained specialist but he can also act as a further screen to prevent the evacuation of patients who really do not need it. It also means that some malingerers will not try it on if they know they will be seen by a specialist (31).

Though these few principles are a good general guide, arrangements do deviate according to circumstances. A special case applies to men at sea. Psychiatric casualties after sea battles are small compared with land battles though hospital ships tend to be more confined than land hospitals and this presents problems for psychiatric patients who, unlike other casualties, can usually move about freely. By and large seriously disturbed patients just have to be kept separately. Another problem with a hospital ship is that men develop symptoms to prolong their stay. It does not work to 'slap them on the back'. Instead some other personal anxiety connected with war, should be looked for. Discussion of this often helps them back to combat. Shelling from other ships, rather than dive bombers, provoked most fear. The closed atmosphere of a ship means that, in general, drugs are used more often to combat disruptive behaviour and the more life can become a continuous group-therapy session the better (32).

During guerrilla warfare, some adaptation of psychiatric services also needs to be made and the Vietnamese War provides a good example. The Americans, Vietnamese and Australians had a total of about thirty psychiatrists (including about half a dozen aboard navy hospital ships). They provided services for about half a million servicemen – the highest ratio of military psychiatric coverage in any war the Americans have ever fought in. They were assigned either to a KO team, a hospital or a division. There were two KO teams in Vietnam most of the time. One was located at Long Binh and was the sole route of evacuation for army psychiatric cases. It provided consultation services for commanders in the field, for the stockade and for other agencies. The psychiatrists who were assigned to the hospitals in Saigon served primarily support troops or infantry troops who got into trouble in the city. The division psychiatrists each catered for about 16 000 men. Towards the end of the war there was a development towards regional psychiatrists in line with the policy of deploying small separate units (33).

One of the advantages during the Vietnam War was the availability of drugs for treatment in psychiatric cases. The skilful use of these, it was found, frequently enabled men to return quickly to useful duty. In many cases they were used by ordinary physicians to counter physical illnesses that could be psychosomatic in origin (34).

The Americans also developed certain general procedures aimed at enabling the soldier to tolerate his situation. One approach was the 'buddy system' in which two 'buddies' looked after each other in combat. This system was based on the crucial role the small unit plays in fostering motivation and morale. Forming ties with other soldiers can, however, be a hazardous experience given the nature of combat. In Vietnam links with home through a weekly phone call and an efficient postal service enabled soldiers to feel they were not cut off and helped them not to get too dependent on their primary units in the field. In addition, the rotation system guaranteed every soldier that he would return to the USA after twelve months (thirteen if he was a Marine). The entire attitude of the men towards the war was governed by this factor (35). The war started the day a man arrived and ended the day he left; his main objective was to stay alive. A side-effect of this was to insulate the men from the opposition to the war in the States: they developed 'combat provincialism' – inured to both political issues and long-term military trends. These factors perhaps contributed to the lower rate of breakdown in Vietnam than in either the Second World War or Korea.

Gary L. Tischler adds to the chronological picture painted by Del Jones. Tischler points out that soldiers in Vietnam presenting during the first three months of their tour 'had been relatively successful in mastering role tasks prior to combat duty', those presenting during the period from the fourth to the ninth month had functioned 'relatively adequately' prior to military service, but those presenting in the final three months had considerable difficulty adjusting both prior to and in the military. Psychosocially, argues Tischler, the different symbolic phases of the war interacted with the men's personalities to produce different types of casualty. His analysis, unfortunately, does not go much further than this (36).

Very little is known about the psychiatric capabilities of the Israelis, but a report by the Israeli Medical Association after the Yom Kippur War said that 9 per cent of casualties were psychiatric, that there had been an unusually high incidence of battle shock, and stressed the need for more rapid treatment of such cases. It was suggested that in future the normal Israeli procedure should be reversed, the most experienced psychiatrists should be attached to field hospitals rather than stationed in the rear (37). In one Israeli experiment a caravan was located near the battle lines. Inside, a shell-shocked soldier would be shown battle films, but instead of immediately hearing the noise of battle, he would hear his favourite music. He would be encouraged to relax, and only then gradually re-introduced to the rigours of battle noise and associated scenes. It was a classic piece of de-sensitization therapy, though whether it worked or not is not known for the accounts so far released have been very garbled and anecdotal and the practice is officially secret. The so-called truth drugs have also been used by the Israelis near the front line to 'abreact' soldiers who are shell-shocked. The drugs induce the men to talk through the circumstances leading to their reaction, an activity which appears to prevent their fears being 'bottled up' and so causing some other, long-term syndrome (38).[3]

THE LONG WITHDRAWAL

The end of a war has its own problems. Soldiers feel that there are few ideological reasons for getting killed and, as anxiety grows, morale becomes brittle. The approaching end of a war will pose problems for psychiatrists, and no war shows this better than Vietnam.

The end of the Vietnam War was as unlike the end of previous wars

3. Dr Niza Yarom, an Israeli psychologist, formerly attached to the University of Haifa, has been collecting material showing the *positive* effects of being in a war. This is the only study I know of which takes this approach.

as the Vietnam War itself was unlike earlier conflicts. By the middle of 1968 there were half a million American soldiers in South Vietnam, plus allied troops – Australians, Filipinos and Thais, as well as the South Vietnamese. In 1967, it is now generally accepted, the allies had gained the initiative, but the infiltration of South Vietnamese troops by the Viet Cong saved the situation for them. By 1968, the Communists were able to start their attack on Saigon in bits and drabs and they also managed to hold Hue for a while. This became known as the Tet offensive, when many thousands of people were murdered by the VC and the massacres such as My Lai began to occur. The significance of Tet was that it forced home to many Americans the view that a military victory was impossible. A period of stalemate (during which the Paris peace talks started) was followed by Nixon's Vietnamization campaign and the beginning of what can now be seen as a long American withdrawal.

The importance of this chain of events is that it began to affect the morale of the American soldiers – especially as the anti-war feeling back home was by now reaching its height. The problem that interests us most is that the marijuana epidemic among the soldiers, which had actually started in 1967, now began to be supplemented by other drugs and by a heroin epidemic. There were several reasons for this: drug abuse was part of the American youth culture even at home, marijuana had always been readily available and the heroin supply routes were not too difficult to establish. But another interesting reason, which Jones brings out, is that Vietnamization relegated the role of American troops to increasingly *support* activities – a milieu conducive to alcohol and drug abuse (39).

During the withdrawal period psychiatric casualties relating to drug abuse came increasingly to dominate the entire medical programme. Out-of-country psychiatric evacuations, which accounted for less than 5 per cent of medical evacuations until the first quarter of 1971, rose sharply to nearly 30 per cent in the third quarter of 1971 and continued to an all-time high of nearly 61 per cent in the last quarter of 1972. These casualties were related to the large-scale drug screening procedures which began about this time.

Moreover, the rate of patients diagnosed as psychotic roughly doubled in mid-1969: since experience in combat and non-combat conditions among troops has long shown that the psychosis rate invariably stays about two per 1000 troops, this probably reflects, as Jones says, the influence of drugs confusing the situation. (When the drug programmes started, the psychosis rates dropped back to roughly two per 1000 troops.) It is important to realize that one of the primary methods of handling the drug-abuse patients in Vietnam, towards the end of the American involvement, was to evacuate them to stateside hospitals and rehabilitation facilities. Thus drug abuse became a kind of evacuation syndrome – a lesson to learn for future wars.[4]

From the studies reported in these chapters it would appear that a kind of plateau has been reached in the study of stress and battle. The main effects with the general body of men are known. Treatment techniques are now understood, and there is less of a fear that in new wars unknown psychiatric illnesses will occur because new and untried weapons will be used. That alone is an advance. The effects of proper psychiatric treatment are also known and in most armies the role of psychiatrists and psychologists is now accepted. That too is an advance, and one hopes that in the event of future wars the old lessons will not have to be re-learned.

The interest of both the military and the scientists now seems to have switched to some of the more special stresses of war, such as capture, torture, isolation, being marooned, and interrogation. This may reflect a general view that guerrilla wars are far more likely in the future than are conventional or nuclear conflicts and that, therefore, these special stresses are more likely to occur.

However, before we examine the study of the special ordeals of capture and imprisonment which a soldier may have to undergo, there is one final example of behaviour under stress while the soldier is still engaged on active duty that we must consider – and that is the commission of atrocities.

4. I have not covered the psychological difficulties faced by returning soldiers when a war is over. I do not minimize these problems but space precludes their discussion. For a useful book on this subject in the Vietnam War, see Lewis Sherman and Engen Coffey (eds.), *The Vietnam Veteran in Contemporary Society* (Veterans Administration, Washington DC, May 1972). This also examines the idea that returning veterans should 'busy themselves in public works' helping the country (like Vietnam) they have just returned from. A sort of 'boot camp' in reverse. After Vietnam, the Pentagon's insistence that returning POWs were, in psychologist Herbert Kelman's words, 'heroes in a war that has no heroes', also created adjustment problems. See 'POW care queries', *Behavior Today*, 16 April 1973.

12 Atrocity Research

SOLDIERS WHO COMMIT ATROCITIES

Atrocities are perhaps the worst type of war crime that there is. Many take the view that atrocity is inevitable in war and that there is little to understand. Here, however, I intend to explore the process whereby individuals or small groups of soldiers come to commit acts which are irrational in the sense that the large number of killings they take part in are unnecessary militarily in so far as the people killed pose no immediate, and possibly not even a long-term, threat to the soldiers' well-being. Where, in fact, the commission of atrocious acts seems more a reaction to extremely stressful events by personalities unfortunately unable to cope.

Neil Sheehan has claimed that the Vietnam War was different to the Second World War in that the belligerents in the 1940s were fighting for survival and therefore enemy cities were legitimate targets even though they contained civilians, unlike in Vietnam where the civilians who were bombed and napalmed were on the side of the bombers, or at least allegedly so (1). One can agree with Sheehan that this makes war crimes more likely in Vietnam almost by definition and that the right way to stop them is through political action and journalistic exposé which Sheehan himself did. But they seem to me to be a different type of atrocity. The purely *military* atrocity, as opposed to the tactical atrocity, occurs when the killing is unnecessary on military – or even psychological warfare – grounds. It also tends to occur on impulse – which suggests that the personalities of the soldiers concerned, or the stressful situation itself, plays a part, quite separately from any general strategic style of war.

Even in the limited context in which I am choosing to regard atrocities, there are a few studies that throw some light on a very murky subject.

SS mass killers

A British psychiatrist, Henry Dicks, studied eight Nazi mass killers who were captured during the Second World War, and although his analysis was not published until 1972, it still makes gripping reading (2).

Dicks interviewed many Nazis, as well as the eight who were involved in mass killings. He was given access to sixty biographies of Nazis in the Wiener Library and also carried out 138 detailed interviews on captured German airmen, sailors and soldiers. He found, as others have found since, that the Germans divided up neatly on a psychological test of fanaticism in a bell-shaped distribution with approximately one in ten falling into the highest, 'very fanatical', segment. He found that the highly fanatical were people likely to describe themselves as 'nihilistic atheists'. They tended to have had a difficult relationship with their father, expressed no feelings towards their mother, were intolerant of tenderness, and showed tendencies to anti-social sadism and to project hostile intent outside themselves. They also displayed extreme neurotic anxiety.

Like others, Dicks identified, in rather more detail perhaps, the personality that is now known as authoritarian, but he went a bit further. For example, he found that these 'high F scorers', as they are known, tended to remain single longer than most, entered the services when they were very young, had often had an elder brother killed in action either earlier in the Second World War, or in the First World War. Most were volunteers, and they came from rural rather than city backgrounds; one third were artisans, small traders or shop assistants. They also showed a tendency towards action rather than thought, were characterized by 'suicidal despair' and were prone to project 'responsibility' for things elsewhere.

We shall see that some of these more sociological findings of Dicks have been echoed in later studies in Vietnam. But his own investigations into the eight mass killers to whom he had access went somewhat further than merely cataloguing background details. What Dicks was able to show was that the men, all eight of them, had in their earlier years had an active fantasy life that fitted the picture painted above – but all had been law-abiding citizens before; and when they went to prison they were some of the best-behaved inmates. The men were never impulsive, being mostly over-controlled people. But they were the type who was likely to be very aggressive when they did get angry, either because the level of instigation had to be high enough to overcome their strong resistance to action, or because they lacked internal cues about 'when to stop'.

Dicks concluded then that atrocities are committed by people who have a history of being mainly mild-mannered, often with a preference for 'things' rather than for 'people' and who, though not usually violent, are extremely so on the few occasions when it happens.

Dicks's study was concerned essentially to provide understanding, rather than to be of predictive value. Since the Vietnam War (and not just My Lai), the search for a psychological method to screen out potential 'atrocity-makers' has, however, assumed greater urgency.[1]

Atrocities in Vietnam

The study of this particular issue first became possible during 1970 and 1971 when psychiatrists in several parts of America began seeing as patients ex-servicemen who had served in Vietnam (3). All the men went to the psychiatrists in order to confess their part in atrocities in the Vietnam War. It appears that none of the men had confessed prior to the psychiatric consultations and they sought help because of serious emotional difficulties, which developed as late as five years after the incident. One psychiatrist in Boston had seen forty such patients by 1971 and another had seen thirty-one. The incidents reported ranged from mistreating prisoners (military and civilian) to mutilating the bodies of dead Vietnamese – cutting off ears, genitals and breasts. Three soldiers at least admitted to attacking their commanders or officers.

These studies suggest that although the conditions of war may make anyone a potential mass murderer, some men *are* more prone to kill indiscriminately than others.

What seems to happen is that when a soldier with a certain type of background and personality finds himself in a certain type of situation,

1. Stanley Milgram, in his well-known book, *Obedience to Authority* (Harper & Row; Tavistock, 1974), argues that his work fits in well with Dr Dicks's. This is not quite so. In Dr Milgram's experiments ordinary people administered electric shocks to innocent victims simply because an 'experimenter' told them to do so. In this his experiments confirm Hannah Arendt's thesis that Eichmann was just an ordinary bureaucrat – not an especially evil man. But Dicks's work, like that reported later in the present chapter, suggests that there are *personalities* who are more likely to commit atrocities although the *circumstances* are important as well.

atrocities are more likely to occur. Combining the information on the patients we find that the great majority of the killers came from rural as opposed to urban backgrounds; none of them was more than nineteen at the time of the killings and most of them had an elder brother who had been killed, though not necessarily in the war; at school the men had varied academic records – intelligence did not seem to be terribly significant – but they had all been fond of, and active in, team sports especially those with a clearly defined single 'enemy team' (football, for example, as opposed to athletics). In both studies it was found that volunteers outnumbered draftees among the killers and that the volunteers had been precipitated into joining up by a specific event in the war or to get away from an unhappy home background rather than for vague reasons like patriotism, money or the excitement of travel. In one study the men were rated on ten psychologically significant aspects of their background such as whether they were from a minority group, whether there was a family history of violence, whether the boy had been a bed-wetter or a fire-raiser, had been arrested and so on. Those who had committed atrocities averaged 5·2 of the symptoms whereas the non-killers averaged only 3·8 (4).

Once in Vietnam these men tended to gravitate to certain positions in the combat units. Whatever their rank, they often found themselves as *second* in command, as radio operator or as point man – the soldier who goes ahead of a platoon on the look-out for the enemy and testing the route for hidden mines and boobytraps. They were certainly looking for action: they were three times as likely as the others to win medals and they tended to have killed legitimately more than those who did not commit atrocities.

What is especially significant about these findings is how well they dovetail in with Peter Bourne's research (see pp. 207–12) into the hormone levels of certain specialist soldiers – officers, radio operators and, in other studies, point men (5). All these, it appears, are highly stressed positions in a platoon. When a high-risk soldier finds himself in a platoon or squad position that is highly stressful for long periods of time, it appears that the chances of atrocities are increased. It is often said, in the anecdotal evidence, that seeing a friend killed is the spur to atrocity but the more systematic studies on this are equivocal. In the one study where atrocity-doers in a platoon could be clearly differentiated from non-killers (that is on-lookers), neither group was more or less likely to have had a close friend killed in combat; nor was either group more or less likely to have been wounded themselves; the only personal difference between these two groups in fact, apart from those already mentioned, was that the killers were far *less* likely to have smoked opiates.

On the other hand, as we have seen, atrocity-doers were far more likely to be volunteers than conscripts, and there were some interesting differences uncovered between volunteers and conscripts by this study (admittedly of small numbers but it is all that is available). Volunteers were about twice as likely as conscripts to have been wounded, three times as likely to have been decorated, exactly twice as likely to have had a close friend killed, were more likely to volunteer for additional duty and, when they were killed, died in a somewhat individual fashion (6). It has also been found in studies that the second-in-command often acts more brutally than average when his platoon or unit has just been given a new commander with less combat experience than himself. This is one reason why, psychiatrists believe, the soldiers who murder innocent civilians were also those who 'fragged' – murdered – officers. Several atrocities were apparently committed after inexperienced officers had given what the men considered to be impossible orders (7).

Knowledge is gradually building up, therefore, to enable us to have some rudimentary ideas about the situations and group dynamics that lead to atrocities. It may not be too much to hope that, on the basis of this, a screening method might be introduced to prevent high-prone individuals being placed in high-risk positions without hindering the military effectiveness of the units. In fact, it may be that such a procedure is already being developed. In order to explain this, however, we have to enter a very dubious field of research – dubious because it is hard to discover the full details and because the intentions behind the research appear suspect.

THE OTHER SIDE OF ATROCITY RESEARCH

In chapter 1 we touched on the political storm surrounding research by Sigmund Streufert for the US Navy into the value people attach to human life (see pp. 36–9). Streufert compiled a questionnaire, part of which is reproduced in Table 24. The preamble to the actual study runs as follows:

> . . . the military decision maker may be concerned with the lives of the enemy, or with the lives of a local population. When is it morally justifiable to kill, and when is it not? What degree of respect or disrespect for the life

of another people can be expected from a soldier who is, or who is not, following orders? What are the motivations, or the belief systems, that are underlying the behavior of military personnel who were involved in the death of civilians at My Lai or at Kent State University? . . . To answer questions of this kind, we must initially learn what 'value of human life' is. Currently we do not know whether value of life is a unitary concept, or whether it consists of a number of components . . . (8)

At this point the account becomes more technical.

What is not clear is whether Streufert intended to use his questionnaire to find out which soldiers do not value other people's lives very much and so would make good killers, or to select soldiers who value their own lives more than most and so would be good at staying alive on a difficult mission. It could also be used by a military commander to assess a local population's attitude towards life so as to be able to predict what would happen if certain military measures were taken – like the introduction of executions and so on. In any event, a spokesman for the Secretary of the US Navy did concede eventually that the

Table 24 Sample questions from the value-of-life questionnaire (the subject has to indicate how much he agrees or disagrees with each statement)

If a terminally ill patient asks to die, one should comply with his wishes.
As long as anyone shows the slightest signs of life, science should prolong his life.
No sane decent person could ever think of hurting a close friend or relative.
Abortion is never justified because the fetus is still a human life.
Obedience and respect for authority are the most important virtues a child should learn.
An international police force should be the only group in the world allowed to have armaments.
Revenge is usually justified; an eye for an eye, a life for a life.
It would be dangerous procedure if every person in the world had equal rights which were guaranteed by an international charter.
Most of our social problems would be solved if we could somehow get rid of the immoral, crooked and feeble-minded people.
A person would be better off dead than leading a consistently immoral life.
Nobody really learned anything important except through suffering.
If a man points something at you through his pocket, assuming you had a gun, you should shoot quick and ask questions later.
Most people don't realize how much our lives are controlled by plots hatched in secret places.
A person's body is inviolable, even after death.
It is best to have some Nazi authorities in German Government to keep order and prevent chaos.

research was meant to be linked to cold and limited war situations, and that in part Streufert's wider work was to predict how men would behave under stress.

In all there are 135 questions. Factor analysis in the preliminary study suggested that the questions tapped four attitude clusters: concern with aggressive punishment versus rehabilitation of criminals; concern with morality; concern with whether a life should be sacrificed for a superordinate cause like a nation; concern about an individual's right to die if circumstances are unfavourable – for example a terminal illness. In general, however, Streutfert was not convinced that these questions did adequately separate the 'efficient' killers from the non-killers.

The question of the exact purpose of the study, of exactly whose attitudes were the real subject of interest, has never been clarified, and Streufert's work in this area has ceased.

There the matter rested for a couple of years. During the course of researching this book, however, I attended a NATO-sponsored conference on stress and anxiety in Oslo in the summer of 1975. A Dr Thomas Narut, from the US Naval Hospital at its southern NATO headquarters in Naples, Italy, addressed the conference on the subject of symbolic modelling – a process whereby anxious people could be taught to cope with certain stresses by watching others (usually on film) cope with these stresses. During the course of his paper Dr Narut let slip one or two comments about the fact that these techniques were being used with 'combat readiness units' to train people to cope with the stress of killing. Intrigued by these comments, I and another psychologist, Dr Alfred Zitani from New Jersey, approached Dr Narut after his paper and questioned him further. Two further surprises were revealed at this exchange. First, Dr Narut said that he was referring to two types of combat readiness unit: the ordinary commando unit and also to naval men inserted into embassies abroad under cover ready to kill. Dr Narut also said, in the presence of Dr Zitani, that men were being sought from military prisons to act as assassins in the overseas embassies. Dr Zitani clearly realized the sensational nature of Dr Narut's claims. His comment to me was: 'Does that guy realize what he has just said?'

Subsequently Dr Narut and I had a somewhat longer interview, during the course of which I took notes, on which the following account is based. Several years ago Dr Narut completed his doctoral thesis on whether certain films could provoke anxiety and whether forcing a man to do tasks irrelevant to the film while watching it might help him to cope with the anxieties the film provoked. Narut's naval work, however, appeared to involve establishing how to induce servicemen who might not be naturally inclined to kill to do so under certain conditions.

The method, according to Narut, was to screen films specially designed to show people being killed or injured in violent ways. By being acclimatized through these films, the men were supposed eventually to become able to disassociate their emotions from such a situation. Dr Narut also added that US naval psychologists specially selected men for these commando tasks from submarine crews, paratroops, and some were convicted murderers from military prisons.

He said that the process had three aspects:

Selection: studies of those given awards for valour in battle have shown that the best killers are men with 'passive aggressive' personalities. They are people with a lot of drive, though they are well-disciplined and do not appear nervous, who periodically experience bursts of explosive energy when they can literally kill without remorse. Dr Narut said that he and his colleagues had therefore been looking for men who had shown themselves capable of killing in this premeditated way. The tests used to select men for these roles include the Minnesota Multiphasic Personality Inventory, especially its sub-scales measuring hostility, depression and psychopathy, and the Rorschach where people have to interpret ambiguous inkblots. Narut says he looks for people who respond particularly to the coloured, as opposed to the black and white, cards, indicating, he says, a preoccupation with violence.

Stress reduction training: men selected are taken for training to the Navy's hospital in Naples or to the Neuropsychiatric Laboratory in San Diego, California. The men are taught to shoot but also given a special type of 'Clockwork Orange' training to quell any qualms they may have had about killing. It works like this. The men are shown a series of gruesome films, which get progressively more horrific. The trainee is forced to watch by having his head bolted in a clamp so he cannot turn away, and a special device keeps his eyelids open. One of the first films shows an African youth being crudely circumcised by fellow members of his tribe. No anaesthetic is used and the knife is obviously blunt. When the film is over the trainee is asked irrelevant questions such as 'What was the motif on the handle of the knife?' or 'How many people were holding the youth down?' In another, the camera follows the movements of a man at work in a saw mill, slicing his way through planks of wood. The film shows his thrusting movements until he slips – and cuts his finger off. Physiological measures are taken to ensure that the men really do learn to remain calm while watching.

Dehumanization of the enemy: In this last phase the idea is to get the men to think of the potential enemies they will have to face as inferior forms of life. They are given lectures and films which portray personalities and customs in foreign countries whose interests may go against the USA. The films are biased to present the enemy as less than human: the stupidity of local customs is ridiculed, local personalities are presented as evil demigods rather than as legitimate political figures.

The total process, according to Dr Narut, takes a few weeks and the men are then passed on. It appears that this technique was particularly employed towards the end of 1973 – at the time of the Yom Kippur War.[2]

Streufert's and Narut's research were separate projects, but they were both undertaken for the US Navy and both were concerned with killing, with selecting and training killers to do a special job. So, far from using research to prevent atrocities occurring in the future, it appears that atrocities are being studied to learn more about killing and to train people to be better at it. A chilling thought.[3]

2. The Pentagon later flatly denied the story, much of which had appeared in the *Sunday Times* and had been widely syndicated. My request to visit the US Naval Neuropsychiatric Research Center at San Diego to question the scientists there was turned down. But I am still in two minds as to what credence to attach to Dr Narut's tale. A few weeks after the episode I received a letter from a psychologist working in the area near San Diego where the Neuropsychiatric Center is located. In it the psychologist said that he had seen the syndicated article and the US Navy's denial and he thought I might be interested to know that, six months previously, psychologists at the San Diego outfit had borrowed from him one of the films – that on circumcision – which Dr Narut had mentioned to me.

3. Dr Chaim Shatan, a New York psychologist who has seen many returning veterans from Vietnam, believes that the brutality inherent in élite corps training may itself prevent easy adjustment back to normal society. As a result disaffected veterans may use their skills (in killing and so forth) to sell to politically disaffected groups. This resolves the personal problems of the veteran alluded to in the preceding chapter. See Shatan's 'Bogus manhood, bogus honor', 'The Hidden Agenda of Basic Marine Training', mimeographed, 1972.

PART THREE *The Determinants of Loyalty and Treason*

13 Captivity

For our purposes we shall consider three aspects of captivity as psychologically distinct. First, there is the capture itself as a result of which the soldier has to live usually in cramped circumstances, undergoing physical hardship and a boring routine existence, always surrounded by the uncertainty of his fate and having to obey the superior position of his captor(s). Next, there is interrogation and possibly torture in which this captor–captive relationship is heightened; when special techniques, some physical, some psychological, are used to try to extract crucial information from him. Third, there is what is called 'brainwashing' – basically the use of interrogation and other psychological techniques not merely to extract information but also to manufacture a confession on the part of the POW for public consumption and which has been claimed at times to 'rework' a man's mind, to create a new set of beliefs.

No one seems to have carried out the obvious study of whether there are any psychological differences between the men who get captured and those who do not. One study by the British in Malaya found that a group of captured *and* defecting guerrilla soldiers were of lower intelligence than the population from which they were drawn, but that is as far as it went (1).

There have been several studies about captivity behaviour itself. These range from A. L. Vischer's *Barbed Wire Disease: a Psychological Study of the Prisoner of War*, published in 1919, through 'The prisoner of war mentality', an article in the *British Medical Journal* in January 1944, to a major book by W. C. Bradbury and A. D. Biderman, *Mass Behavior in Battle and Captivity*, published in 1962. The latter is probably the most comprehensive account for the interested reader (2). Here I want simply to describe captivity in four situations: under the

Germans in the Second World War; under the Japanese in the Second World War; under the Koreans and Chinese in the Korean War; and later under the Koreans in 1968, when the crew of the *Pueblo* spy ship was captured and held for eleven months. These provide a range of experiences which enable us to see how much captivity varies and what its salient features are. Finally, we examine an attempt to reproduce experimentally the experience of captivity, and how it affects the behaviour of both captives and captors.

INTERNMENT: A GERMAN CONCENTRATION CAMP

The camp of Terezin was situated in Theresienstadt in Northern Bohemia. It served as an internment camp for Jews from November 1941 until the end of the war – 139 666 people being sent there, 33 468 of whom died. The average population was 30 000–40 000. Dr V. A. Kral, a psychiatrist, observed that the normal reaction to the camp was initial shock, 'the severity and duration of which was rather uniform in the average adult of both sexes' (3). This was characterized by depression and retardation, loss of initiative (even for washing and eating), and accompanied by anorexia, sleeplessness, constipation and, in most of the women, loss of the menstruation period. Suicide or panic, however, were *not* observed. Many experienced feelings of unreality, as if surrounded by ghosts.

This initial phase usually subsided after about two weeks without any treatment. Young children did not go through this phase, and old people were often unable to overcome it. Women, it seems, became adjusted earlier and better than men. Level of intelligence did not appear as decisive as other personality factors: people 'educated to fairness . . . and trained to control their emotions . . . became better adjusted than those with higher intelligence and not so trained.' Religion also helped, as did being allowed to practise a profession, such as medicine. Some were unable to accept the reality of camp life, were given to boasting of past achievements and to day-dreaming; they adjusted less well than those who accepted the camp for what it was and set to to make the best of it.

Regardless of this, however, the degree of starvation did produce psychological sequelae. The vitamin deficiency, for example, caused memory deficits for recent events, most marked in the aged, but present even in adolescents. Orientation at night was affected, as were attention and the ability to concentrate. In the longer term, indifference and instability set in, leading to occasional temper outbursts. It became characteristic that such things as the death of a relative met only superficial sympathy, whereas trifles grew in importance. Thousands could be shipped east by ten SS officers without any trouble.

Food governed everyone's thoughts, day-dreams and night-dreams; sex became less necessary. The only thing to interfere with this was political discussion, which was sometimes capable of breaking the otherwise chronic indifference and apathy.

Dr Kral's observations were in general duplicated by others: for example, by Dr Bruno Bettelheim and by Dr Bandy (4). But Kral also looked at the fate of the mentally ill at Theresienstadt. Hundreds of mentally ill patients were sent to the camp's psychiatric unit from all over Europe. Dr Kral's finding was that the camp exerted no influence over endogenous psychoses, but that senile dementia and other organic psychoses were quickly exacerbated by the camp. No new cases of psychoneuroses originated in Theresienstadt. But after the liberation, neurotic reactions could be observed in some of the former internees.

JAPANESE POW CAMPS

Reactions in German POW camps are reported as roughly similar in most studies. With the Japanese one difference was that they were often wholly unprepared to manage a large body of prisoners of war. As a result the individual guards seemed to be given free rein in the treatment of prisoners. They were capricious and unpredictable. When orders were not promptly carried out they could be vindictive and cruel.

A study by Major Stewart Wolf and Lt Col Herbert Ripley considered soldiers from several nationalities, including British, Americans and Dutch, who had survived three years of imprisonment and torture by the Japanese (5). The Japanese camps were just as crowded, dirty and as short of food as the German ones, but they were also far more disorganized, according to Wolf and Ripley. The Japanese, according to this study, divided the men into groups of ten, and indicated that should any of the ten escape, the remainder would be killed. This effectively stopped escapes and the fact that officers had to enforce this policy also drove a wedge between the officers and the men.

Tortures used included beatings, standing to attention in the sun for hours, having hair or finger nails pulled out and the 'water treatment' in which the prisoner had his stomach distended with water via a long tube down his throat. In the 'sun treatment' prisoners had their eyelids kept open with sticks and were then left to look into the sun for hours on end.

Reactions were more extreme than in Germany. Some prisoners developed cravings for tobacco, others gave it up and traded what they had for food. None thought of home – just of work. To have thought of home, report Wolf and Ripley, would have meant the end.

During captivity most of the men were 'seclusive and taciturn'. There were few group activities though many cliques held mock 'classes' in which someone would describe in detail the preparation of some especially delicious dish. Sex was a topic of conversation for the first few months only and when extra food was available. At such times homosexuality and masturbation were common, giving rise to guilt feelings. The Americans took it out on their officers more than the British, no doubt due, say Wolf and Ripley, to cultural differences: the Americans were more prone to 'speak their mind' and needed to stress their masculinity. Yet they could not do this and survive. So it was re-directed onto their own officers.

There was a lowering of moral standards: food was often stolen. Depression was common in the early months, plus anxiety arising from the unpredictable behaviour of the guards. These disappeared with time to be replaced, say Wolf and Ripley, by 'hysterical suppression of resentments and conflicts with loss of function . . . . Some eventually became unable to cry or laugh.' Several later reported that they could stand the torture because they could 'turn off the pain'. Hysterical blindness or deafness was reported, in most cases similar to those seen in people with gross vitamin deficiency.

Wolf and Ripley report that the survivors said that the men who succumbed usually did so because they allowed themselves to be dominated by thoughts about home, developed a distaste for food, and died. Difficulty in concentration and loss of memory were also reported as widespread. Other studies on captivity under the Japanese (for example, that by Nardini) confirm Wolf and Ripley's findings though there is some doubt about the effects of self-image on survival. Nardini found that people who persistently told themselves that they were 'a soldier, a Texan, a father' survived. But how one does this *without* thinking of home is not made clear (6). Nevertheless, there emerges a basic picture of captivity reactions as due to two things: physical hardships and the unpredictability and cruelty of the captors.

CAPTIVITY IN THE KOREAN WAR

The North Koreans and Chinese took the process one step further in the Korean War. There they put their POWs through extreme physical hardship, including forced marches, and developed their own unpredictability to high degree; but they also applied chronic pressures on the men to collaborate and to give up existing loyalties in favour of new ones. This was the brainwashing process. We look at it in detail in chapter 15, and also consider the psychological differences between those who did collaborate and those who did not. For the moment, however, we are concerned with captivity and the effect of forced compliance of this nature on the men in the camps.

The Chinese systematically destroyed the prisoner's formal and informal group structure. Men were segregated by race, nationality and rank (7). Any group meeting was banned, the lower ranks were put in charge of higher ranks. Harvey Strassman, an American psychiatrist, reports that the only way the men could keep from either collaborating or resisting to the point of eliciting punishment 'was to withdraw as much as possible from any but routine interactions with either the Chinese or *other* POWs'. This was not the 'apathy' that characterized concentration camp victims but a more conscious attempt not to collaborate (8). There was no element of resignation: 'The men were waiting and watching rather than hoping and planning', says Strassman.

We shall see in chapter 15 how these studies have been fitted into training techniques to help men survive prison camps and brainwashing. But before that we consider the *Pueblo* incident–which occurred in peacetime, making it psychologically different from what has been considered so far.

THE PUEBLO INCIDENT

The US spy ship, the *Pueblo*, was seized off the Korean coast on 23 January 1968. A long-term follow-up of the crew is still being carried out by the US Navy, so severe were the psychological sequelae of this ordeal: but what is already available is very interesting (9). The *Pueblo* was a small intelligence ship and when she was attacked, one crew member was killed and two others sustained substantial injuries. The ship was then boarded, captured and sailed into Wanson Harbour. The crew of eighty-two was transferred to Pyongyang by rail, where they were imprisoned in a building known to them as 'the barn'. The commanding officer was separated from the others who were quartered three or four to a room. They were threatened with death and beaten. Forced 'confessions' about criminal aggressive acts were obtained from all crew members as a result. Subsequently, they were transferred nearer to military installations, where officers got their own rooms and the men were eight to a room. They were later given lectures, field trips and written material by the Koreans in an effort to convince them that the US government was imperialist and unjust. Despite this and despite the 'confessions', the crew attempted to communicate to the world their lack of sincerity. In letters home they referred to dead relatives and used uncharacteristic phrasing: and in the propaganda photographs which they were forced to take part in they could be seen making obscene gestures.

After eleven months they were suddenly released, arriving in California on Christmas Eve. They were immediately examined both physically and psychiatrically. A psychiatric investigation was carried out not only to assess current functioning at the time of release but also to explore how the men had got on as captives. The psychiatrists probed the men's psychological defences, their ability to contribute to group support and morale and to provide realistic resistance to their captors.

Half the men admitted they had experienced serious worries but all said this was due to the unpredictability of their treatment; sixteen reported they had been chronically depressed, most saying that separation from their families upset them most of all, though four were upset by the fact they had 'confessed'. Suicide ideas were uncommon though three said they had made attempts. On average they had lost about 30 lbs, twenty-seven complaining of anorexia during their confinement. Disturbances of sleep had been common and related to mood; dreams were common and included wish-fulfilment dreams of wanting to be back home and of escape. Many men fantasized that the US government would seek retribution by killing the *Pueblo* crew for its breach of discipline and loss of ship.

It was clear that some men had acted as informal group leaders in each room, where talk was found to resemble group therapy – there was ample time for discussion of lives, hopes and so forth. Sex was also a popular discussion topic though there was no reported homosexuality.

Chuck Ford, one of the psychiatrists studying the crew, was able to divide the men into three groups according to how well they had coped with the captivity experience. Comparing the upper third (twenty-four) men with the bottom third (twenty-seven) he found no differences in age, military service, educational level or psychiatric or legal history. There was little difference in marital status, though among the bad captives there were three who were divorced or separated as opposed to none in the upper group (10).

There were, though, definite differences in personality. More than a third of the bottom group had passive–dependent personalities, whereas two thirds of the good captives were either healthy or 'schizoid'. As regards ego defence mechanisms, Ford says, the higher group used an average of 5 whereas the poorer group managed only 2·4. The better captives also used different types of mechanism – reality testing, rationalization, faith, denial and humour, most noticeably, whereas the poorer captives were more likely to defend themselves with 'obsessive ideation'.

So some neat results appear to be emerging from the *Pueblo* study (Ford revealed the latest details at a NATO conference in 1975) and from which counter-capture training could benefit quite handsomely. We shall see, however, that it is by no means certain that those men who stand up well to the experience of capture are those who adjust back to normal life most readily when the conflict is over. Before we move on to that, however, let us look at a study of the captive–captor relationship that has been carried out in the laboratory. It lacks some of the credibility of war-time studies, yet it does enable more systematic manipulation of the captor–captive situation. And this can provide some equally striking insights.

ZIMBARDO'S PRISON STUDY

The foremost researcher in the area of captivity has been Dr Philip Zimbardo. In his work for the US Office of Naval Research, Dr Zimbardo and his colleagues were especially interested in captivity behaviour. Their most colourful study was one carried out in 1972–3, the methods and results of which are highly controversial. Let us look first at the experiment itself and then come back to the implications (11).

Ostensibly, Zimbardo wanted to study the group dynamics of a prison: he was interested in whether abuses in prison arise because, for example, the prisoners are a group who have a long record of disregard for society, or whether the power inherent in the prison officer's job made him especially vindictive. Accordingly, Zimbardo advertised in his local newspapers for some healthy males to help him in an experiment into 'prison life'. Out of seventy-five people who applied, the twenty-four most emotionally stable men were picked; none of them had a previous history of either criminal or any other deviant activity – they could not be seen as disadvantaged in any way. Half of the subjects were arbitrarily assigned the role of 'guards' and the other half were designated the 'prisoners'. The prisoners were told to be available on a certain Sunday when the experiment would begin. The guards were called to a meeting before the experiment started where they met the 'prison superintendent' (Zimbardo) and a 'warden' (one of his assistants). They were told that within the limits of moral and pragmatic considerations, it would be their job to maintain a degree of reasonable order within the prison, although how this specifically might be done was deliberately not explored. They were made aware of the fact that although there might be escape attempts and so on, their job was to make sure that these did not come off. They were to work in shifts, and had to fill out shift reports and critical incident reports. They were intentionally given only minimal guidelines on their treatment of the prisoners though it was made clear that all forms of physical punishment were ruled out. To make it even more realistic the guards helped to construct the final part of the prison.

This was designed to be a functional prison but not necessarily life-like in every detail (it was in fact built in the basement of Stanford University psychology department). There were three 6 × 9 ft cells with black, barred doors. There was a strong door at the entrance to the cell block and an observation door. A small room represented the enclosed prison exercise yard and a much smaller room (2 × 2 × 7 ft), which was unlit, was used for solitary confinement. The entire prison was fitted out for video and audio surveillance.

The experiment proper began without warning when the subjects who had been assigned the role of prisoner were 'arrested'. Zimbardo was successful in getting the cooperation of the Palo Alto City police department and a police car called at their homes. The officer said that he was arresting them on suspicion of armed robbery or burglary, advised them on their legal rights, handcuffed them and gave them a thorough search (often in the presence of curious neighbours). They were then driven to the police station where they were photographed and their finger-prints were taken. Then they were placed in a detention cell before being blindfolded and driven by one of the experimenters to the prison. Throughout the entire arrest sequence the police officers maintained a restrained attitude; they gave no indication that the arrest was anything to do with the mock prison study. On arrival, the prisoner had all his clothes taken from him, was sprayed with what was supposed to be a de-lousing chemical (but was in fact merely a spray deodorant), and had to remain alone and naked in the 'yard' for a while. He was again photographed and then given his prison tunic.

The guards wore khaki shirts and trousers, and had a police whistle, a night stick, and reflecting sunglasses which made eye-contact impossible. Each prisoner wore a loose-fitting muslin smock with an identification number stamped front and back, no underclothes (which forced him to sit in feminine poses), a light chain and lock around one ankle, rubber sandals and a cap made from a nylon stocking to remove differences in hair length (in some prisons, inmates' heads are shaved). When all the cells had been filled, the warden greeted the prisoners and read them the rules of the institution (developed together by the guards and the warden) which they had to memorize and follow. From then on prisoners were only referred to by number.

The prisoners were fed three bland meals a day, were allowed supervised toilet visits and two hours daily for the privilege of letter writing and reading. They were given work and received an hourly wage to make up the $15 a day that they had been promised. Two visiting periods a week were to be allowed, as were film nights and exercise periods. Three times a day prisoners were lined up for a count – these parades coinciding with the beginning of a new shift for the guards.

The results of this experiment were shattering. Right from the start the prisoners were found to adopt a passive attitude whereas the guards became more and more aggressive. Forbidden to give physical punishment, they came increasingly to rely on the use of verbal aggression.

Extreme signs of stress emerged in five of the prisoners as early as the second day. The pattern was similar in four of these and included depression, fits of crying, bursts of anger and acute anxiety. The fifth subject had to be released prematurely after he had developed a skin rash due to his harsh treatment. Of the remaining prisoners, only two said they would not forgo their pay for being let out early. When the experiment was in fact terminated a few days before schedule, all remaining prisoners were delighted. In contrast the guards seemed distressed by the decision to stop. None of them had ever failed to come to work on time and indeed, on several occasions, guards remained on duty and uncomplaining for hours extra without any additional pay.

There can be little doubt that the simulation became a very real experience for much of the time (and a very unpleasant one for the prisoners). Zimbardo and his colleagues monitored guards and the prisoners without them being aware of it. Ninety per cent of the prisoners' private conversation was directly related to immediate prison conditions – food, privileges, harassment. Only one tenth of the time did they talk about their lives outside the prison. As a result the prisoners got to know surprisingly little about one another, says Zimbardo, and the excessive concentration on the vicissitudes of their current situation helped to make the prison experiences more oppressive for them, because instead of escaping from it when they had a chance to do so in the privacy of their cells, they continued to allow it to dominate their thoughts and social relations. And when prisoners were interviewed they spent a lot of the time deprecating one another.

Similarly, the guards spent a lot of their free time talking about 'problem' prisoners. They also had the tendency to be even more cruel when they thought they were not being watched by the experimenters. They continued to intensify their aggressive behaviour even when it had become obvious to everyone that the prisoners' deterioration was marked and visible.

A measure of how real the experiment had become to most of the participants is that when the prisoners were visited by a real priest, many referred to themselves by their number and not by their names; some even asked him to get them a lawyer to get them out. And when some of them appeared before a 'parole board', three said they would be willing to forgo their pay in order to be released. Yet when the chairman of the board (Zimbardo himself) said that their requests would be considered, the prisoners all allowed themselves to be led quietly back to their cells even though they were in reality able to leave just for the asking.

Undoubtedly a powerful study. The question is, however, what was its real aim? At the end of his paper, Zimbardo rightly says that his experiment shows the powerful dynamic built into the prison situation and that it might provide a basis for a new type of prison guard training – presumably to allow guards to see the dangers of arbitrary power. This is plausible: but it is by no means the only interpretation that can be placed on this study – nor is it the one which fits most of the facts of the experiment.

Starting from the premise that the US Navy does not really need to spend a lot of money in training its stockade guards in how to understand their role, let us first remember that the study is part of a Navy inquiry into the basic psychological principles underlying aggression. We can be forgiven perhaps for assuming that the Navy is more interested in the warlike uses of aggression than the behaviour of prison guards. We can then note that there were some crucial psychological aspects to the experiment which made it much *less* like a prison than it could have been. The prisoners had no record: yet it surely is a psychological fact of life that prisoners by and large end up in jail because they deserve to. To make the prison experience realistic, there must be some measure of feeling in the prisoners that they took a risk, and failed. A consequence of this should be that a real prisoner would feel much less outraged at the arbitrary nature of his arrest than was the case in the experiment. Next it would have been relatively easy to make the Stanford basement physically much more like a prison yet the experimenters seemed to prefer the idea of a compromise: it was a temporary prison, something which at other times had other uses. Then there was the symbolic shaving of the heads of the men: in truth this is done very rarely in western prisons, but – and this is perhaps the most overt clue to the experiment's purpose – it *is* done to prisoners of war. Next there was the admission by Zimbardo that his guards were given khaki uniforms – to make them look as much like military men as possible?

Whatever the ostensible purpose of this research, its implications for prisoners of war are clear. The psychological situation facing the 'prisoners' was similar in many respects to that facing a prisoner of war: the surprise of the arrest, the blindfolding on the way to prison, the corporate aspect of the greeting. In prison men are still known by their name *and* number; only prisoners of war usually are referred to by their number.

Many armies now prepare their soldiers to be able to withstand interrogation. Zimbardo has in effect developed a method to teach people how to cope with the experiences of capture and imprisonment. The fact that 90 per cent of the conversation by prisoners was about prison, and that it made matters worse, suggests that POWs should not talk about prison, but should use every opportunity to escape from it mentally. The tendency to deprecate fellow prisoners and its security hazards should be pointed out. Continually dwelling on their immediate circumstances prevented the men from becoming a group and suggests that POWs should attempt to get talking and form themselves into a group straight away. Zimbardo's prisoners were strangers to one another, paralleling the POW situation. The fact that small guards can make large prisoners act foolishly and childishly is dramatically brought home; Zimbardo found that subjects mainly felt that the selection of guards had been on the basis of height – bigger men became guards – but in fact there were no physical differences between the two groups. At times in the experiment prisoners were encouraged to belittle each other publicly, during the counts, perhaps an experimental version of public confessions which were so infamous during the time of brainwashing.

On many matters, therefore, the Zimbardo experiment seems as much an exploration for a course in how to train soldiers or sailors to cope with the stresses of captivity as an experiment to find out how prison guards and prisoners behave, and perhaps that was Zimbardo's initial intention.

Zimbardo's experiment is but one of many that, since the Korean War in particular, have begun to examine captivity, torture and interrogation on a more scientific basis. The Korean War itself spawned a whole series of studies into who collaborated and into the thorny question of treason and loyalty. But laboratory experiments have not been lacking either, including many studies on forced compliance, how to threaten, the psychology of suffering and much else, which we now examine.

14 Interrogation

In his book, *Beyond Breaking Point,* which is a study of interrogation techniques, Peter Deeley devotes as much space to mental as to physical techniques (1). And a bibliography prepared by the Center for Research in Social Systems (CRESS) in 1967 listed 122 separate psychological studies into interrogation. This gives some idea of the enormous interest, both public and military, in psychological aspects of interrogation and torture (2).

For our purposes, interrogation methods are perhaps best divided into those techniques which employ torture and those which employ more subtle psychological pressures to extract information.

PHYSICAL AND MENTAL TORTURE

The most obviously psychological aspect of physical torture, of course, is that everyone fears pain and the job of the successful torturer is to manipulate his threats so that fear in the person being interrogated is maximized. This can be done by building up the amount of pain gradually, by making it come at regular intervals and so on. However, people differ greatly in the amount of pain they can tolerate. There also appears to be a difference between short sharp bursts of pain and longer, duller ones. Consider, for example, being held under water until you nearly suffocate and then being brought up to confess. Most of us probably feel that we would cave in under these circumstances. But the fact remains that many do not. The situation is such that some people prefer to die rather than submit. Also the interrogee knows that if he is killed his captors are certain not to get the information he has, so they will always pull him out when he gets unconscious. Rather than terrify a man into confession therefore, repeated short sharp ducking works more often than not because eventually the man is exhausted and cannot resist any more (3). The only problem from the interrogator's point of view is that this method takes time.

The next psychological move therefore is to use a technique that is less acute in the pain it presents. It does not make the prisoner unconscious – but he knows he will not escape the dull, exhausting pain *until* he has given away whatever information he is thought to have. For this various electrical methods have been used – wires fixed to sensitive parts of the body, or gradually moved from one part of the body to a more sensitive part, like the genitals or the anus, leaving the subject fully conscious, but in pain, and faced with yet more acute pain in the future. This method enables the interrogator to vary his threats and to tailor them to his prisoner's mood (4).

All effective interrogation/torture is a mixture of mental and physical methods. Usually, interrogation begins with a softening up process that is itself a cross between the mental and physical. The skilful interrogator will use the subject's psychology to turn it against him. He will manipulate the interrogee's pride and dignity, and sense of self. He will remove all his clothing, refuse him access to a lavatory so that he has to soil his own cell; rats or other animals symbolic for humans of filth and degradation will be allowed to roam the cell at will – exploiting a freedom the prisoner does not have. Meals and other privileges will come irregularly and infrequently, symbolizing not only the low worth of the prisoner but also the complete power over him of the guards; this will also add to his disorientation of not knowing where he is, why he is there, what is going to happen to him and what time it is. Man fears uncertainty probably as much as he fears anything. Threats to relatives or friends may also be used. After the revolution in Portugal in April 1974, it was revealed that PIDE (the political police) had made regular, and apparently effective, use of recordings of women's screams to 'persuade' prisoners that their wives/girl friends were being tortured in nearby cells (5).

More often than not, however, when interrogators interfere with a prisoner's sense of place, time and self, this signals the beginning of torture by sensory deprivation – a process which many think is the most effective form of all.

Sensory deprivation: the pathology of boredom

Sensory deprivation (SD) has been seen as the worst form of torture because it is relatively new, provokes more anxiety among the interrogees than more traditional tortures, leaves no visible scars and, therefore, is harder to prove, and produces longer lasting effects after it is all over. Compared with other traditional tortures, it is probably more difficult to prove that SD has taken place. Whether its effects are longer lasting – well, these claims are probably correct but they have not been completely substantiated by scientific results.

My purpose in saying this is not to belittle the effects of SD (there are several effects which are never mentioned in the literature – for example, many ordinary students start to masturbate after about 16 hours within full view of male and female experimenters). Rather, it is to point up two things: first, that it is not in this area that the most bloodthirsty psychological methods of interrogation have taken hold; and, second, to argue that the military do have a legitimate interest in the effects of both social isolation and sleep deprivation for situations other than interrogation – in the Arctic, in nuclear submarines, and in space flight.

The Canadians were the first to carry out major experiments in sensory deprivation, notably Donald Hebb of McGill University for the Canadian Defense Research Board (6, 7, 8), and John Lilly and Jack Vernon, all of whom were working in this area in the early 1950s. For some reason the Canadians were interested in manipulating attitudes and were in the brainwashing business as early as the Koreans and Chinese.

Throughout the late 1950s and 1960s, as the nature of warfare altered, SD research (and particularly social isolation research) took on a new significance. The military interest in remote aspects of our environment provided a need to find out the effect on group dynamics of social isolation and of sleep deprivation. There are, then, three aspects of the general subject of SD. First, the more experimental studies researching the basic effects of sensory deprivation and sleep deprivation; second, the use of these techniques in interrogation; third, (considered in chapter 16 on survival) their significance in special warfare conditions.

The experimental basis. The early experiments reported seemingly bizarre effects of isolation. The experimental techniques included fitting men with cuffs, goggles, semi-immersing them in tepid water, and playing monotones and meaningless noise (9). Hallucinations occurred, and many of the experimental subjects were too disturbed by the procedure of doing nothing all day to finish the experiment (even though they were being paid for it). Such findings attracted a lot of publicity, particularly what appeared to be parallels with the effects of brain-washing in Korea.

In the rumpus, several things happened. Most importantly, the general public came to believe, and *still* believes, that the effects of SD are invariably bizarre and terrifying. But in fact these early experiments were rather milder in their effects than expected (and produced *no* effects on some); and many follow-up experiments failed to duplicate their more exotic results.

In 1966 HumRRO released a report that covered a whole series of experiments in SD carried out during the previous ten years (10). The senior scientist in that project, Thomas Myers, carried on his research and published another detailed account in 1971 (11). These two reports comprise one of the fullest accounts of SD work during that time. Myers emphasizes that in a sense his research has been into *monotony*, rather than SD, because in practice it is very difficult to reduce stimulation to zero. What usually happens is that all change in stimulation is minimized – monotony.

At the Presidio of Monterey, an army base in California, subjects were confined to soundproofed cubicles which were kept at 72°, lit with diffuse light at all times, and contained a rubber bed, a chemical toilet and a fridge with as much food as the men wanted. Subjects wore loose-fitting pyjamas to reduce sensation; their watches and cigarettes were taken away.

The subjects were soldiers, ranging in age from seventeen to twenty-seven, of above-average intelligence and had been screened for any psychiatric abnormality. They were given a series of psychological tests before they went into their isolation period, similar tests at various points during the isolation, and finally after they came out.

There was no compulsion on the men to stay in the experiment, just the incentive of extra pay, but the experiments were terminated after four days. The first things which Myers and his colleagues measured were the movements of the men around the cubicle and on the bed; Figure 47 shows that daytime restlessness increased gradually as the days went by. The feeding and eating habits of the men, however, stayed pretty much the same.

The men reported that they spent a great deal of time thinking and dreaming about the past – for some they were pleasant dreams and helped to pass the time, but for many they were unpleasant and frightening. The men were on occasions frightened that they could not

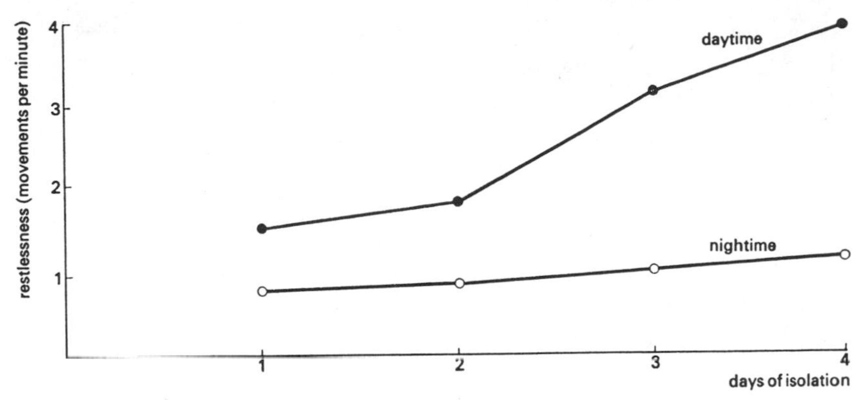

Figure 47 Increase in restlessness as a function of period in isolation.

stop these thoughts. Several became fearful that something terrible might happen to them or that they might be affected adversely by the experience. Still others were bothered because they could picture clearly what they were thinking in front of them and some reported later that they had been unable at times to distinguish wakefulness from sleep, feeling that the room was closing in on them or that their bodies were different from normal. Most ate less than they wanted, even when hungry, and as a result there was an average weight loss of 4·5 lbs.

An hour after the end of the experiment all men were given a subjective stress scale (first developed to measure how frightened in battle men are – see p. 204). The men in the cubicles rated themselves as considerably more frightened than a control group, who had spent the same time in similar cubicles but in the company of others, and with TV available, and with an intercom by means of which they could talk to the experimenters. The isolated men averaged a score of 50 compared to the controls' 24. But – and this is important – this figure of 50 on the scale translates only into feelings of 'indifference' or 'timidity'. It was *not*, on average, a 'terrifying' experience.

On the other hand, Myers and his team found that 37 per cent of their subjects withdrew before the four days were up: most complained of tedium and boredom and almost all said that they could not sleep any more. Many said they thought time had gone more slowly than they reckoned, they were disturbed at not knowing how much longer they would have to do and that their thinking had become jumbled. Interestingly, many were surprised when they were let out, claiming that they must have requested release during their sleep, since they could not remember saying the words. Early-releasers, Myers also found, were more likely to estimate that more time had elapsed over a given period and were more restless early on the experiment. Overall they were younger, higher on the psychopathic deviance and hypermanic scales of the Minnesota Multiphasic Personality Inventory and more likely to smoke than the long-staying subjects.

On release, after ninety-six hours in the dark and quiet, ordinary visual objects were seen as exceptionally rich in detail and saturated in hue, with more sounds apparently crowding in on them. Many also felt light-headed, a bit dizzy, possibly weak. Some were excited and garrulous (important possibly for interrogation) but though many spent the first day taking things easy, these effects lasted for no more than a few hours. Much later, however, memories of the event could still be evoked – for example, in one subject who was beset by insomnia and another who was by chance engaged in testing food products – which he found reminiscent of the cubicle diet.

Anxiety scales showed no change with isolation: on various psychomotor and arithmetical tests (dotting, tracing, copying, pursuit and so on) the isolated men were somewhat slower. On intellectual tests, varied results were found. Compared with control subjects, the men in isolated cubicles were worse on two more difficult tasks, much the same on another, but actually performed better on the simpler tasks. This was in spite of the fact that the subjects in the isolated cubicles did report that they had much greater difficulty in thinking than did the controls. Greatest difficulty was reported among the isolated subjects who did best on the tests, which suggests that those who tried hardest experienced more subjective difficulty.

These were the more basic psychological changes measured in the early phases of Myers' work and it can be seen that though there were definite changes, they were not as dramatic as often thought, nor were they all negative.

But Myers also looked at more applied changes. He found that on vigilance tasks (spotting and reacting to a random buzz), the isolated people actually made only half the errors of the controls. They remained more awake, and less drowsy than the controls. This would appear to have some implications for military situations like the monitoring of radar screens, say in submarines which are themselves confined and isolated environments.

Myers found that hallucinations were affected by the instructions he gave the men before they went into the cubicles. Those whom he told it was normal to have visual sensations reported more than those who did not receive these instructions. So he began to classify hallucinations into: amorphous light or contrast in shade of darkness, geometric forms, single objects, and integrated scenes. He found *no* difference in the amount of visual sensation reported between the isolated men and the control men. In both cases most reported hallucinations but what seems necessary for this, according to Myers, is not sensory deprivation or isolation, but a period of darkness. Nor was there any difference in the complexity of the hallucinations experienced by the two groups. On the other hand, after the isolation was over, the isolated men did report more visual sensations than the controls.

If stimulation was so important, Myers reasoned, then if the isolated men were given the opportunity to stimulate themselves, they would do so more than the controls. So they were told that if they pulled a lever they would receive in some cases pleasant music and in others a fairly unpleasant droning sound. There was no difference between the groups in the amount of pleasant stimulation they wanted. Some in the isolated cubicles were resentful of the outside world of which it reminded them. But those in the isolated cubicles wanted *less* of the white noise; it seems, says Myers, that SD reduced the threshold for irritability – another possibly important result from the point of view of interrogation (12).

Myers also explored the effect of SD on an individual's ability to maintain independence of judgement in the face of contrary group pressures (13) – again of interest in interrogation. Results showed that the isolated men of lower intelligence (remember though they were still above average for the army) were about twice as conforming as the controls. But the men in isolated cubicles who were of higher intelligence were, if anything, *less* conforming than the controls.

Next, Myers tackled the subject of propaganda direct. He gave the men a set of tapes which were either favourable or unfavourable to one of America's allies in NATO – the Turks (14). Myers found that the isolated men requested more playing of the tapes, but that their attitudes did not change any more than did the control groups (and what change did occur was minimal). Once again there was a slight tendency for the more intelligent isolated subjects to change less than the others – that is, they were *more* resistant to the propaganda.

Finally, in an experiment to try to condition unfavourable adjectives to certain nationalities – by pairing the names of the nationalities, say Greek or Italian, with such adjectives – Myers found again that there were no differences between isolated and control groups. What learning did occur did not last beyond isolation (15).

These experiments, therefore, suggest that sensory deprivation is not the monster it has been portrayed. Some psychological changes do occur but they are minimal, and some are apparently beneficial. Hardly any of them last.

But this was only after four days' isolation: what would happen if it went on longer? By now Myers had moved from California to the Naval Medical Research Institute in Bethesda, Maryland, and the project was now dignified with the codeword 'COMONOT', standing for 'comparative monotony'. In a further set of experiments, in which the deprivation consisted of confinement in total darkness and silence lasting seven days, Myers looked at brain states, sleep patterns and hormone levels, as well as the aspects explored in his earlier experiments (16).

His most important findings were of two clusters of effects which he called 'distress' and 'deactivation'. He found that the people who lasted four days were extremely likely to last seven. On the other hand, those who were extremely restless on the first day were unlikely to last even four. On the mood questionnaire and other tests, the long-stayers in the experimental group were indistinguishable from a confined control group and an unrestrained control group. It was only the people who showed the distress symptoms who made the SD group different (17).

Distress turned out, on statistical analysis, to have three components: a tedium factor related to boredom, time disorientation and restlessness; an unreality stress factor embraced hallucinations, strange bodily sensations and the like; and a third factor, which Myers had not anticipated, he labelled 'positive contemplation' – this denotes constructive thinking efforts, clearness of memories and the pleasant tone of some cubicle activities. Early-releasers were higher than long-stayers on the first two factors and lower on the third.

Only the early-leavers were most likely to seek out stimulation; the isolated long-stayers were no more inclined to want it than were the control groups. This Myers called the 'deactivation syndrome': certain people prefer just to get their heads down for the duration of their isolation without interruption. Men who did not want further stimulation were also very low on motility – they hardly ever left their beds, and this was constant over the seven days. They also performed better on the mood questionnaires: they were less angry, less frightened, less depressed – some even said that they enjoyed the experience. On closer examination, Myers also found that these men had slower EEG alpha rhythms as the experiment proceeded (18). The confined control group also had slower EEG rhythms, which Myers thought was due to decreased motility as subjects confined in the smaller of two rooms had slower alpha rhythms than those in the larger. 'Kinesthetic activity', concluded Myers, 'appears to be an important source of sensory input, one that is "necessary" for the maintenance of "normal" cortical tone' (19).

Thus confinement by itself produces scant evidence of distress. It appears instead that confinement produces or helps in deactivation, which is, therefore, mainly determined by conditions. But distress is rooted in personality differences.

Using psychoanalytic methods, Myers is now able to predict the men who can withstand SD. For example, he has been using the complex scoring of inkblot measures – reactions to the Rorschach inkblot cards. Particularly useful are those responses which indicate the intrusion into consciousness of aggressive, sexual and peculiarly non-logical content – what he calls 'primary thinking' (20). This though, he says, is normally anxiety-provoking; consequently he then examines secondary responses to see how these anxieties are defended against. Those who can defend well cope. Using this test he was able to predict seven out of nine men who actually remained in the cubicles for the seven days. All six poor defenders were early leavers. Using the Fitzgerald Experience Inquiry, which is a measure of flexibility of thought – the extent to which a person can draw on artistic inspiration, brainstorm, indulge in divergent thinking and so forth – Myers again found that performance on this is related to withstanding SD. Seven out of eight high scorers (very flexible people) weathered the SD treatment for seven days; only one of nine low scorers did so (21).

Myers' work is consistent over a number of years. Furthermore, it dispels some of the myths that have grown up about sensory deprivation and at the same time has relevance for many military situations – both for survival in special circumstances, and in interrogation. But it would be wrong to give the impression that Dr Myers is the only person to argue that SD is not necessarily as horrific as is generally thought. In *Man in Isolation and Confinement*, which arose from a NATO seminar on the subject in Rome, many topics were explored, including the effects of SD on resistance to pain, the perception of time, 'planned' SD versus 'accidental' SD (as in capture) and so on. There too it was reported that in many cases SD can have a beneficial effect and that the selection of people who responded well to it is perfectly possible (22).

So far as controlled laboratory experiments are concerned, therefore, there can be little doubt that, by itself, SD is not as horrendous an experience as it has been painted. But of course, there is a large difference between the laboratory situation and SD when used as part of 'hostilities'. The captor–captive relationship is entirely different, the role of uncertainty and fear is magnified and attempts to get people to talk will be far more assiduously pursued. Let us look at a real-life case where SD has been used, so as to understand the differences.

The Ulster hybrid

As Tim Shallice, a British psychologist with a long-standing interest in the subject, has pointed out, the normal methods of SD take some time to produce 'confessions' or usable information (23). In Northern Ireland, when internment was introduced, a hybrid technique appears to have been used which produced its effect more quickly and so elicited information sooner. Shallice reports on twelve men who were especially subject to these procedures. When not being actively interrogated these men had their heads hooded in a tightly woven black bag; they were subjected to noise of 85–87 dB – 'like the whir of helicopter blades'; and were forced to stand with their hands above their heads against the wall (similar, Shallice says, to the KGB Stoika position). If a man moved he was beaten and periods of wall-standing lasted for up to sixteen hours at a stretch. Internees wore loose-fitting boiler suits; were deprived of sleep for the first two or three days; and their diet was restricted to the occasional dry bread and water. Temperature 'appears', says Shallice, either to have been very hot or, when the men were allowed to sleep, too cold. The total time men spent at the wall varied from nine to forty-three hours and the entire interrogation experience lasted for six days.

Undoubtedly the whole procedure was a terrifying experience for the men. Three men later seen by Professor Daly, an Irish psychiatrist at University College, Cork, were reported to have become 'psychotic' within twenty-four hours of the beginning of the interrogation. The symptoms were loss of the sense of time, perceptual disturbance leading to hallucinations, profound apprehension and depression, and delusional beliefs (24). One man is said to have heard and seen a choir conducted by the Protestant leader, Ian Paisley; another could not stop himself from urinating in his trousers and on his mattress; a number had suicidal fantasies. Shallice reports that most other cases of hooding suffered the severe mental injuries after it was over and that almost all of Daly's other patients (twenty in all) had overt psychiatric illness. Anxiety, fear, dread, as well as insomnia, nightmares and startle response were common; depression was almost universal. Psychosomatic symptoms such as peptic ulcers had developed rapidly.

The picture revealed here is much more complex than the straightforward SD one. Only three aspects of the treatment – hooding, noise, wall-standing – can be called sensory or perceptual constancy. The boiler suit and temperature variation both provide a certain amount of sensory stimulation. All these factors, when combined with sleep deprivation, make the Ulster interrogation methods closer to Zimbardo's fake prison than to Myers' SD cubicles. The British Army, in fact, seems to have arrived at a highly effective compromise between two pieces of research, finding that together they are far more powerful in breaking people down than either method used alone. In Ulster, psychiatric symptoms began to appear on the second day; many people survived seven days' SD, whereas no one remained unaffected by Zimbardo's prison, with the bouts of crying, anger and resignation settling in during the second day. The psychosomatic symptoms which developed in Ulster did not always occur in the pure SD experiments; but they *were* seen in Zimbardo's prison. Finally, there seems to have been *no* evidence of the deactivation syndrome among the prisoners in Ulster; if they had been subjected to SD alone, one or two 'survivors' might have been expected. But no one survived Zimbardo's experiments intact, and this fact also supports the idea that Ulster was closer to the practices used in his experiment.

It is clear, then, from a comparison of the work by Zimbardo and by Myers in relation to the Ulster hybrid, that the experience of captivity is far more disorientating than the experience of sensory deprivation, but together they are lethal.[1] To counter them soldiers have to have

1. The same can be said of the special 'control units' introduced into British prisons in 1974. Two new units at Wakefield and Wormwood Scrubs prisons were designed only for 'intractable troublemakers'. Their chief features were isolation (all the men were in solitary confinement), guards who had been trained to be 'cool' (the Home Office's word), and a system whereby each day was numbered 1 to 90; should any prisoner transgress the strict rules he automatically went back to day 1, even if he were on day 89. As a result of articles I wrote in the *Sunday Times* and the many adverse reactions, the control

training in both. The world's armies are probably more aware of this distinction than the non-military scientists. We often see newspaper reports of soldiers playing military manoeuvres in which they get 'caught' and are 'interrogated' for long periods under rather brutal conditions (25).

The Russians, the West Germans and the South Africans have all tried variants. For example, following the appearance of the Baader-Meinhof group, the Germans developed long-term solitary confinement units, known as *Tote Trakt* ('silent floor') in which all walls and furniture were painted white, the light was always left on, and windows were dark and difficult to see through (26). This system led to Holger Meinz starving himself to death and to Ulrike Meinhof's suicide by hanging in May 1976. According to L. E. Hinkle and H. G. Wolff, the Russians use similar methods, but also keep the prisoners short of food (27). The treatment is shorter but more severe, and produces a state in which prisoners, reportedly, become totally dejected, depressed, obsessed with nightmares (the deactivation syndrome?). In South Africa conditions of prolonged solitary confinement are less extreme than in Germany or Russia but last for a longer time under the ninety-day detention law. According to two South African lawyers, this has led to two suicides and the committal of others to mental hospitals on their release (28).

Drugs in interrogation

As early as 1957 the use of drugs in interrogation was outlined by Joseph Kubis. He discussed scopolamine, sodium amytal, sodium pentothal and various barbiturates. He reported that 'under slight sedation (with narcotics) . . . individuals feel impelled to talk and often reveal personal matters they would otherwise conceal'. From 1958 to 1965 the US Army was involved in a series of trials with meprobamate and its effects on mood, emotion and motivation. It was found to

---

unit at Wakefield was eventually closed, and the one under construction at Wormwood Scrubs never opened. However, several scientists criticized the units for their use of sensory deprivation. Yet the men in the units were not really deprived in this way: they could move about, had an hour's exercise each day, could talk to the guards from time to time. More horrific was the manipulation by the prison authorities of the prisoner-guard relationship. The deliberate 'coolness' and the ninety-day ordeal were much more powerful, it seems to me, in breaking the men. And it further shows that we should not be misled into thinking that SD is responsible for *all* the horrors of psychological torture (29).

produce a 'drowsy, washed-out, bored' feeling useful in interrogation (30).

Subsequently, however, Lionel Haward, a British forensic psychologist, has claimed that the 'so-called truth drug' has been shown to be 'ineffective', but that 'some of the newer tranquillizing anxiolytic drugs induce a sense of well-being which . . . lowers the prisoner's mental defences and enables information to be obtained more easily' (31).

Recently, the use of drugs has come closer to direct physical torture. About thirty cases of alleged administration of drugs by security forces in Ulster in 1972 have been documented (32), although the information on some of these is scanty. Three cases are particularly well recorded: Francis McBridge, after drinking a cup of tea at Ballymoney police station where he was being held for questioning, became dizzy and began to have hallucinations. Similar symptoms – suggesting an overdose of amphetamines – were experienced by two Newry men: they could not urinate, experienced dryness of the mouth, and disorientation. When one of the men was eventually able to urinate, a sample showed traces of amphetamines.

The Russians are reported to have used succinyl choline on Israeli prisoners in Syria after the Yom Kippur War in 1973. The drug, administered by injection, causes convulsive muscular spasms, leaving the victim totally paralysed, in agonizing pain, unable to breathe properly, but conscious. The victim feels himself dying for lack of oxygen; the effect is transitory, but he is then threatened with a repeat experience. By all accounts the use of this drug was very effective (33).

Thus, the trend seems to be towards using drugs which produce extreme physical effects to terrify a captive into talking, rather than merely making him garrulous.

Bluebird, Artichoke and Often-Chickweed

But of course most of this pales in comparison to the series of revelations, made in the summer of 1977, about the activities of the CIA in the field of 'mind control'. Following the assiduous digging of John Marks, a former State Department employee and co-author of *The CIA and the Cult of Intelligence*, and owing to the extraordinary provisions of the USA's Freedom of Information Act, eventually some 8000 pages of secret CIA documents were made available to the world (sanitized as to names and places, but otherwise apparently intact) (34).

These showed that, between 1953 and 1963, the CIA had authorized 149 projects attempting to control man's mind in one way or another, involving eighty institutions and 183 research workers, a good many of them in independent colleges and universities. Special organizations, like the Society for the Investigation of Human Ecology and the Geshikter Fund for Medical Research, had been set up to fund and carry out much of the research. Before it was all to end, an estimated $25 million would have been spent as Project Bluebird became Project Artichoke, which became MK/ULTRA, which turned into MK/DELTA. After 1963, one project, 'Often-Chickweed', lasted until 1973.

What had initially provoked such public interest in this story was the revelation, in 1976 before a Congressional Committee on US intelligence activities, that a Dr Frank Olson had been given LSD in a drink (unknown to himself, of course) and as a result had thrown himself from a tenth-storey window in 1953. Not until then was his family aware that he had not committed an insane suicide. John Marks eventually showed that the CIA had tried all sorts of behaviour-manipulation exercises – altering sex patterns, discrediting potential adversaries by causing 'aberrant behaviour', creating disturbances of memory, creating dependency in people, creating suggestibility. In 1954 forays were made into Communist countries by special interrogation teams (consisting of a psychiatrist, a hypnotist and an interrogator) to try out the new methods. As well as interrogation attempts, there were attempts to induce hypnotic amnesia in spies, so that they could carry information but not consciously remember either the information or the fact that they had even carried it.

Up to now, many of the experiments, except that on Dr Olson, have not been described in detail, so in spite of the publicity, a full assessment is not possible. Two or three points, however, are worth making in the context of this book. In the first place, it is important to note that the CIA's aim, initially defensive, was later acknowledged to have become offensive. It was initially felt that the Chinese and Russians had made startling advances in the field of brainwashing. As we have seen elsewhere in this book, however, and as a memo from Richard Helms, the deputy director of the CIA in charge of these projects, to the Committee on the Assassination of President Kennedy makes clear, this did not last long. It was soon apparent that the USA was ahead. As we have also seen in this book, the ability to do something almost invariably means that an attempt will be made to actually do it.

Secondly, the secrecy surrounding the projects undoubtedly affected the nature of the science carried out. Overwhelmingly, the ideas did not work. The attempts to induce amnesia or suggestibility, to discredit enemies by inducing aberrant behaviour, were to the CIA on a par – as the documents made available to Marks make clear – with attempts to dissolve the Berlin wall, to devise a pill to make a drunk sober, or to devise a knockout drop. As such they catch the imagination; but – and this is important – as has been acknowledged, most if not all the projects came to naught. That is my third point. They failed – not only because the CIA became worried lest a San Francisco brothel maintained on its payroll to provide a source of subjects for experiment would be found out; they failed as science also.

It is, I must confess, too early to judge these CIA disclosures. *Operation Mind Control*, by Walter Bowart, is due to appear at roughly the same time as this book and contains material which, Mr Bowart says, shows that, for example, attempts to induce amnesia were successful. He has been kind enough to show me the manuscript of some of his chapters, but I remain less than totally convinced. Perhaps his book, and the one planned by Mr Marks on a similar subject, will change my mind. But at the moment it seems to me that the CIA behaviour-control experiments are significant mainly for the publicity they have attracted as a kind of public bogey. But in the longer term they tell us less about the ways to change (or 'control') behaviour than other sections in this book.

PSYCHOLOGICAL INTERROGATION

We may now turn to those methods in which pain is not necessarily present in the interrogation but in which psychological principles are employed to extract information. These techniques can be roughly divided into those which extract information from the interrogee either without his consent or without his knowledge. In a military context it is highly unlikely that someone being interrogated would consent to give away information easily, but it is still a useful distinction.

Perhaps the most purely psychological interrogation technique is that in which the interrogator uses no aids other than his understanding of human behaviour. He comes to know how to exert influence on his captives so that they comply with his wishes. So his first aim is to exploit the compliance of the interrogee.

Albert Biderman, in an account for the US Air Force on interrogation during the Korean War, concluded that many prisoners of war did not conform to the international convention (of giving names, rank, date of birth and military number *only)* because of the psychological need to maintain 'a viable social role and an esteemed self-image' (35). Biderman interviewed 220 air force men repatriated by the Chinese Communists and coded the interrogations according to the amount of interaction between interrogator and prisoner, the demeanour of the interrogator, the amount of duress used, and the significance of the statements made by prisoners. He found that over half the interrogations lasted more than twenty-four hours, 10 per cent for more than a month. 'Coercive' methods (pain) were actually less 'effective' than non-coercive methods, unless the pain was debilitating and self-inflicted (for example, prolonged standing). Whatever their instructions, nearly all the air force men disobeyed them and conversed (in general terms) with their interrogators. Biderman concluded that, for many Americans, silence in an interaction is far more stressful than verbal banter, for example. In this situation, the interrogator asks short, simple questions quickly, leaving the prisoner in the role of a silent but *compliant* listener, since the interrogator already knows the answers to his questions. 'You are from the 351st, right?' 'Based in Okinawa, right?' and so on. In addition, the assumption by the interrogator that silence is incriminating also puts pressure on the prisoner to speak to defend himself. A further technique exploiting the need to comply is to spend hours asking, shouting, questions to which a prisoner cannot possibly know the answer – details of atomic weapons, lists of leading military and political figures. Many prisoners spoke of '. . . the tremendous feeling of relief you get when he finally asks you something you can answer'. This is one of the more obvious examples, says Biderman, of how the interrogator manipulates the situation so that it is frustrating for the prisoner and his only relaxation is to 'please' the interrogator. Others yielded, Biderman found, 'while I still had my wits about me'. To Biderman this was not necessarily giving in prematurely, but a way of avoiding full compliance in the situation and of evading guilt after it – it reassures the prisoner that he is still in full control of the situation.

The interrogator will often appear to 'accept' a fabricated story. This forces the prisoner into a compliant role and later he can be made to show how 'insignificant' his statement was by giving away more significant material. The artful interrogator will also dodge the prisoner's hostility by never getting angry in return. This could be very frustrating for the prisoners in Korea and was often deflected onto their own colleagues. Biderman also found – and we return to this later in discussing resistance training – that men who had a rigid notion of their own behaviour, that they had, for example, a 'breaking point' beyond which they would collapse, gave in more than men with less rigid attitudes who could 'soak up' an indefinite amount of treatment.

Biderman explored the nature of the threats used. He found death threats remarkably unproductive – they produced information on only 5 per cent of occasions. Far more effective were threats that were vague about time and the degree of physical harm: this disorganized the prisoners more and made them more compliant.

Later, classified USAPRO work in simulated interrogations found that officers who gave away secrets differed from those who did not. The 'confessors' were those who were bothered by being alone; suffered drowsiness or fatigue, especially during the mission; felt 'battered' by the dangers of the mission and by events between capture and interrogation (36).

Political interrogation

Interrogation techniques, according to the evidence, appear to work less well with politically motivated individuals than with others. Police and other security forces have, therefore, developed new ways to break political suspects, for instance, the 'interrogation area', which ideally consists of a reception area and the interrogation room proper which can only be reached via the reception area. The area is usually selected for its remoteness, suspects or prisoners being taken there by the most roundabout way possible to increase the sense of isolation. The prisoner will normally be taken to the area by a uniformed guard who will then leave. This is to highlight the sense that the episode is out of the ordinary. The suspect is not left alone to 'sweat it out'; the interrogator is usually in the room when the prisoner/suspect arrives, so gaining the psychological advantage. The interrogator will endeavour to look as business-like as possible: neat, clean, composed, in control. The suspect/prisoner will be given a chair that will make him sit upright, but not too comfortably, and most probably fixed to the floor, just far enough away from the desk for him *not* to be able to rest his elbows on it. The interrogator, in contrast, will have a comfortable chair, well padded, and with a swivel, so that his movements can continually emphasize his freedom.

The interrogator will work to a three-stage plan. First comes the friendly interviewer stage, used partly to get basic information about the suspect and his movements and to establish in the mind of the suspect that the interrogator is no ordinary policeman/guard and that he does not have the common attitude to criminals/prisoners. At first any mention of the alleged crime or sensitive information is general enough. The aim is to get across the personality of the interrogator, his professionalism, his humanity. If the suspect does not clam up the interrogator is doing his job properly.

After a while, however, and when the suspect/prisoner is talking readily enough, the situation is dramatically and suddenly changed. The interrogator himself may suddenly change his manner. He may have a hidden buzzer with which he can signal to the reception area so that someone will enter with, say, some official-looking papers which the suspect/prisoner is not allowed to see. The interrogator reads the papers and his attitude changes – he acts as though the whole thing is sewn up, an eyewitness has been found, one of the suspect's accomplices has talked. The suspect may as well give himself up. It is a bluff, but often it works. If it does not work, then the interrogator will be changed: he will be 'called away' to be replaced by someone who is a totally different figure – a verbal bully. This sudden-change tactic works often enough for it to be the most common technique.

When this technique does not work the third phase is put into operation. The suspect/prisoner is removed from the room and when he returns the desk has gone and the original interrogator is back with just two chairs. The interrogator moves in close, changes regularly from a friendly to an unfriendly attitude. In this phase he may make use of psychological findings which show that many people have a body 'buffer zone' – an area inside which we do not like other people to come. In most of us it is a circle around us, two or three feet in radius. Some criminals, however, have a differently shaped zone. They can stand people coming closer to them at the front but do not like anyone to come at all close to them when they stand behind, out of sight (38). So rather than the conventional eyeball-to-eyeball interrogation, the modern interrogator may ask many questions sitting or standing behind the suspect, out of sight. All this, of course, is part of a series of 'interviews', not a one-off affair.

A technique used in Korea that applied perhaps more to brainwashing than interrogation *per se* is called by Cyril Cunningham, the psychologist who examined the British captives in Korea, 'ideational bullets' (39). By this he means the insertion into the interrogation of ideological assumptions designed to re-align the interrogator and prisoner so that they appear to be on the same side, working against a common enemy. It must be, says Cunningham, at the level of *assumption* not explicit statement if it is to work. The interrogator will, for example, 'let slip' his feeling towards the prison camp commander or his feelings that neither capitalism nor communism works. Over time the suspect does the same and starts revealing things he would not otherwise.

The interrogator's skill lies in the degree to which he can trick the prisoner into giving away information, while the prisoner is usually fully aware that these tricks are likely to be played. There are, however, two methods in which it may not be obvious to the prisoner that he is being manipulated. In one, the interrogee's behaviour is 'conditioned'; in the other a 'stool pigeon' is used to elicit information.

Conditioning. Using the technique of instrumental conditioning, a prisoner's behaviour may be modified by rewarding or punishing him according to principles laid down by the captors. These principles are not necessarily readily apparent to the captive. For example, a prisoner may be fed only when he talks about the general area in which the interrogators are interested. This way the prisoner 'finds it easier' to talk about some topics rather than others. Every time the prisoner says something of which the guard approves, the guard says 'right' or offers some form of encouragement. Gradually, imperceptibly, the field is narrowed down so that the prisoner is talking about what the guard wants. Even if successful, however, this method is a slow one (40).

Stool pigeons. The use of a stool pigeon is a variant of the 'hidden microphone' technique used, for example, by the Portuguese political police (41). Someone, a confederate, even a fellow prisoner, may appear to the interrogee to be a sympathetic listener in whom the interrogee can confide in an unguarded moment. Haward recalls the story of a German pilot in the Second World War who baled out over the Channel while flying a new type of fighter aircraft. The usual interrogation methods failed to elicit any information about the aircraft; but when the pilot was later admitted to hospital, he was put into a bed next to a British pilot who was able to discuss technical details of various aircraft. It was not long before the German pilot was discussing the technical merits of his own plane (42).

A similar technique was used in Korea. It is based on *folie à deux*, a rare psychiatric abnormality in which two people come to share the same mental illness, usually the same set of delusional beliefs. A stool pigeon is placed in the compound with the prisoners and he tries to 'infect' their thoughts with well-placed suggestions. In time, and if well done, this can apparently so confuse a prisoner that he may come to believe the stool pigeon and give information away without realizing it (43).

THE PSYCHOLOGY OF LYING

The good interrogator naturally needs to know when his captive is lying. There are two ways that a liar may be spotted. The most well-known is the lie detector, but that often takes time to set up and under certain conditions its reliability may be in doubt. Before we consider that, then, let us look at the other method which stems from a study of the way guilty people respond to certain questions as compared with honest people.

Frank Horsvath found that on several general questions, people who were later found to be liars gave different answers compared with the answers honest people gave (44). Horsvath's work was based on a five-year study of people taking polygraph tests by Fred Inbau and John Reid (who runs a large lie-detector agency in the USA) (45). Their combined efforts now enable any interrogator (police more often than the military so far) to ask a series of about half-a-dozen questions and, even without a lie-detector, to have a very good idea of who is telling the truth and who is not. For example, when asked if he has any suspicions as to who might have committed a crime, a typical truth teller will usually try to identify people by name and give reasons for his suspicions. A person who is lying will refuse to blame anyone or suggest people who could not possibly be implicated. Another useful question is to ask the person/suspect if there is anyone who could *not* have done it. Truth tellers tend to name two or three, sometimes including themselves; a liar on the other hand will make excuses, for example, he doesn't know anyone well enough to vouch for them.

Horsvath lists about half a dozen tell-tale questions but points out that this type of analysis is a long way from being 100 per cent reliable as a guide to the guilty and no one suggests that it can ever be used as evidence in court. On the other hand, police work as well as military work often depends on speed so all clues that are available to the interrogator are helpful.

The lie-detector

The lie-detector has a celebrated history and a controversial one. Some security forces seem only too ready to use it; others – the British, for example – remain politely sceptical. Nevertheless, its use is spreading: the USA pioneered its development, but Japan and Israel at least now make regular use of this machine.

The theory of the lie-detector is basically simple: there are some aspects of our bodily functioning over which we have no conscious control – this is our autonomic system and among the changes we cannot influence are our heart rate, and the conductance of the skin. It is these changes which the polygraph (the technical term for the lie-detector) measures. These processes change their rate of activity when we are stressed. Lying is a stressful process. Therefore, if we ask someone a question the answers to which are stressful only for the guilty person because he or she is lying, this should show on the lie-detector.

In practice there are all sorts of complications. A relative of a murdered person, for example, is going to be highly stressed immediately after the death whether or not he or she committed an offence. And it can often be extremely difficult to ask a suspect a question in such a way that only the guilty will show a response. There are also ethnic differences. One American 'intelligence operative' who had worked in the Far East, is quoted by Peter Deeley: 'The orientals are so impassive, so capable of hiding their feelings, that we had to pack the lie-detector in; it was a complete failure' (46). Blood pressure changes can be brought on artificially by muscular contraction; there are instances of criminal suspects doing things like wriggling their toes to throw the machine off (47). In certain cases the machines show *more* stress when the average subject tells the truth than when he tells a lie. For example, in an experiment by Jan Berkhout and his colleagues, subjects were asked: 'Do you masturbate?' The majority of subjects showed more stress in admitting that they did than those who denied it. And once someone has lied on one answer, the physiological effect of this lie may run over the next few answers. When an interrogator is leading up to an all-important question, therefore, this process could throw his whole approach out of kilter. A suspect may lie on an early inconsequential question and this then confounds the important matter. Mentally subnormal suspects may not understand a question and therefore lie without knowing it; or, conversely, they may be so upset by not understanding the question that their distress may show as lying (48).

On balance all this means that false positives (indications that people are guilty when they are not) are much more likely than false negatives (indications that people are innocent when they are not). This can pose ethical problems about whom to accuse and when. As many detectives in America rightly say, the polygraph can only ever be used as an aid to interrogation. It is never more certain than that.

The Israelis have recently developed a lie-detector which, they claim, is more reliable than the polygraph (49). It measures the changes in breathing rate, for example, by using radar which is bounced off the subject's chest. In time it may be possible to use this at quite large distances so that, for example, police could question a criminal in a siege about his ammunition and monitor his replies for honesty.

Another lie-detector gaining popularity is the Dektor voice print analyser (50). This works on the principle that our vocal chords produce different frequencies when we are stressed than when we are not.

Hypnosis

Hypnosis of witnesses has been used in America, but the real pioneers in this appear to be the Israelis. Meyer Kaplan, the detective who commands the Jerusalem CID, has found that witnesses can recall far more under hypnosis than they can consciously. He thinks that this makes the method particularly useful in terrorist-type crimes when people may not, for example, remember someone putting down an ordinary-looking carrier bag. 'Under hypnosis the witnesses get a second chance to look at the scene of the incident immediately prior to the blast and describe it to us,' says Kaplan. 'In this way we get much fuller and more accurate descriptions of who was carrying what, who sat near whom and so on.' Hypnosis has recently been introduced as a routine police measure in all terrorist bombings in Israel where there might have been witnesses (51).

LIMITED LANGUAGES

Last in our consideration of methods of interrogation is one that has been specifically developed in the military: the concept of a limited language.

A combat soldier at the front line may capture a prisoner who in fear may be ready to talk, but the captor may not speak the same language. In the 1960s HumRRO developed a 'limited language' to enable front line combat soldiers to grill captives in the field before any information they might have of tactical value becomes out of date (52).

Limited language ignores rudimentary grammar but contains all words and phrases suitable for extracting information. The language developed by HumRRO consisted of sixteen commands (e.g. 'Surrender', 'Hands up', 'Speak slowly'); ten question frames ('Have you there . . . ?'; 'How many . . . ?'); and 100 inserts such as 'tanks', 'guns' or 'men'. One hundred basic words which a prisoner might use in answering were then included. Among these were 'yes' and 'no', cardinal and ordinal numbers, compass headings, time divisions, plus connecting words. The entire language consisted of about 450 words – 150 in the questioning vocabulary and 300 more in the answers. The question frames, when put together with inserts, were not necessarily grammatically correct though all answers were.

In the pilot study the language of interrogation, for the English-speaking American soldiers, was Russian. HumRRO found that it took soldiers about twenty days to learn the language. Then in tests, it was found that, with about forty questions relating to tactical situation, between 85 and 97 per cent of the information available in a given situation could be obtained.

The researchers emphasize that the system can be used with any enemy language – and can even be used between allies to improve tactical understanding on common operations. Since then the entire system has been put on tape so that the method can be taught in the field. The tapes are available in all 'enemy' languages.

15 Hsi nao: 'Brainwashing'

Brainwashing differs from interrogation in two important respects. One, it seeks not merely to extract information from a captive but to actually change his beliefs, his attitudes, his thoughts; and two, it has, in the past, resulted in the captive, as well as parting with information and changing his views, actively collaborating with the enemy by indulging in such things as broadcasts and false confessions which can be used as propaganda by the enemy captors.

As such, the whole process has achieved a notoriety in recent psychological and political history. A great number of extravagant claims have been made on its behalf. Since there is an enormous literature on the subject, no attempt will be made here to explore the subject exhaustively; instead, we will confine ourselves to an examination of the methods used, and the psychological principles involved, take a critical look at what has actually happened – and leave it at that. It seems to me that brainwashing was a stage in military psychology, effective – so far as it *was* effective – because it had not been used before, but unlikely to be as effective ever again precisely because of the notoriety that resulted from the first and only major use of it, in Korea. Interested readers are referred to an exhaustive bibliography on brainwashing, *Brainwashing: A Guide to the Literature* (1).

ORIGINS OF BRAINWASHING

Edward Hunter claims to have been the first to use the term 'brainwash', a translation of the Chinese 'hsi nao' (wash-brain). In 1951 he published *Brain-washing in Red China* in which he described Communist indoctrination techniques (2). Hunter established that brainwashing of enemy soldiers grew out of indoctrination techniques developed in

China immediately after the revolution in 1948,[1] which in turn drew on the expertise of the Stalinist purges of the thirties.[2] A CRESS bibliography on interrogation techniques (3) gives four historical roots to the science of thought reform: the Russian purges of the 1930s, the Second World War, the Korean War, and the Communist Chinese thought-reform programme.

However, Dr William Sargant, a British psychiatrist, puts these roots back even further. In *Battle for the Mind* he links all sorts of belief changes – religious conversion, political indoctrination, the changing of delusional beliefs among the mentally ill by psychiatrists – to certain common changes in the physiology of the brain (4). And these, he says, have been exploited since the times of ancient Greece, at least. In this context, Sargant discusses the work of the Russian physiologist, Ivan Pavlov. Pavlov studied the way in which reflexive responding, e.g. an eye-blink or salivation, could be conditioned. He found that if a stimulus that naturally led to a response – a puff of air causing an eye-blink, for example – was presented at almost the same time as a neutral stimulus – a sound – the neutral stimulus would eventually lead to the same response. By suitable manipulation of conditioned responses marked changes in behaviour could be induced. Sargant has attempted to explain the nature of the progressive dysfunctioning in brain-washing by reference to the work of Pavlov. He lists three stages. First, in what Pavlov called the 'equivalent' stage, all variations and strengths of outside stimulation produce the same responses, so that, in Sargant's words, 'the individual gets the same amount of feeling whether he is given a thousand pounds or sixpence'. Next comes the 'paradoxical' phase when he gets more satisfaction from small stimu-

1. For a somewhat different account of brainwashing and 'thought reform' to that given here, readers are referred to Dr Robert Jay Lifton's *Thought Reform and the Psychology of Totalism* (Penguin Books, 1967). Dr Lifton was in Hong Kong in 1954–5 and arrived there shortly after many US 'brainwashed' soldiers had been released from North Korea and returned home. He examined twenty-five Western and Chinese *civilians* who had been prisoners in China and who had gone through various techniques in prison similar to those the soldiers had experienced. Dr Lifton's concern is more with the way totalitarian regimes exploit fear and guilt, and the public confessions of any-one who deviates from accepted ideology. In dwelling on civilians, placing the episode in historical context, and raising the wider issues of conformity in societies, his account does much to de-sensationalize brainwashing and to defuse some of the more extravagant claims that have been made for it.

2. See Augustin Bonnard, 'The metapsychology of the Russian trials confessions', *Int. J. Psychoanal.*, vol. 35, 1954, pp. 208–13, for an account of the interrogation techniques and confession used around the time of Stalin's death. Some useful parallels with techniques described in this book may be seen.

lation than from a large one – people get very worked up about minor things while remaining impassive to more important events. Then there is the 'ultra-paradoxical' phase in which many things which we usually like (or believe in) we cease to like (or believe) and those things we do not like (and do not believe in) suddenly become those things we prefer or believe. It is in this phase that new convictions are said to really start to take hold on the person.

Together with this, Pavlov said that a person may also enter a hypnoidal phase in which he stops being critical and receives suggestions, commands and so on without question. Pavlov found that prolonged excitation of the nervous sytem, the induction of insoluble mental conflicts, and extreme fatigue brought about these four phases more quickly.

Dr Raymond Bauer has roundly criticized this type of explanation (5). He has argued that Communist brainwashing was characterized not so much by its rigidity as by its great variety. Mostly, he says, it exploited the pressures of group conformity rather than simple conditioning. As Khrushchev once said, Stalinism's psychology, Bauer reminds us, was 'beat, beat, beat'. Brainwashing, concludes Bauer, is often little more than common sense, was not new in the 1950s, and – more often – did not work. Many allegedly brainwashed in fact ended up merely confused. He makes a powerful case.

This, then, is a brief account of the early history of brainwashing. Even if what Sargant says is true, and the techniques had been used under other names for centuries, the whole process seems not to have finally coalesced in such a striking way until the Korean War.

METHODS IN KOREA

During the Korean War some 7000 Americans were captured. In every war in American history, some men have managed to escape. But it is generally conceded that the Korean conflict was the exception. Roughly one in three of the American prisoners collaborated with the Communists in some way, either as informers or as propagandists (13 per cent were subsequently charged with offences). In the twenty prison camps, 2730 (roughly 30 per cent) died, the highest mortality rate among prisoners in US history. For the first time in American history, twenty-one men chose not to return to America (6).

Moreover, the unusual sight of American soldiers 'confessing' that the USA had a germ-warfare programme against China and taking part in other propaganda broadcasts directed against the interests of the USA naturally attracted a great deal of attention. What were the methods used to achieve these remarkable changes?

According to Major William Mayer, a US Army psychiatrist who served in Korea and researched almost 1000 cases of POWs forcibly indoctrinated in North Korea, the Chinese chose as directors of the brainwashing programmes persons who had been educated in the United States during the previous ten to fifteen years and so understood the American mind and way of life (7). To begin with, squads of captured American personnel would be placed in a general camp. The first efforts were directed at destroying the unity of the squad; leaders would be discouraged from acting as leaders and if they persisted would be moved to a special camp for 'reactionaries'. (It has to be remembered that all the time the prisoners were uncertain as to their fate: the dread of being killed no doubt made the personal situation much worse than it must appear to the reader.) Another tactic was the development of the informer system. The Communists understood enough group dynamics to know that in any group someone would turn informer from his own personal needs – to gain satisfaction, to feel superior and so forth. But they also ensured that, in the early stages at least, no one suffered as a result of informing. All they would do was have a chat with the man who had been informed upon and say that they realized he was really a 'victim' of society and sympathize to an extent. This ensured two things: first, that the informer was not picked upon by his fellow group members and thus information kept coming in; second, longer-term prisoners ceased to trust anyone. This helped to break up the group, added to the soldiers' insecurity and anxiety, and made thought reform that much easier.

Then came the indoctrination sessions. These were made up of both formal lessons and informal talk sessions. After capture the prisoners, expecting torture and possible death, were instead told that they were not held by their Communist captors to be responsible in any way for the war. That responsibility was laid at the door of 'Imperialist Wall Street Warmongers'. Although this may sound unconvincing to us, to people who had just been reprieved from what they thought was certain death, it could easily have had a very different ring. They could not help but be grateful. Capitalizing on this feeling, the formal lessons did not try, unlike so much propaganda, to foster communistic feelings in the POWs, but rather to 'unsell' America. No Marxist tracts were used: instead articles from the American press and American literature were used – *Time*, *Wall Street Journal*, *Fortune*, and so on. Items were selected to show up the contradictions in American life, and to show how some people were still making a lot of money back home while 'these boys' were being captured. Post was censored so that they received only 'Dear John' letters, bills or bad news. In the informal self-criticism sessions, the soldiers stood up and confessed the error of their ways, criticized their own conduct and apologized to their fellow prisoners for this straying. It was, says Mayer, a form of self-informing to the enemy and it succeeded in creating yet more feelings of guilt and anxiety. It was the first step in collaboration.

In 1956 the US Group for the Advancement of Psychiatry held two special symposia on forceful indoctrination (8). In the first of these Dr Harold Wolff presented research to show that pain is most effective in altering attitudes when it succeeds in convincing prisoners of their isolation and the hostile attitude of their captors, when it makes the captive feel humilitated, degraded and cut off. Eventually 'these individuals then became especially susceptible to the contact of a friendly person who might approach them'.

The symposia also heard that, contrary to popular belief, isolation was only one of *eight* common Communist coercive methods for eliciting individual compliance. The eight are shown in Table 25.

Table 25 Communist coercive methods for eliciting individual compliance

Method	*Variants*	*Effects*
1. *Enforcing trivial demands*	Enforcement of minute rules and schedules Forced writing	Develops habit of compliance
2. *Demonstrating 'omnipotence' and 'omniscience'*	Confrontations Pretending to take cooperation for granted Demonstrating complete control over victim's fate Tantalizing with possible favours	Suggests futility of resistance
3. *Occasional indulgences*	Unpredictable favours Rewards for partial compliance Promises of better treatment Fluctuation of captor's attitude Unexpected kindness	Provides positive motivation for compliance Reinforces learning Impairs adjustment to deprivation

4. *Threats*	Of death or torture Of non-repatriation Of endless isolation and interrogation Against family or comrades Mysterious changes of treatment Vague but ominous threats	Cultivates anxiety, dread and despair
5. *Degradation*	Prevention of personal hygiene Filthy, infested surroundings Demeaning punishments Various humiliations Taunts and insults Denial of privacy	Makes continued resistance seem more threatening to self-esteem than compliance Reduces prisoner to concern with 'animal' values
6. *Control of perceptions*	Darkness or bright light No books or recreations Barren environment Monotonous food Restricted movement Absence of normal stimuli	Fixes attention on predicament Fosters introspection Frustrates all actions not consistent with compliance Eliminates distractions
7. *Isolation*	Complete physical isolation Solitary confinement Semi-isolation Isolation of small groups	Develops intense concern with self Deprives victim of social support Makes victim dependent on interrogator
8. *Induced debilitation and exhaustion*	Semi-starvation Exposure Exploitation of wounds Induced illness Prolonged constraint Prolonged standing Sleep deprivation Prolonged interrogation or forced writing Over-exertion Sustained tensions	Weakens physical and mental ability to resist

What, for the Chinese, did brainwashing achieve? According to Mayer, they obtained quite a bit of military information. Another important achievement, of course, was the propaganda value of the converts. Another, more subtle, change was that even in the minds of those who were not converted, the methods created doubts and confusions about themselves and their country and turned them into a very docile batch of prisoners – they rarely tried to escape and, according to Mayer, 'apparently learned nothing from the Communists that would be of strategical value to the USA'. Many prisoners died in captivity, evidently from the severe living conditions rather than as a result of torture. Survival depends in part on psychological factors. For example, several hundred Turkish soldiers held under similar conditions survived almost to a man. They maintained military discipline and their self-discipline better than the Americans; when a man became ill a detail of soldiers was assigned to care for him and ensure his recovery, bathing him and spoon-feeding him if necessary. No matter how much the Communists segregated the leaders, one of the remaining men always assumed command (9). (See also chapter 16, Survival.)

Mayer concluded that American soldiers had defects in three basic areas: in character development and self-discipline; in general education, particularly about the operation of a democracy and the multicultural role of the world; and in military preparedness – while good on weapons and tactics and so on, they were weak on the psychological and moral side of war. These points have implications for the training of resistance to interrogation and brainwashing as we shall see in just a moment.

INVESTIGATING THE BRAINWASHED

Immediately the brainwashed returned home the relevant authorities had to concern themselves with two problems. First, they had to undertake a proper scientific investigation to establish what had happened and, in the long term, they had to devise methods to counteract interrogation and brainwashing.

Investigation was also necessary to determine whether or not POWs had been guilty of war crimes. In America, for example, a Colonel Schwable of the Marine Corps was one of those who confessed to using germ warfare and along with many others was sentenced to a long term of imprisonment for these actions, although many soldiers claimed that they had been under mental duress (10).

British POWs alleging torture were all examined by Cyril Cunningham, a forensic psychologist employed by the Ministry of Defence. The military authorities in Britain thought that the allegations of mental torture were being used to excuse war crimes such as giving away information. During three years of intense investigations, Cunningham examined thousands of case histories and the evidence he uncovered prevented the premature trials of many British servicemen (11). Cunningham turned up only sixty cases 'in which psychologically significant techniques had been clearly used'.

The British method was to 'fingerprint' the various techniques and interrogators so that if a crime had been committed it could be carefully described and attributed. Cunningham used accepted forensic techniques to establish whether a violent or unnatural offence had actually occurred. For example, the claims of some men that they had been drugged were disproved because they gave inconsistent and inaccurate ingestion periods, or because witnesses contradicted one another. It became clear that certain techniques were used more often in some places than in others; sleep deprivation or stool pigeons appear to have been used in specific prison camps. By cross-checking individual accounts, it became possible to establish that some prisoners *had* lied about their treatment. It also emerged, for example, that six men in one camp, who were widely regarded as informers, did not have access to the information attributed to them. It became clear to Cunningham and his colleagues that others were the real informers. Clearly, the Chinese had broken down the group cohesion. The guilty in this case could not be brought to court as the evidence was not strong enough, but Cunningham did prevent the six innocent men being prosecuted.

He also found that a knowledge of psychosomatic symptoms was useful; there was evidence that behaviour normally regarded as psychotic had been deliberately induced by manipulation of events. In another case a man betrayed in attempting to escape had been brutally maltreated by his captors. He became paranoid and, after release, he committed suicide in the mistaken belief that he had caught VD on the journey home. As a result of the investigation, his family were able to sue for his pension.

Roughly half the American soldiers taken captive during the Korean War were repatriated and their personal histories and prison-camp conduct were recorded. Julius Segal, of HumRRO, working from Army documents, divided the men into three categories:

Participators: 15 per cent of the men were recommended for court martial or dishonourable discharge.

Resisters: 5 per cent recommended for decoration who committed at least two acts of resistance in internment.

Middlemen: 80 per cent on whom there was little or no information.

In general Segal found that background characteristics did not serve to differentiate resisters and collaborators nearly so well as did traits of personality shown during internment (12). The resister group had fewer men with low IQ – 35 per cent as opposed to 45 per cent of the collaborators. Contrary to popular view, the amount of pressure applied by the captors was *negatively* linked to giving in – the more it was needed, the less likely it was to succeed. About 12 per cent of all POWs accepted the Koreans' ideological teaching to any degree, but little relationship was found between the degree to which a man accepted the captor's ideology as his own and the extent to which he collaborated.

The main difference, according to Segal, was the way a man responded to promises of preferential treatment. In other words, men who were influenced by, and accepted, material favours were those who collaborated: 91 per cent of the resisters were not in the least swayed by the enemy's promise of rewards; but the same can only be said of 8 per cent of the collaborators. They were opportunistic. The bag of prizes included everything from better food to money. No conditions were ever plush – perhaps a fur-lined jacket for petition signatures or a position of prestige for informing on a colleague. One indication of the real meaning of the difference, says Segal, is that the collaborators were repatriated in better physical shape than the resisters but with more neurotic symptoms: 'acceptance of the enemy's bribes was not made, evidently, without some degree of guilt and anxiety.'

Incidentally, only 13 per cent showed great concern for their fellows; 49 per cent showed moderate concern and 38 per cent showed none. An estimated 10 per cent of people informed on others and cliques of resisters sprang up to search these out and beat or murder them.

Segal's conclusions are prosaic – and realistic. Collaboration was achieved in Korea by the old technique of blackmail. What *was* new – and, says Segal, 'grossly misunderstood by many of our soldiers – were the conditions the enemy placed on the pleasures of preferential treatment and the elimination of threat. The propaganda petitions were not just innocuous pieces of paper . . . [they] were potent weapons in the enemy's psychological warfare arsenal.'

Segal noted that 80 per cent of the middlemen (who neither collaborated extensively nor resisted particularly well) were characterized by apathy: they did not collaborate but neither did they carry the fight to the enemy. This must count as a victory for the Koreans and an indictment of the American's own indoctrination of its men.

It is a situation which the Americans have since worked hard to eradicate. Besides greater indoctrination of its soldiers and the development of specific techniques to resist interrogation, many studies have been initiated to gain a fuller understanding of the captor–captive relationship.

Countless studies, for example, have been pursued into the question of how best to give and how best to counter threats. Even though the Office of Naval Research and the Disarmament Agency have funded several of them, much of what has been done has been very artificial. Nevertheless, certain findings are worth noting. One would not expect to learn, for example, that if you wish to change someone's opinions you should pay him a small amount of money rather than a large amount. Yet this is precisely what you should do *if*, as in brainwashing, the person's changed position is to be made public. It is as if paying a man too much smacks of bribery whereas a small amount counts as legitimate earnings (13). This is presumably what the Koreans practised when they only gave *small* privileges for collaboration. Other experiments have shown that the prisoner is more persuaded by a captor who leaves his threats vague than by one who makes them explicit (14). Conversely, a captive who holds information which the captors want should make it clear that he knows what they are after. He is likely to be treated with more respect. The more the captive indicates that he understands the psychological reasoning behind threats the more the wind is taken out of the threateners' sails (15).

Morris Friedell found that people who tend to retaliate can be distinguished by psychological tests and clearly this could be useful in screening men for hazardous missions (16). The most effective strategy in reducing aggression directed towards oneself has been found to be where the 'victim' matches his aggression to the 'aggressor'. In a prison camp this option will often be unrealistic yet, as with other research, this does encourage people to use whatever power they have, rather than adopt a completely passive role (17).

Other military research in this area has shown, for example, that people with a low height–weight ratio stand up better to threats than do others (18); and that nations differ in the role of underdog (19) – Danes, for example, are more submissive than Americans; that torturers *do* ease up the more pain the victim appears to be in (20); that people are *more* likely to devalue, and therefore *less* likely to help, someone who bleeds and complains rather than someone who does not (21).

RESISTANCE TRAINING: HOW TO BE LOYAL

Clearly, if interrogation or brainwashing is going to be an aspect of capture, then it is the duty of the soldier to resist it. Any training he can be given in this thus makes a lot of sense. At the lowest level, merely making the soldier aware of the techniques available is a help, on a 'forewarned is forearmed' principle. But it is even more effective to give the soldier *experience* of these techniques: recent reports indicate that British naval officers in training on Dartmoor, Belgian paratroopers and US Special Forces personnel in Germany have been doing just this (22).

More systematically, studies of people who have resisted interrogations and brainwashing in the past (and how they differ from collaborators) help the military select people in the future for high-risk specialities (such as reconnaissance flights) where capture and interrogation are extremely likely.

Dr Edgar Schein concluded from his study of resisters and collaborators in Korea that the best way to look at this problem is to think of interrogation and brainwashing as a stressful situation where the people either remove the stress by withdrawing psychologically from the situation (the resisters) or by taking some action to remove it (the collaborators). He based this conclusion on a whole series of comparative studies of the background characteristics of resisters, collaborators, and neutrals (23). On all measures he obtained the same intriguing shape to his graph (see Figure 48). In other words, resisters and collaborators have more in common with each other psychologically than with neutrals. He also makes the point that in many cases collaboration should be seen more in a group context than an individual one. Where the Communists succeeded in breaking down the structure of the captives' groups, their hierarchy and so on, the ensuing group disorganization contributed substantially to the lack of resistance.

In general it would seem that the collaborator and the resister are

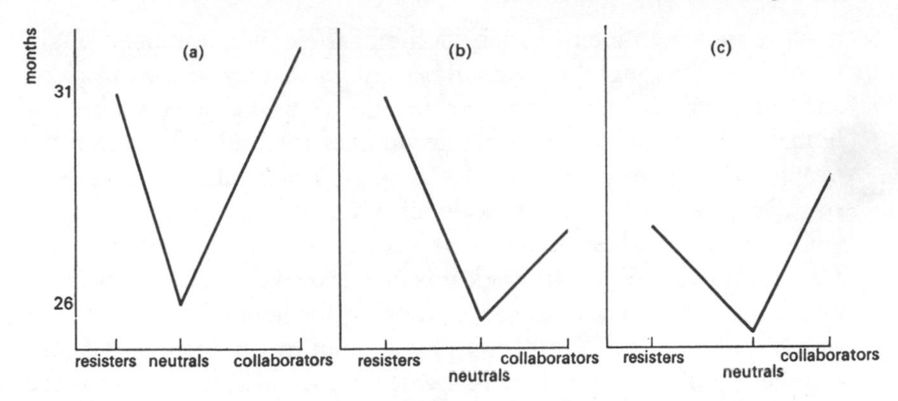

Figure 48 Comparison of collaborators, neutrals and resisters in Korean War as a function of: (a) length of internment; (b) intelligence; (c) personality as measured by the psychopathic deviance score of the Minnesota Multiphasic Personality Inventory.

not to be easily distinguished from one another on the usual background variables of psychology. It is true that, as was mentioned earlier, soldiers with a better political grasp appeared to be better resisters in Korea than those who were less knowledgeable politically. But whether this was specifically due to their political knowledge or to their natural interest in politics or to their higher general level of education is an open question. Since the Korean conflict, the military have, however, taken the liberal arts and the social sciences a lot more seriously – or at least have devoted more time to them in training their men. To that extent it can be said that they had taken the lesson of Korea to heart. At the same time, it is also true that the military still seem to favour a more specific approach – teaching resistance to the specific aspects of interrogation and torture rather than more general matters. This seems to mean that they are more interested in how men stand up to pain, or isolation than how they stand up to persuasion. And it means, of course, that they are more worried about men giving away information than repeating what happened in Korea and confessing to war crimes they could not have done. (In so far as the confessions of Korean POWs were later shown to be largely false, the military are now right: any power publicizing the confessions of a POW would have to contend with the Korean history – and it would have to be good propaganda to overcome that.)

We can thus proceed to a description of the more sophisticated special techniques under consideration or in use by the military to teach resistance to interrogation. Some of these were certainly taught to soldiers who fought in Vietnam because, as that war drew to a close, and many of the returning veterans grew disillusioned with their homeland, they passed on these techniques to some of the terrorist groups then spawning in the USA. California police were at one stage very worried about the scale on which this was happening (24).

One of the earliest accounts of resistance training was given by Dr Louis Jolyon West. His work was undertaken for the US Air Force and entitled 'Psychiatric aspects of training for honorable survival as a prisoner of war' (25). West begins his account by questioning the concept of 'breaking point' beyond which a man will invariably break down. He quotes evidence to show that people vary over time in their ability to withstand stress and it is clearly helpful for trainees to be told this, to give them faith in their ability to 'soak up' punishment. A second point is that Communist techniques tend to be fairly rigid and therefore the various stages of brainwashing may be anticipated by the captives and this also forearms them to some extent.

Thirdly, 'It is vital,' Dr West continues, 'to educate the combat crew member with regard to the overall nature of the conflict, the significance of the role of his organization, and the importance of the behavior of each member thereof after capture.' This should include, says West, a clear-cut account of the enemy's activities in the past – especially, if possible, from someone who has actually experienced them. Didactic lectures, films, demonstrations and personal participation are all recommended by West. 'The endurance of some physical discomfort may actually be helpful', though physical pain or injury need not occur. Training that familiarizes the trainees with the fear reaction is also a help and need not be linked too closely with any particular method. It should be pointed out that a good show of resistance in the early stages means that prisoners are likely to be left alone in the future. The experience of objectionable food is also a help and may make difficult diets easier to bear in the future. The men are specifically told that it is possible to go ten days without food, that they can travel 150 miles or more during that period with proper pacing, that hunger pangs often diminish after seventy-two hours and that they can exist for six months on half their usual diet – all of which, says West, can be made doubly meaningful if they train on long treks with short rations. The men are also told that a large percentage of enemy threats will be bluff, that the enemy will impose a demanding schedule, and use isolation to try to make the captive more suggestible. Men are taught ways of 'not hearing', to expect intensive interrogation after an initial softening up, and that they can accept some of what the interrogator says without losing their overall judgement. They should know that they can go without sleep for up to five days without harm, that they may have to stand for forty-eight hours, that a delirious reaction usually achieves some release and that this can be simulated. So far as indoctrination is concerned, says West, it is rare for physical coercion to be used; invariably it is psychological, and usually group-oriented. What counts here is that the soldier is well briefed on the war he is fighting and who and what his enemy is.

Dr West's account concerns pilots particularly and appeared after the Korean conflict. Since then other sophisticated techniques have been developed or are in the process of development based on recent psychological and physiological research. We can, in fact, now expand on Dr West's account to conceive of modern interrogation resistance training as having four stages (26).

First, for selection, background differences between good and bad resisters are of minimal use. A small proportion of people are extremely resistant to pain (and so are more likely to withstand torture) whereas others seem to be very good at withstanding sensory deprivation – and so can presumably stand up well to isolation cells and similar treatment.

A simple method can be used to select people who will be resistant to pain. To begin with, the men are shown a screen on which a symbol or word appears for a fraction of a second; the duration of the exposure is gradually increased until the man can just see what is shown. If this time is relatively long in comparison to other people, then the chances are that the man's nervous system is 'turned down' and stimulation has a job getting through. This type should be able to stand up to pain well. But to double check the experiment is repeated using another sense – say, hearing. The man's life style is also checked: does he smoke? drink a lot? wear bright clothes? or lead a hectic social life? If the answer to all these is 'yes' then the chances are that stimulation *does* indeed have a job getting through to this man and that he enjoys stimulation and works hard to get it. He should stand up to pain best of all.

If, on the other hand, the soldier has a low threshold for the senses – he picks up what is on the screen very quickly – and on top of this leads a quiet life, this may mean that the quiet life provides him with enough stimulation to satisfy him. He should be able to withstand solitary confinement much better than most, but not pain. A dilemma.

The second stage, once the men have been screened, is to deliberately give them a plausible *alter ego* or alibi. In Korea, POWs had the greatest difficulty in maintaining that they were ignorant of important military details. Yet a plausible but fairly bland alibi did enable those who tried (which was against the rules then) to claim a realistic ignorance of much information. The soldiers themselves have to choose the alibi: for example, one used successfully in Vietnam by a major in a reconnaissance company was to say that he was an engineer building roads. He knew enough engineering to keep his end up and this also enabled him to deny convincingly that he had access to any intelligence.

The troops also undergo self-criticism sessions in some cases, this being the third stage of the counter-interrogation process. These are not dissimilar to the self-criticism sessions of the Chinese. Each man is trained just to sit there and take deliberately brutal and personal criticism without responding in any way. He has to carry out irrelevant tasks, say mental arithmetic, designed to help him 'switch off' and so avoid confrontations which an interrogator will invariably try to provoke.

The last method is basically the design of ways to counter specific interrogation aids: the lie-detector, for instance, or hypnosis (27). Biofeedback machines, which convert changes in skin resistance to audible buzzes of different frequency, enable a subject to control a process which is supposed not to be controllable. In time he might learn to control his skin conductance to such an extent at least to render his record chaotic. He can also change his pulse rate by wriggling his toes and so on. Dr M. T. Orne at the Institute of Pennsylvania Hospital in Philadelphia (who became well-known as one of the experts defending Patty Hearst) has been investigating the use of hypnosis in a number of military situations including those where the soldier has to resist pain, interrogation, isolation, sea-sickness and so on (28). His main interest has been in how soldiers can hypnotize *themselves* in order to do certain military tasks and particularly how subjects can successfully *pretend* that they have been hypnotized so as to fool a potential interrogator using this method.

Very little has been published on this project as yet but it appears that one avenue being followed through is to see whether soldiers can be hypnotized *before* they go on a mission to resist interrogation, or trained to simulate hypnosis without actually being hypnotized. This seems mainly to be in the area of producing nystagmus (rapid eye movement) to give the effect of being hypnotized. Another possibility is to give soldiers who have to go on dangerous missions information of a classified nature under hypnosis and then to induce post-hypnotic amnesia. The subject is given a code by the hypnotist and can only recall the material in response to the code (29).

The Korean brainwashing episode, though it provided the most notorious chapter in the recent history of interrogation, is, therefore, for the reasons outlined here, unlikely to occur in the same form again. Interrogation, however, crops up again and again as a controversial political and scientific issue. If any study is the keynote, it seems to me to be that by Zimbardo, described in chapter 13. For the soldier, it demonstrates that the enemy in uniform, if he has power, is almost powerless to stop himself from using it. And for the rest of us, it should help us to understand the nature of cruelty. The role a man has – even a normal healthy man who is merely a subject in an experiment – can exert an influence so strong that he may seem like a psychopath. This experiment, above all others, teaches us that treason and loyalty are slippery words; and should be used with great care.

PART FOUR *Survival*

16 Living Where Others Would Die

We have already encountered survival as an aspect of our discussions of battlefield casualties in chapter 11 and in prison camps in chapter 13. In this chapter we shall range much wider to look at the human factors which affect man's survival below the sea, well below freezing point at the north or south poles, high above the atmosphere in space, and after nuclear warfare. Given the definite possibility of nuclear warfare, armies have not been slow to explore the remoter regions of the world – areas where post nuclear survival seems most likely. A post-nuclear age is bound to be highly stressful for all concerned: so psychological factors seem likely to loom very large.

ADVERSE CONDITIONS

Isolation – to come back to a topic we explored in chapter 14 – is a feature of many of these situations: the Arctic, space, submarines. Here we can see that the military do have a very real need to study why some men can withstand it and some cannot. Isolation research need not involve sinister explorations of sensory deprivation for interrogation. Consider first, for example, its applications in research with submarines.

Submarines

Two aspects of underwater psychology have been explored: the effects of confinement and isolation on submarine crews; and the effects of water pressure and temperature on dexterity of divers and on other abilities of military significance. The early phases of the crew research we have already come across – they were carried out by Dr Tom Myers (see chapter 14) (1). In those experiments the effects of SD were explored and it was found that certain people could survive seven days' darkness whereas others could not. Some responded with distress and requested to leave the experiment; others went into what Myers called the 'deactivation syndrome' and survived well over the period.

The next phase of this project was more applicable to submarine work in that it explored the ways in which small groups of people responded to isolation. This was largely carried out under the direction of William Haythorn (2). Eighteen pairs of men were selected in such a manner that in each experimental condition a third of the pairs were either both high, both low, or heterogeneous on each of four personality variables: dogmatism, the need for achievement, the need for affiliation, and the need for dominance. Each pair in the experimental group had to spend ten days in a small room, 12 × 12 ft, carrying out selected tasks with no contact with the outside world. They were compared with control pairs who did the tasks in similar rooms but were free to leave them at any other time and were able to lead a normal life on the naval base.

Haythorn found that two of the nine pairs were unable to complete the ten days' isolation and that two others showed a very high level of 'interpersonal hostility', so much so that the experimenters had to intervene to avoid 'serious physical conflict'. None of the non-isolated pairs had any such difficulty. Responses on subjective stress scales (see chapter 9, p. 204 ff.) showed that the isolated groups experienced far more stress than did the non-isolated groups.

On the other hand, Haythorn found that although the isolated crews experienced more subjective stress, they also performed better on several of the cooperative tasks that he gave them to do. The apparently compatible men performed best and reported less stress – 'compatible' meaning where the crew consisted of one man low on a certain personality trait and the other man high on the same trait.

Haythorn also looked at the use which the various subjects made of the space they had available. He found that isolated crews tended to withdraw from each other and to establish clear-cut territorial preferences to a much greater degree than did the controls. This was especially true for the incompatible groups where they were incompatible on social aspects such as the need for affiliation. Both the establishment of territorial claims and withdrawal appeared to be methods of reducing stress. Haythorn neatly demonstrated, therefore, just how crew incompatibility might influence cooperation: it followed that by modifying crew composition he should be able to improve performance. His next series of experiments – with a further eight crews – was to see whether he could indeed manipulate various aspects of the physical environment so as to make crew life easier and thereby more productive (3). For some crews, the men had their own rooms in which to retire if they wanted, once the crucial tasks had been performed, while in others they were able to talk with the outside world, play records and so forth; and they were also divided according to whether their mission was definite in length or indefinite. Haythorn found that the crews did worst where there was no privacy, no communication and where the length of their mission was unknown.

The next stage was to try similar experiments in an actual submarine: on board a laboratory vessel, Sealab II. Here the men's movements and interactions were monitored by television as well as by ordinary questionnaire methods (4).

The Sealab research used ten-man crews, who had to work in the laboratory vessel on the sea bottom 200 feet down, maintaining the submarine and carrying out excavations on underwater wrecks for between two and four weeks. The experimenters looked at the aquanauts' physiological responses as well, but only their psychological results will be considered here. Virtually all aspects of their behaviour were recorded; these ranged from regular checks on their mood, their ability to cope well with the jobs they had to do, their choice of friends, their dependence or independence with regard to their leaders and even their requests for telephone calls to wives, friends or other relatives back home. In addition, many background details on each man were available to the experimenters such as the size of their family, the amount of education each had received and even the type of town (rural or urban) they had been brought up in.

The two main investigators were Roland Radloff and Robert Helmreich and their main findings were fairly consistent. They found that the men did much better than anyone had expected them to do. As one naval officer put it, although the conditions were crowded and noisy and the diet was restricted, 'Everyone seemed to go out of their way to be nice.' Even so there were measurable differences between the men in how well they performed. Radloff and Helmreich found it useful to look at three psychological aspects – these three being based on the research by Eric Gunderson who had for years been looking at the psychological factors which predispose people to stand up well to the rigours of cold and isolation in the Antarctic (see later in this chapter for an account of this research).

Gunderson had found (5) that a man's attitude to his job, his emotional stability and his social compatibility all affected the success of a mission. Radloff and Helmreich found that this way of looking at aquanauts helped too. The best-adjusted divers were highly ranked by their leaders; they were 'sociometric stars' – that is, they were popular with everyone else, good at their job and liked being 'one of the boys'. The Sealab experiments also showed that standard personality classifications are not much good at predicting who will do well. Mood scales helped, though; the divers who reported that they were frightened and were low on a sense of well-being were poorer performers and less well-adjusted in general. Demographic variables were also important – later borns and aquanauts from smaller towns for some reason reported less fear and were better adjusted. The preference an aquanaut had for his leader was found to be *negatively* related to performance – as if the more stressed a man found the experience, the more dependent he became on his leader. Analysis of the characteristics of the preferred leaders showed only that older and more experienced men were preferred: there appeared to be no one style of leadership that worked better than others.

To all intents and purposes, then, the field studies are in general agreement with the laboratory experiments. It would seem that, up to a point, the successful aquanaut can be predicted. In later research, and in a later project, called Tektite 2, Helmreich expanded some of these ideas and found that a life-history questionnaire, concerned with *events* in a diver's earlier life, was a better predictor of performance and adjustment than attitude questionnaires (6). A feature of the questionnaire is that the *same* twelve questions about school, quality of clothes, fights with brothers and sisters, clashes with authority and so on, are each asked *nineteen* times, once for every year in the diver's life prior to military service. Patterns of behaviour during those years are then analysed. Receiving a lot of physical affection, trying hard in school (without necessarily being highly intelligent), having long periods of illness forcing someone to lead a restricted life for a while, were some of the factors associated with being a good diver. Since Sealab and Tektite, Haythorn has also been investigating more sophisticated psychological concepts. For instance, he has been looking at how fear in one man may 'infect' the thinking of another, the importance of generosity in enclosed spaces and so on (7). These are intended to help captains understand the dynamics of groups under danger. Other investigations have studied the relationships between territorial preference and interpersonal compatibility so that the navy might better understand who to billet with whom on long missions (8); and finally the way that men on a long, isolated mission evaluate their own professional actions so as to see whether they can continue to tell when they are doing a good job or not (9).[1]

Divers

In some respects, the underwater environment is more hostile than any other, even space, because of the high pressure, the lack of light, the cold, and dangerous fish. 'Nitrogen narcosis' – breathing nitrogen under pressure – can cause 'rapture of the deep': people vary in their susceptibility to this but most are affected below 30 metres (10). Divers become euphoric, stupid, forgetful and careless, as if under the influence of drink. Below 130 metres a man may become unconscious. Narcosis does appear to have an effect on dexterity too, though this may be a secondary effect of the way it alters reasoning (11).

The effects of anxiety have been studied by Dr A. D. Baddeley, from Cambridge in England. He has found that divers' performance near the shore is quite similar to that in the laboratory under the same water pressure. But in the open ocean performance deteriorates – narcosis and anxiety appear to interact to produce an exaggerated decrement in performance (12). Perceptual narrowing also occurs – the diver who is anxious tends not to notice things happening at the edge of his field of vision. This can of course be highly dangerous, the more so as target detection also drops off when divers are anxious. There is some evidence that in cold water divers' memories are impaired too, but this is inconclusive (13). The effects of cold can be in part diminished by breathing more slowly – a third of heat loss is due to breathing in cold air or oxygen.

Time perception is affected in deep water diving – time seems to pass more quickly. Visual and hearing problems are more difficult to offset – hearing possibly being the most important because, with

1. It should be noted that British naval psychologists think that the psychological problems associated with life in a submarine – even a nuclear submarine at sea for months on end – have been overstressed. Edward Elliott, one of the Royal Navy's top psychologists, has spent much time aboard subs to see for himself. His conclusion is that the men are too highly motivated for the isolation to have much effect.

exploding shells, etc., it is more difficult to adjust to than visual distortion. (Sound travels four and a half times faster underwater than in the air and so its source may appear closer than it actually is (14).)

Military research on diving currently appears to centre on two areas: the perception of danger by divers and communication. Most navies have produced statistical curves which relate the amount of time a diver may stay down at certain depths (the French Navy usually being rather more generous than either the Royal Navy or the US Navy). But the search for divers who respond well to danger goes on. On communication, military efforts seem to be directed towards training divers to decipher the 'Donald Duck' speech which occurs when they breathe helium at high pressure, and a system of skin electrodes (developed for use in infantry combat; see chapter 5) which can transmit a series of simple messages and commands in pulsed code (15).

Space

Astronauts are always in touch with mission control and in that sense are not isolated. However, space could easily become the most isolated military situation; moreover spacecraft are at present more confined than the machines on the seabed and space flight lasts for much longer periods than the average conventional flight. It is therefore appropriate to consider the psychology of space flight and its military implications here.

The most complete survey of the research in the area to date is that carried out in 1972 by the the US National Academy of Sciences and entitled *Human Factors in Long-Duration Spaceflight* (16). (It lists 324 references for anyone who is interested.) The report makes three points of substance. First, space flights to date have been too short for psychological problems to really show themselves. The problems so far have been those concerned with reacting quickly to a stressful stiuation (re-entry and splash-down) and here space flight is not substantially different to other military activities that are highly technical and dangerous. Second, the confinement inherent in space flight so far has had consequences that are physiological rather than directly psychological, in producing a 'deconditioning' of the physical state of some astronauts. This has affected several functions including those of the central nervous system. Longer flights can be expected to be more spacious than those to date, but the report recommends more research on the phenomenon of deconditioning since this could affect the way men cope with the stress of re-entry and landing after long periods of relatively straightforward flight.

The report's third point is that the short duration of space flights so far has also meant that many of the anticipated problems of extended flight have yet to be encountered. It is only when space flight involves a long middle period of 'empty flight', when there is little to do except wait for arrival, that the major psychological problems of monotony and boredom are expected to arise. The research carried out by Haythorn and Radloff and Helmreich on submarine crews has relevance here, and by careful crew selection most of the psychological difficulties that might be expected in space flight should be eliminated.

The Arctic

The Americans rather more than anyone else seem convinced that the north and south poles could one day be a battlefield of the future (possibly they will be one of the few areas of the world relatively uncontaminated after a nuclear war). Accordingly, a surprisingly large number of studies have been carried out into what type of man or group of men survives on the polar ice cap. We shall review these first, since they are essentially studies of groups in isolation. Then we can discuss the effects of cold upon military performance, as the first part of the next section on climatic influences.

Interest in the Arctic seems to have blossomed in the early 1950s. The US Army undertook one study of nearly 1000 men who had *not* been nominated for Arctic duty, finding that, while the single largest cause was being physically unfit, a conglomeration of psychological disabilities (emotionally unstable, lazy, careless, low aptitude, 'doesn't get along well with others') accounted for about a third of those refused nomination (17). The quartermaster branch of the army also had problems selecting personnel to carry out tests of food products in Northern Canada (18). Neither physical fitness nor how good a man was at his home base appeared to be much of a help in predicting performance in isolation in the cold. Throughout the fifties and early sixties much effort went into the study of the psychology of survival in the Arctic. Much of it was carried out from the Navy's Neuropsychiatric base at San Diego in California, notably by Eric Gunderson who was interested both in individual characteristics that help a man adjust to the remote cold areas and in the groups that get on best and perform well.

In the Arctic winter temperatures drop to between 20° and 30° below freezing, winds reach 140 miles per hour, and it is dark continuously for up to six months. There is, on the stations Gunderson uses, no way to evacuate in case of emergency or to receive additional supplies until summer comes round; the only contact with the outside world is the radio.

Over the years, a few hundred men have completed a whole range of attitude questionnaires given to them before the winter experience, then again half way through and finally at the end (19). The questions cover everything from living conditions and mood to what the men think of one another. Gunderson has found that there are ten attitude clusters which are important to survival. These are: physical adjustment; expedition motivation; trust in the organization; personal usefulness; boredom; compatibility of group members; group teamwork; group efficiency; group achievement; and the egalitarian atmosphere within the group. In his early experiments Gunderson varied the composition of the groups: some had more officers, some less, others had a few civilians, still others none at all. What he found was that the groups which adjusted well early on remained well adjusted throughout – they were stable in this respect. But he also found that the groups with the *smaller* percentage of officers and the *larger* proportion of civilians were happier with their experiences and accomplished more than did other groups (20). The great variability which arose from having civilians present kept up motivation (but did not affect boredom) and the lack of many officers appeared to give the groups a more egalitarian feeling which Gunderson found was important. When men did show psychological symptoms as a result of their isolation these usually took the form of either sleep disturbance, depression or irritability.

Using official records, supervisors' reports and so on, in later research Gunderson has elaborated his work, and has shown, for example, that cooperation is linked to an ability to bring conflicts into the open; this is even related to the successful maintenance of basic equipment. So with a little selection, to maintain variability in the groups and a limit on the proportion of officers and, with a little preparation (warning the teams about the avoidance of festering grievances), performance of the Arctic teams can now be significantly improved.

Navy psychiatrists also screen men who are candidates for winter duty in the Antarctic on four other factors now: motivation; the man's history of past effectiveness; his ego strength – that is the adequacy of his defence mechanisms; and his ability to work with others. Anyone who shows evidence of not being able to cope in the past – showing reliance on drink, or obsessive worries, for example – is ruled out. At one point, five out of six men who had to be put on the sick list in the Antarctic for psychiatric reasons were last-minute replacements for others who had dropped out, the replacements *not* being screened (21). The psychiatrists confirmed Gunderson's findings that depression was common, often sparked by events over which the men had no control, and that problems were more common in the smaller groups (with between fifteen and forty members). So Gunderson's administrative solutions outlined in the last paragraph also receive support from the psychiatric research.

Climate

Away now from the remoteness of the Antarctic and on to its other main feature – the cold. Studies into the effects of cold have mainly been concerned with specific skills: rifle firing, target detection, manual dexterity, rather than the way temperature affects, for example, the workings of a small group. In fact, surprisingly little has been done in this area. The most wide-ranging survey was carried out by HumRRO as part of a background study for the development, in the late sixties, for a new Tactical Individual Transporter (TINT) for cold areas (22).

Before the Second World War it had been found that, compared with normal tactical efficiency at warm temperatures, aircraft pilots got worse at progressively lower temperatures: 80 per cent efficiency at –2° C; 68 per cent at –18° C, for example. L. J. Peacock studied rifle aiming in the cold after vigorous exercise and found it to be severely affected, even when the effects of shivering were accounted for (23). What matters is not how cold a person is in general but the skin temperature of the relevant part of the anatomy (hand, foot, etc.). For both skin sensitivity (which might be used in new communication techniques) and for manual dexterity (used, of course, in everything from typewriting to operating switches and tying knots) there is a critical hand skin temperature above which performance is relatively unaffected but below which there is a 'precipitous decline' (24). For tactile sensitivity, this appears to be near 8° C and for dexterity somewhat higher – at 12° to 16° C. Clear guidelines for the quartermaster.

So far as overcoming the effects of cold is concerned the HumRRO survey found that many people *can* adapt to it, that breathing pure oxygen helps, and that on some tasks hypnosis is beneficial. A. T. Kissen and his colleagues found that hypnotized soldiers shivered less, had a slower heart rate, but a higher rate of body heat loss. Hypnotism did not affect skin temperature but it did affect vigilance – performance under hypnotism being much better (25). On more personal matters larger men report feeling warmer than small men, and though there are no specific effects of personality on resistance to cold, men who differ from the norm on such questionnaires as the Minnesota Multiphasic Personality Inventory have been found to take longer to recover from the effects of very low temperatures.

The HumRRO survey and several other strands of research gave rise in the early seventies to a new line of thought at ARPA – the Advanced Research Projects Agency in Washington. One of the more interesting areas of development in psychology during this time has been biofeedback. By devising special instruments which measure variations in a person's biology – such as the heart rate, skin temperature, muscle activity – and then producing these measures in a form which the person himself can assess (feedback), it has been found possible for someone to control aspects of his bodily activity that hitherto had been assumed beyond such control. ARPA studied skin temperature and by 1976 had managed to train men to change their skin temperature by up to 16° F (26). At present it is not an operational technique but ultimately it may be possible for men working in the cold to 'direct' their blood to – say – their hands so as to keep their fingers nimble for delicate operations.[2]

The studies carried out into the effects of heat are fairly prosaic. In the 1960s the US Quartermaster General carried out a survey into the attitudes of troops in the tropics (27). In general it emerged that what the troops wanted above all in such circumstances was *comfortable* equipment, but they also tended to be somewhat more hostile to their officers compared with troops in a temperate zone and more dissatisfied with the entertainment they were offered. They also appeared to take the view that the tropics are fit only for serving soldiers and for this reason seemed not to mind the absence of families in this particular habitat.[3]

2. The US Navy is also sponsoring work by Dr Robert Meyers at Purdue University into the hypothalamus and the effect that calcium and sodium ions have on body temperature. By changing this composition Dr Meyers has so far been able to vary body temperature of some animals by up to 10° F. 'It is anticipated . . . that this [research] . . . might permit a physician to control a patient's temperature by pills or injections' (28).

3. The US Army Medical Service Research and Development Command com-

Continuous operations: going without sleep

Continuous, or extended, operations in military parlance usually applies to those situations in which men are required to be on duty and therefore have to go without proper sleep for forty-eight hours or more. Laboratory studies on sleep loss have usually been followed up by field and combat simulation exercises. It thus becomes a useful area in which to compare these two types of result.

Early laboratory studies suggested that sleep loss would occasion a significant drop in military performance (29). It has since been found that not only total loss of sleep but loss of any *one* part of the sleep cycle can be upsetting (30). Sleep has been found to consist of five phases, varying from low-voltage brain wave patterns seen with the onset of sleep to high-voltage one-cycle-per-second waves during 'slow wave' sleep. If a subject is wakened during a particular phase on one night, then, as he enters the same phase the next time he sleeps, it is more difficult to wake him. This resistance progresses until it becomes necessary to give the subject continuous electrical shocks in order to wake him, demonstrating, it would seem, the need not merely for sleep in general but for the various stages of it. Later research has shown that deprivation of certain parts of the cycle have different effects: for example, going without rapid-eye-movement sleep (associated with dreaming) makes a person excited and irritable, whereas deprivation of slow wave sleep makes him depressed.

Better technology and better methods, incidentally, have modified the view, gained during research on interrogation methods, that sleep loss invariably produces hallucinations and delusions. Sleep loss does produce changes in mood, but so far as performance is concerned it appears to exert its influence mainly in the form of lapses of attention. Tasks with a strong incentive for success and which are of brief duration suffer less than boring long-term tasks (31). Performance in any case follows the diurnal cycle: the best performance is usually in the evening and the poorest around dawn. On viligance, Robert Wilkinson, working with British sailors, found that up to three hours' sleep loss had to occur before there was a change in vigilance performance. Over

---

missioned Dr R. W. Russell to explore the use of chemicals to control appetite. The research was to find whether, on short missions, men could use pills to inhibit appetite, rather than have to carry more bulky food which might interfere with their fighting capacity (32).

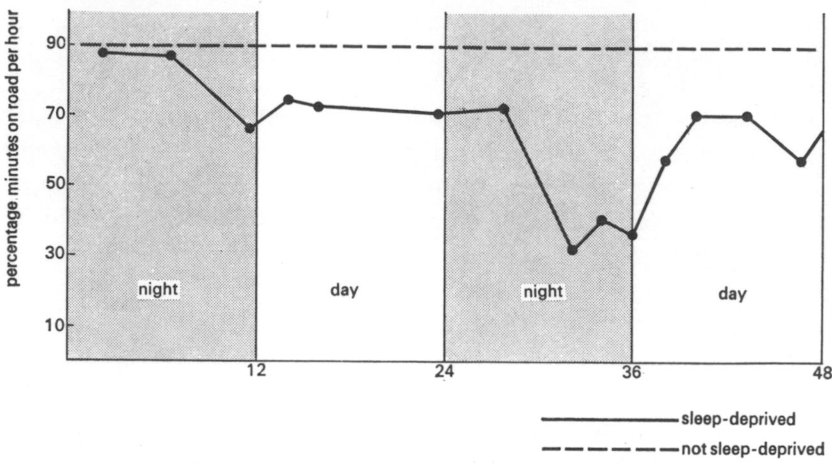

Figure 49 Simulated tank driving: the effects of continuous operations.

five hours of lost sleep, however, made no further difference (33).[4]

The effects of sleep loss on various military tasks have been studied in the laboratory. In the late sixties, a series of men and tank crews were tested on three types of combat behaviour – driving a tank (using a simulator), looking for targets, and assembling and dis-assembling a machine gun (34). The crews had to perform these jobs at intervals over forty-eight hours. They started at different times in the day. It was found that performance dropped off significantly during the first night compared with soldiers doing the same thing but getting normal sleep, was impaired somewhat during the second day and dropped drastically during the second night (see Figure 49). It was also found that the time of day in which the experiment started had no effect. Whether the experiment started in the morning or in the evening, the low point of performance for virtually everyone was the

4. A military conference in San Diego, though it was clearly concerned with the effects of sleep loss on continuous operations, showed how easily a military mind slips from one type of military problem to another. For in considering the future of sleep deprivation research, the uses of sleep in interrogation procedures was clearly seen as a factor of major importance (35). Dr Harold Williamson, for example, from the University of Oklahoma, was interested in the following: to carry out repeated periods of forty-eight hour sleep loss separated by two or three weeks of normal sleep to see whether losses have a cumulative effect – this is likely to simulate the normal battle pattern of short, sharp bursts of activity followed by much longer periods of inactivity; to find out whether the recovery period was a linear function of the length of deprivation; to investigate whether selective sleep-stage deprivation could be used to motivate behaviour – in other words, could offering sleep, or various stages of it, be used to induce men to confess?

early morning – between five and seven a.m. It thus seems that diurnal rhythms exert a strong influence and it should follow that if an engagement starts in the morning the men will be able to go on longer before their performance starts to wane.

However, this was in the laboratory. When the *same* group of people tried similar experiments in the field rather different results were obtained (36). In the field five basic types of military behaviour were tested – gunnery, surveillance, communications, driving, and maintenance. For example, the crew had to drive for forty-eight hours across a test landscape that included offensive, defensive and retrograde operations, making a total of 143 miles travelled in the tank during the allotted time. The results showed that, in fact, there were *few, if any*, performance decrements that could be attributed to the effects of continuous operations. The researchers concluded that the motivational aspects of the situation in the field (and presumably even more so in real combat) outweigh any effects of sleep loss. Morale, for example, remained very high.

Much the same results were obtained by a further US Army study released in 1970 (37). This looked at other military skills, notably rifle shooting and grenade throwing accuracy. This study found that, despite clear evidence of fatigue on the second day of a forty-four-hour exercise, there was no degradation of performance, morale again remaining high.

In the previous study, the combat realism was fairly good except that no sleep was allowed at all. In this later study, however, the operations were continuous in the sense that they are in the real battlefield – the men were allowed to take catnaps between bouts of activity as in actual battles, in a realistic series of day and night manoeuvres (they were even ambushed at one point). Most 'players' reported that, on the second day, they were very tired but could keep going if they had to. On all objective measures – rifle firing, grenade-throwing, night target spotting – performance remained unaffected by the amount of sleep deprivation the men had suffered. No attempt was made to link performance to whether or not a soldier took catnaps, and in this sense the experiment cannot be called completely methodologically sound. However, it confirms the earlier studies that sleep loss does not really affect military performance in the field over twenty-four to forty-eight hours (it generally being reasoned that in most situations the average soldier could get some sleep after such a period) (38). There is some evidence that the sleep deprived can 'plug' straight into the phase of sleep that they need, and that sleep carries more redundancy from this point of view than we think. Therefore, it may have been that the men were actually getting a significant kind of sleep during the catnaps. But that of course does not explain the similar performance in the earlier studies in the field where catnaps were not allowed. It certainly seems that, in so far as forty-eight hours can be called continuous operations, the average soldier can lift himself psychologically to overcome the effects of sleep loss.

Research on sleep loss which lasts longer than forty-eight hours is thin on the ground, presumably because most military authorities think that the average soldier would be able to get some sleep after that. But, of course, in interrogation that may not be so. The psychological effects of several days' sleep loss on a soldier who thinks that, in battle and on manoeuvres of forty-eight hours or so, he has proved to himself that he can carry on unaffected may be that he will crack all the quicker when he finds that his well-proved abilities are starting to wilt.

The Berlin airlift

There are no properly controlled experiments about sieges or blockades, perhaps the most critical type of continuous operation there is. However, the Rand Corporation did carry out a survey of the Berlin airlift which took place in 1948–9 (39). The Russians sealed off the land links between West Germany and Berlin and almost overnight food and other essential supplies had to be flown in, all available aircraft being commandeered by both the British and Americans.

To begin with, there was a great deal of enthusiasm: the importance of the mission was clear-cut, the moral issue sharp, and the mission, though arduous, was not exceptionally dangerous. However, most of the personnel had expected the blockade to last at the most a few weeks and as it passed the 120-day mark some problems of fatigue and morale did start to show themselves.

According to the Rand study, some of the morale problems were brought on by the sheer physical and mental fatigue, others by inadequate base facilities, uncertainty about how long the operation would continue, and domestic worries. A British doctor, R. H. Stanbridge, studied the major causes of fatigue in British aircrews (see Table 26). He also found that although there was some mental fatigue resulting from the unaccustomed pressures, the main symptoms which the men began to suffer were physical, caused mainly, it was felt, by the lack of

Table 26 Major causes of fatigue reported by British aircrews

Problem	*Percentage mentioning*
Lack of sleep or undisturbed sleep	57
Waiting about between trips	46
Unsatisfactory living conditions	40
Unsatisfactory ground crew organization	28
Long working hours	28
Aircraft design	26
Irregular meals; poor food	23
Extra flying	20
Domestic worries	20
Lack of recreation	10

sleep (40). The Americans, on the other hand, reported more difficulty over domestic problems till at one stage the crews could think of nothing else and performance was greatly impaired. In part this was due to the fact that the Americans were further from home, but also because the British operated a rotation scheme which gave crews fewer – but longer – breaks, rather than frequent short breaks. This appears to have enabled more relaxation and also the opportunity for men to return home and sort out any domestic problems. The Rand study also reports that there were many compensating factors – not least the judicious use of military cartoonists and a vigorous daily newspaper where grievances, experiences and jokes could be shared. However, the overall seriousness of the problem in the siege can be gauged from the fact that, after a while, every month 10 per cent or more of the air-crews on the airlift bases had to be removed from flying, compared with 2·5 per cent at non-airlift bases. Respiratory diseases were given as the most frequent causes by Colonel Moseley, the American doctor looking into these questions. But he took the view that very often this was just an acceptable excuse for 'some type of sub-clinical fatigue' (41). He concluded that since there were plenty of people to help out at the time of the Berlin crisis none of this mattered, but had there not been, as could easily happen in a similar situation in the future, the blockade could have been a much more hazardous experience for all involved.

NUCLEAR WAR

Though nuclear conflict is arguably a less likely variant of war than the other special examples considered so far in this chapter, studies into its psychological aspects – perhaps because our experience is so limited –

has been more diverse. In general two aspects have been distinguished: these are the psychological screening of military (and other personnel) with access to nuclear missiles either at their launch sites or else in the aircraft or submarines that carry them as part of the defence system; and behaviour after the attack – both immediately and long-term. Necessarily, much of the study is speculative in nature.

Security risks

The problem of selection is a highly specialized one. Not all soldiers have access to nuclear weapons and many nuclear weapons cannot be fired by one man. Screening thus amounts to ensuring that unstable persons do not get their hands on weapons that can be fired by single individuals and that nuclear submarine crews, who spend a long time on their own together, do not develop a collective personality that may result in 'accidents' of a catastrophic nature.

About three per 1000 soldiers develop psychosis each year in war and peace. Though this sounds a small figure, it is worth recording that according to one document in the Pentagon, there are 17 500 men in the US armed forces alone with access to nuclear weapons (42). So without screening we may expect that roughly fifty to fifty-five of those develop psychosis each year. (For nuclear submarines, a 4 per cent referral rate was reported in the *American Journal of Psychiatry* in 1967.) The US Army treats this under a policy set out in document AR 611/15. This document has four annexes on the possible extension of psychiatric techniques for better security screening following April 1964 but it started with a study of sixty US individuals who had defected to Communist countries plus another of fifty people all of whom had had their top security clearance 'voided for cause' – that is, they had been exposed or sacked. It was felt at that time that selection for nuclear access should be tougher even than that for spies because of the risks involved. This meant that all personnel with access to nuclear positions must have: a positive attitude to assignments involving nuclear weapons; a good military record; a good education; good social adjustment; and be physically fit.

Conduct which disqualified them included court martial, unit punishment, heavy drinking, delinquency, financial irresponsibility and a poor attitude towards a nuclear job. Recognizing the dangers, both the US Navy and the Air Force now re-examine their men in sensitive positions every six months as part of a 'Human Reliability Program' (43). The first screening under this programme is believed to have resulted in the replacement of two men per 1000 tested. According to Dr Jerome Frank, however, there has been at least one documented case where a demented individual with access to nuclear weapons has threatened to exploit that access (44)[5].

At the same time, the documents point out, psychiatrists and psychologists are much better at predicting successes than failures (which is what security risks essentially are) and that the true traitor is not so much emotionally disturbed as a stable ideological convert; so psychiatric screening as such is pretty useless in this context. The document concludes that the *behaviour* of people should be watched and that this is best done by someone close to the job, *not* a psychiatrist. So now the policy is for each person with access to nuclear bombs to have a 'target' to watch.

Behaviour after an attack

In December 1963, the US Army Combat Developments Command requested HumRRO to undertake a study designed to provide information on the 'probable effectiveness' of soldiers in tactical nuclear battle and on the 'probable behavior' of civilians exposed to nuclear tactical warfare (45). In addition, HumRRO was asked, if possible, to devise a method by which 'human factor considerations could be introduced into the conduct of war games'. HumRRO surveyed the literature on battle stress and applied it to the nuclear field; a model was drawn up which was intended to allow for the adjustment of casualty rates, 'based on psychological factors' for use in nuclear war gaming.

The actual experience we have, of course, comes from the blasts, in 1945, at Hiroshima and Nagasaki. The most comprehensive survey of the psychological effects of those events has been written by Dr

5. Another method advocated (but never used) in the prevention of nuclear war is to use the weaknesses of major public figures to assess whether the intention exists to use nuclear weapons or at least to circumvent existing treaties. In the early 1960s, for example, when there was much discussion about how to enforce any nuclear test-ban treaties that might be agreed, the Rand Corporation came up with the concept of 'psychological inspection' or P I. The idea was to give such things as lie-detector tests, truth drugs, hypnosis and cross-examinations of top people – politicians, scientists, people on exchange visits (Khrushchev was mentioned in the document). The basis of the idea was that since missiles bases or weapons stockpiles could be hidden almost anywhere it was more efficient to question people who must know about such developments, to see whether treaties and the like were being violated. Lewis Bohn, of the Social Science Division of the Rand Corporation, was the author of the document in question. It is not known what the fate of his ideas was (46).

Robert Jay Lifton, a Yale psychiatrist (47). More statistical accounts of disasters such as Hiroshima are Irving Janis's study based on the US Strategic Bombing Survey, and John Hersey's *Hiroshima* (48). Though not as moving or as well written as Dr Lifton's study, these are more useful from a policy-maker's viewpoint – conserving life in any future nuclear war.

They show that the great difference between the survivors of a nuclear attack and a conventional attack is that, besides the lack of warning, almost all the survivors 'were personally exposed to physical dangers and were aware of the direct threat to their own survival, although there was no realization of the extent of the disaster until the evacuations began' (49). The other main point was that the incidence of narrow escapes – or near-misses – was very high. The near-miss phenomenon is important. Studies of conventional warfare and natural disasters show that being involved in a narrow escape from danger gives rise to acute and persistent anxiety reactions. It is thought that these experiences tend to disrupt the psychological defences which control emotional response in war (50).

The three main events reported by the majority of survivors of Hiroshima and Nagasaki were the flash and blast of the explosion and the experience of a large number of casualties. The impact brought on acute fear; blast effects produced high arousal and the large number of burned, cut and otherwise maimed individuals was what brought on emotional trauma (the trauma itself appeared no different from that occurring in 'normal' bombing raids in Britain, Germany or Japan but of course the scale was much bigger). A few terrified people behaved in an impulsive or disorganized manner for a short time but there was no evidence of overt panic or anti-social behaviour on a mass scale. Fear reactions persisted among a sizeable proportion of the population for many days, but the incidence of neurotic responses and severe types of psychiatric disorder, such as depression and apathy, was small; transient emotional effects predominated just as in conventional war, but again on a larger scale. Perhaps the most surprising result was that the morale of people in and around the bomb sites fell no more after the blasts than it would have done after a conventional raid (though Janis concedes that morale had been higher in these cities before the bombing). On the other hand, Leighton, in his studies for the Foreign Morale Analysis Division of the Office of War Information (51), found that some aspects of morale were less resistant to change than others. For example, the Japanese retained their faith in the Emperor, his way of life, the rightness of the war, but they did lose faith with Japanese medicine and health-care systems, Japanese weapons, their communications system and the Japanese productive capacity.

Military sociologists and others have a long history of over-reacting to potential effects of new weapons. The point to realize about nuclear warfare, therefore, is that it is not necessarily different, qualitatively, to conventional warfare, in its impact, except in three aspects:

(a) With the much greater killing power of nuclear weapons, the average soldier (or civilian) is more likely to perceive death from an attack as inevitable and thus his way of escape diminishing. Therefore it seems likely that the *threat* of an attack is likely to provoke panic more than the blast itself. This may be precipitated by the wish to search out loved ones and so forth.

(b) A nuclear blast is more likely than conventional blast to disrupt the arrangement of the primary group for a soldier (in terms of greater loss of personnel and leaders, breaks in communication, supplies and medical care) and thus isolation is more likely. Studies of the Second World War showed that isolated soldiers, even when comparatively safe physically, were much more likely to surrender than those who, though in greater danger, could still maintain contact with their unit. In nuclear war, therefore, surrenders and defections may become more common.

(c) It seems more likely that soldiers will fight on their home ground and that their families will be more involved. One study found that soldiers find it more rewarding to fight on their home territory; on the other hand, Edward Shils reported that when these conditions existed in Germany in 1945, soldiers exhibited a strong tendency to desert homewards – the primary group of the military had a less strong influence. The same author also pointed out that interrogation of Germany POWs indicated that they discussed the possibilities of many alternative courses of action when talking about their families rather than when discussing the war. It is thus possible that desertion, as well as surrender and defection, may rise in a nuclear war (52).

Results like these have provoked several armies into studies designed to prepare their soldiers for nuclear emergencies. It is one of the more intriguing areas of investigation precisely because it is so speculative.

Worry and warnings: nuclear warfare training

Most of the studies draw their ideas from schemes to cope with natural disasters which cause havoc and loss of lives on a scale more similar to nuclear warfare than do conventional weapons. In the USA, for example, HumRRO started with an indoctrination process about nuclear warfare, dealing with specific information relating to the whole range of medical and social effects. This included explaining to the soldiers what reactions they were likely to have themselves *after* the blast. It was done so that they would know that they were not alone in feeling what they were feeling, would regard themselves as normal and so be better able to cope (53).

In Irving Janis's view the main problem to be faced after the Hiroshima experience was that in nuclear warfare it may be very difficult for soldiers' feelings of invulnerability to be maintained in the face of so many near-misses. It has repeatedly been found in wars that those soldiers who regard themselves as invulnerable and extremely unlikely to get killed, even though this is untrue, are thc ones most likely to cope with the stresses of battle. Another psychological fact to get across to the soldiers is that worrying – the work of worry – is not only a normal reaction but that it is in fact essential to their emotional health when there are so many near-misses. It helps the soldier maintain his grip on reality.

Worry is important. But not unnecessary worry. One of the problems of nuclear warfare – a problem it shares with natural disasters – is that for many people it is an inconceivable situation. The problem, as the threat approaches, is to warn people that the situation is changing but so as not to create panic; to get people to behave according to instruction; and to minimize any false-alarm effects.

In general it was felt by HumRRO that no real purpose is gained by telling soldiers about remote dangers. As specific crises approach, however, the HumRRO people advised that more and more soldiers be prepared for the nuclear possibility in the way outlined above (54).

Thus, nuclear warfare training, says Janis, should arouse in the soldier a moderate degree of fear – by the evocation of facts of nuclear life, followed by statements showing how these fear-provoking events can be controlled or minimized, which rub in for the soldiers that they alone among the population know how to survive and are therefore in some way special (55). Several studies have also shown that soldiers, no less than the general population, are extremely poorly informed about, for example, radiation and its effects – making their own reactions worse than they might be if they understood the situation properly. It is important then not to overlook the relevance of accurate information. In Korea, for example, after an indoctrination course was introduced to explain certain aspects of the war to the soldiers, the nature of the risks and so on, the incidence of self-inflicted wounds dropped dramatically (56).

As Janis has said, anticipatory fear needs to be aroused, but kept to a moderate level. The point of this, the HumRRO studies show, is that the 'work of worrying' about nuclear warfare may be stimulated in this way. The soldier may apply himself to learning the facts about a nuclear future more assiduously; he can then be encouraged to think of various scenarios and how he would act in them. Here at least the army has the man facing up to the situation, even if only in his own mind. Once he has done that he should be better able to cope when the real thing comes along.

This preparation and the work of worrying is also regarded as being important in ensuring that the soldier remains loyal during the worst periods of the blast and immediately after. In moments of extreme danger, for example, it is known that people may tend to blame 'the authorities' – in the army this may mean higher command (and has been observed in other, more conventional, battles). Information presented to the soldier should be personal to him and no information should be given for its own sake. Thus he is told about epilation and the significance of losing one's hair after exposure to radiation, but he is *not* told abstract nuclear physics about how the bomb works, or physiological details that he cannot readily apply to himself (these have been found to be disturbing and can give rise to quite unnecessary fears). The training, however, should cover things the soldier *might* encounter because, in a post-blast world, ambiguous situations can be among the most fearful. It follows that highly fearful information that cannot be resolved should not be told to the soldier either. Military authorities are clearly very worried that the number of people killed and the number of people experiencing a near-miss will be so large in a nuclear war as to make the physically wounded and the psychiatrically crippled the main post-war problem (quite apart from the effects of mass bereavement). This accounts for the second of the HumRRO studies (57), which we can dwell on briefly here: the amount of neuro-psychiatric casualties there would be in a nuclear war.

The HumRRO study adopts a quite straightforward model. It

makes three points: first, that different types of soldier are at different risks (and it quantifies this); second, that different types of unit have different risks; and third, and by far the most important, it shows that psychiatric risks are related directly to the number of people killed and that therefore you can calculate the psychiatric casualty rate from the known killing power of a bomb, and the area in which it is dropped (58).

On the first question, for example, the psychiatric attrition of various types of personnel has been calculated (see Table 27).

Table 27 Psychiatric attrition rate for various types of personnel (average division risk: 100)

Ammunition handler	132
Gunner	134
Platoon sergeant	155
Squad leader	212
Rifleman	432

The studies also show that armoured personnel experience a higher rate of neuropsychiatric attrition than the infantry but that the rate for the airborne forces is much lower.

The psychiatric attrition rate has also been calculated for length of time in combat (see Table 28).

Table 28 Psychiatric attrition rate for length of time in combat

Cumulative days in combat	*Psychiatric casualty rate per 1000 men per 10 day period*
1–5	5·8
6–10	20·3
11–20	42·2
21–40	66·7
41–80	88·9

On average for every 1000 men killed, something like 130 are lost for psychiatric reasons. Nuclear psychiatric casualties are a function of the total number killed, of the time in battle and the type of soldier. HumRRO calculations have been made for the various types of troops; in Figure 50 we show just one, for the infantry, to give some idea of orders of magnitude. If, therefore, the bomb was dropped at the outbreak of hostilities we should expect to see the bottom line apply

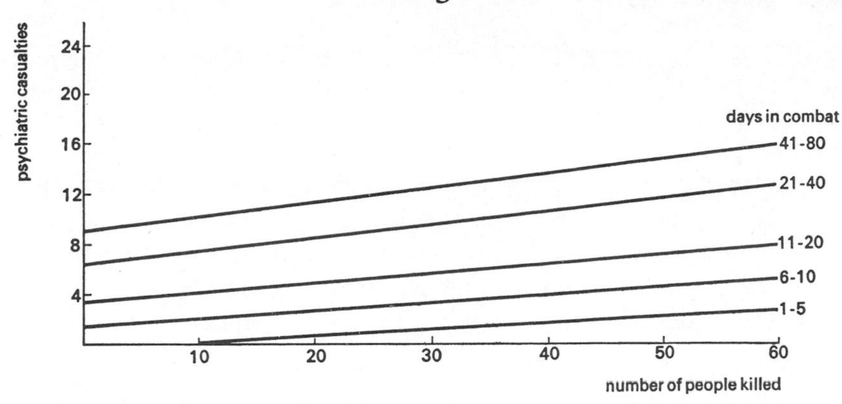

Figure 50 Psychiatric casualties among infantry troops as a function of combat intensity, by cumulative time in combat.

(though much further along it). If, on the other hand, it was dropped part-way through hostilities which had already been going on for some time then one of the other lines would, accordingly, be nearer the mark.[6]

THE FIVE ENEMIES OF INDIVIDUAL SURVIVAL

My basic source for this section is a paper prepared by Paul Torrance for the US Air Force and entitled 'Psychological aspects of survival'. Although prepared some time ago, in 1953, the principles it raises are still valid. I have up-dated the examples given in that paper with more recent research (59).

Torrance simply but effectively defined survival as 'living where others would die'. He reviewed psychological literature and rescue reports and interviewed Korean survivors.

First, he deals with fear and how to minimize it. As we have seen, for nuclear alerts it is important to train the soldier to know what to expect and to know that fear is the normal reaction. Prior to a mission, men usually experience most fear immediately before take-off and perhaps again over the enemy coast. But action normally dispels fear. If, however, they are downed, fear returns and a man may feel that he is abnormal. It is important to realize that this reaction is perfectly normal. (During the war, in air flights, 33 per cent of the officers and 42 per cent of the airmen, according to one study, were afraid on every or almost every mission.)

6. The HumRRO people also thought that civilians would be affected in the same way – if not more so.

Men who can busy themselves with technical things manage their fear better than most, Torrance found – important this for the leader of a small group that is isolated; also a blind belief, as we have elsewhere seen, helps people get through survival situations. It is always important to explain to airmen particularly that there are *patterns* of fear: some men will fly into strafing with apparent delight but still be scared to fly over water; some hate to fly above a certain height but are nevertheless happy at lower – sometimes more dangerous – heights.

Torrance then goes into what he calls the 'five enemies of survival':

Pain: the soldier should be told that the movie idea of the injured soldier letting the others go on without him to save all their skins at his expense rings true less often than the opposite state of affairs. Figures show that about 60 per cent of all survival cases receive some type of injury (but still make it). Among those who suffer injuries, the ability to endure seems to be largely a matter of the time perspective of the individual; persistence in trying to escape is in turn related to survival. Definite goals, whether distant or immediate, are important and necessary, based on the sound psychological principal that, as you get nearer a goal, the tendency to move towards that goal gets stronger.

Cold: we have explored this earlier in this chapter. All we need to add here is that studies of Arctic explorers show that what counts is to accept the practices of the natives of the cold regions so far as living is concerned; a rigid adherence to more 'civilized' ways of life – our concern with personal hygiene, for example – seems to have caused more trouble than any other single factor.

Hunger and thirst: these can be severe problems. It is important to spell out to the soldier what they entail, for two reasons. First, the explanations help the soldier to gain insight into his own behaviour and to minimize conflicts within his group which arise from hunger or thirst; second, such details break down frightening single words like 'starvation' and 'thirst' into more manageable steps. This means that the soldier's anxiety rises in gradual steps and acts as a warning sign. They also prevent him from panicking or drying up and not trying when he thinks he is going to starve or die of thirst.

The soldier is therefore told, for example, that he will lose his sex drive when hungry; he will, of course, begin to feel tired and weak. He may be surprised to learn that hungry people do not like to be touched; are very susceptible to the cold; get bored and take less interest in others; lose their sense of humour; and eventually take on a 'don't care' attitude, the most dangerous sign of all.

In a group suffering from hunger friendships can become very severely strained and sub-group divisions grow in importance; discipline works best here it seems. Also groups may have extreme prejudices with regard to strange and novel foods. It has been found that individuals are much more adventurous in eating new foods than are groups.

The findings about thirst are less plentiful than about hunger, probably for the grim reason that the really thirsty did not survive to be studied. But again the idea is to break it down into steps: the very thirsty man gets very tired, his temperature goes up and by the time he feels sick he has lost 5 per cent of his body weight; dizziness means the loss is getting on for 10 per cent; from then on down, there is difficulty in breathing, tingling in the arms and legs, the body gets bluish, and speech indistinct; after that the symptoms become even more horrible but are not spelled out because they arouse fear and by that time survival is probably academic anyway.

Fatigue: another enemy of survival which has been explored earlier in this chapter (see pp. 317–21).

Boredom and loneliness: these may lead to suicide when evasive action is not possible and where the survivor is alone. Suicide thoughts have been found to be more common among single survivors than in groups, though if a man is made a scapegoat by a group he may well feel like killing himself. A psychological gulf often develops between a survivor and the main unit. This may be so great that he begins to identify more with enemy soldiers in the same isolated situation as himself. While this may do no harm if the enemy soldiers *are* in the same situation, if they are not it can, of course, be disastrous and lead to capture.

Leadership is important. However good the leader may be, he may die or be lost, so responsibility should be shared as soon as possible. Behavioural contagion also occurs: in aircraft crews, for example, where everyone can hear everyone else talking over the intercom, if the captain shows his fear, it spreads. It is more important than ever then for the commander to behave calmly, speak calmly and crack a few jokes. The leader, more than anyone, must curb his feelings of fear and panic. He can also benefit from systematic research which has shown, for example, that the touchiest time is when the rescue plane or helicopter becomes overdue. This is when to keep the men busy (60).

The aggressive individuals in the platoon, who may be a nuisance in ordinary circumstances, have been found to stand up better in survival situations; it may be that they become informal leaders of the group under these circumstances. The things to look out for in someone who may be cracking up include irritability, insomnia and depression. Hypochondriacs also do not stand up well to this kind of stress.

Many conflicts within a group or between an individual and his surroundings can threaten survival. In Korea, for example, it was found that many men were modest in toilet functions in front of others and this was a source of great concern impairing unnecessarily the health of the men.

The will to survive, then, is more than a 'mystic' quality given to a chosen few. It means establishing a goal to be striven for, thinking up gradual steps to go through on the way to that goal, acquiring in training specific skills and the removal of fear through knowledge.

PART FIVE *The Psychology of Counter-Insurgency*

The 'special' war that has occurred most frequently since the Second World War is guerrilla conflict. There have been innumerable books on the theories employed by guerrillas, and countless descriptions of individual encounters. Most are agreed, however, that the guerrilla war, above all others, is a supremely psychological battle, a fight for the 'hearts and minds' of the civilians in the middle. Britain's most notable expert on counter insurgency operations, General (then Brigadier) Frank Kitson, has roundly condemned British apathy in organizing its psychological operations: '... the British seem to persist in thinking of psychological operations as being something from the realms of science fiction, [but] it has for many years been regarded as a necessary and respectable form of war by most of our allies as well as virtually all of our potential enemies' (1). General Kitson goes on to list the psychological operations units maintained by other NATO forces: West Germany (3000 men in the regular army); Greeks and Turks (300 men each); Italians (one full-strength company) and of course, the enormous American contingent described in chapter 20. General Kitson concludes: 'Undoubtedly the British are "bringing up the rear" in this important aspect of contemporary war.' Even the Rhodesian PSYAC (Psychological Action) troops numbered 'nearly 1000', according to one report early in 1977 (2).

Except perhaps in Britain, therefore, it should come as no surprise that many psychologists have studied guerrilla war. Indeed, the following chapters testify to the many techniques which have been thought up to deal with guerrillas – or insurgents as the rebels are also known.

Yet most of these techniques have had very little attention. In many cases they have not been published at all; in others they have been released in only a limited way. Put together in coherent form, as I try to do in the following chapters, they amount to a formidable battery of bizarre weaponry. The techniques, however, arise out of an analysis of the guerrilla situation – the human aspects of an insurgency make the explicitly psychological approach more important here than anywhere else. In chapter 17, therefore, we chronicle some of the major psychological studies that have been done on underground organizations and rebels in an attempt to show what makes them tick. Then we look at the techniques themselves and discuss their implications.

17 Human Factors in Guerrilla War

The most explicit psychological analysis of insurgencies which I have come across is that produced by Andrew Molnar, Jerry Tinkler and John Leitoir at CRESS in 1966 entitled 'Human factors considerations of undergrounds in insurgencies'. It consists of a detailed analysis of twenty-four insurgencies which had taken place since 1946 (1). It is a contentious document. Right at the beginning, for example, Molnar says that the analysis revealed that neither the state of a nation's economic development, its rural–urban composition, its rate of illiteracy nor its educational level affected either the occurrence of insurgencies, or the result – whether they were successful or not. (The insurgencies studied included Algeria, the Congo, Cuba, Haiti, Hungary, Malaya, Vietnam, the Yemen and Venezuela.) In other words, the accepted wisdom and commonsensical notion that socioeconomic factors cause political conflict was regarded as largely irrelevant to insurgencies, according to Molnar and his colleagues. (He even says that, looking at individual guerrillas, the number who joined a guerrilla unit for political reasons was in a tiny minority.) We shall return in a moment to other studies which disagree with this one but, with socioeconomic matters neatly out of the way, Molnar was left free to explore the individual or psychological factors in undergrounds. Without necessarily accepting his argument, it is useful for us to follow the organization of Molnar's document in this chapter, since he does cover all the areas we shall be interested in. From time to time we shall be referring to other studies which either add to or contradict his.

THE PSYCHOLOGY OF THE CELL: GROWTH AND STRUCTURE OF INSURGENCIES

First, Molnar tackles the question of how widespread insurgencies are in terms of the population they tie up. In one small-scale study of the undergrounds of seven insurgencies, he concludes that only a very small percentage of the population ever took part – the range was from 0·6 per cent to 11 per cent, the average being 6 per cent. (But, as he concedes: '. . . there may have been sympathizers within the rest of the population'.) The ratio of underground to guerrilla population ranged from 3:1 to 27:1 with an average of 9:1 – 'indicating,' he says, 'that a large proportion of insurgents work at everyday jobs . . .'. In another small-scale study which he made of three insurgencies in the Far East, he found that 60 per cent of the rebels were native to the areas in which they were most active and that although men constituted the majority there were sizeable minorities of women also. Active guerrillas, contrary to the impressions sometimes given, are actually older than underground members. Few active guerrillas are under twenty, the average age of the Viet Cong, for instance, being 23·4 years according to Molnar. The younger and older men tend to be members of the underground but not of the active service units. The breakdown by occupational structure produces few surprises (see Table 29).

Table 29 Occupational background of Korean insurgents

	%
Labourers and artisans	40
Farmers and peasants	30
Students	20
Former municipal employees	5
Former policemen	5

So what Molnar's analysis reveals is that the guerrilla comes from a fairly straightforward cross-section of the population. There appear to be no crude sociological indicators which drive one type of individual underground rather than another. It will be seen that as Molnar's account progresses, the role of individual psychology in the motives of guerrillas assumes an increasing importance as the weight of other explanations fades.

Molnar also discusses the structure and organization of underground movements. The basic cell structure of clandestine organizations usually consists of five to eight members, this being ideal for securing maximum conformity (experiments show, he says, that larger groups fragment, whereas in smaller groups rival leaders may spend most of the time locked in disagreement). However, it appears that the cell size also depends to some extent on the phase of the insurgency. If a legal political party exists, then the cells may be larger because their main activity will be recruitment. In such a case they will have a leadership of three: the political leader, the administrative organizer and the agitprop leader. The more the organization goes underground the smaller the cells become. Many undergrounds, according to Molnar, also have auxiliary cells: these are cells where people whose loyalty has not been tested get some – though not very risky – jobs to do. At the other extreme, the intelligence cells and sabotage cells rarely act in units of more than three and are specially chosen, says Molnar, not just for technical skills but because they are good at being 'left out in the cold'; that is, members do not mind not hearing regularly from the others in the clandestine organization.

As an insurgency moves from the clandestine phase, where cells are on the small size in view of the need for security, and into the 'psychological offensive phase', more communication is needed. This is, therefore, a risky period. It is when the police and the army of a country become targets for attack, when underground newspapers are formed, all of which makes it necessary to have cells in series. For example, one cell will act as reporters for a newspaper, another as an equipment procurer, a third as editors and so on. Further expansion of the movement is not much safer either. The spreading of unrest and uncertainty continues but now there is a greater number of auxiliary cells formed so that more members can be accommodated. From being highly selective, recruitment has to develop until at the end there is the open recruitment of almost anybody.

The process of infiltration by (and of) underground movements and the activities of *agents provocateurs* are relevant here. Gary Marx of the Massachusetts Institute of Technology and Harvard Center for Criminal Justice has studied both (2). His study was done mainly on the infiltration of subversive and protest groups by police or the FBI and his investigations (of thirty-four cases of known *agents provocateurs* and informants) show that in the police the men selected have recently joined the force and have often deliberately been excused academy training so that they can avoid the 'smell' of being police. Where they are civilians people are often chosen who already lead double lives, such as homosexuals and drug dealers, because they have 'practice'. Motives are obvious, Marx says – patriotism, coercion, money, disaffection from the group under suspicion – but he notes that about 20 per cent of agents actually defect *to* the group they are infiltrating. For this reason every FBI official, says Marx, must have twelve informants in such organizations at any one time.[1]

The final stage in development of an insurgency is the militarization phase, when the most important cells are the active guerrilla units whose job it is to inconvenience the police and troops – and through them the government.

There are various phases of the militarization stage, according to Molnar, which employ different types of guerrilla. In the beginning there are sabotage units – made up of men highly qualified in technical things like explosives. Later they give way, as the battle continues, to more conventional soldiers as the guerrilla units expand, become more regular and begin to tackle the government forces head on.

Most governments, however, would much rather be able to stop insurgencies right at the beginning and this is the main purpose of Molnar's study: to see what factors accounted for the occurrence of insurgencies and what affected their outcome. He devotes less than a dozen pages of a 334-page document to the wider socioeconomic influences on rebellion; he and his colleagues seem happy enough to conclude that the real answer lies in the psychological backgrounds of the guerrillas themselves. He certainly devotes a large segment of his analysis to these factors.

1. Before we leave this topic of infiltration, just consider briefly the most remarkable example ever conceived. God knows whether it was ever implemented, but it shows what some psychologists get up to. According to Mr Kenneth Goff, testifying to the House Un-American Activities Committee, a sixty-four page booklet, the *Communist Manual of Instructions of Psycho-political Warfare*, advocates infiltration of the American health profession by Communists to make America 'more crazy' (3). The document apparently recommends recruiting doctors and psychiatrists who should then attempt to disrupt people's lives by 'treatment' which raises the neurosis rate, encourages people to commit suicide, and 'keeps up the mystique of insanity'. This manual is supposed to be used in at least one labour school – the Eugene Debs Labor School (113E Wells St, Milwaukee, Wisconsin). It is difficult to believe that such a campaign could ever be more than marginally successful. Nor is it clear whether this is a real 'manual' or a piece of black propaganda surreptitiously put out by the right wing in America (or the CIA). But either way, it shows just how far thoughts on infiltration have gone in some quarters.

RECRUITING REBELS, TRAINING TERRORISTS

Recruitment into an underground depends on the insurgency's stage of development. To begin with, primary attention it given to the development of carefully chosen, well-disciplined cadre. Later, mass support is the object. First, says Molnar, the recruiter identifies talented people who have personal grievances – though they are seldom placed in a position where they must immediately decide whether or not to join the underground. 'Instead, through a series of seemingly innocent or slightly illegal acts which, when viewed by an outsider, appear subversive or illegal, the recruit is led to believe that an overt commitment to the underground is his only alternative.' Grievances and vulnerabilities are played upon but the role of coercion is never neglected, Molnar says. In Malaya, in some of the Communist-led unions, lectures were given to party members several times a week while agents in the audience observed the reactions of those attending. The topics discussed were usually broad social issues – *not* ideological material – and those who seemed interested were followed up. At first, the individual is not asked whether he will join but whether he is 'free' to distribute leaflets, collect funds or carry messages. From then on more weighty things are demanded of him until, says Molnar, he is given a test. Invariably this is some minor illegal act, often carrying a message to a certain address. The message, inevitably, is a blank piece of paper in an envelope with a hidden seal so if he defects to the government, nothing is lost. Then comes the oath, at which the recruit is sponsored by two members who have an interest in making sure he stays in. He is then placed in a probationary cell, where he is tried at various tasks to assess his particular talents.

The era of mass recruitment, when the movement is on the psychological offensive, comes after cells of hard-core cadre have been set up in most areas of a country. To begin with popular support is sought for some particular grievance and only later is this channelled into active insurgency. The methods exploited are indebtedness, coercion, suggestion and alienation from the government. The local men look for problems which the farmers may have and help is given. Locals who are short of cash are lent money but have to give a compromising receipt. Suggestion is particularly effective if, at meetings, the young can be separated from the old. The young, it has been found, are particularly susceptible to recruitment at rallies, celebrations and so forth where stooges 'volunteer', taking fresh members with them. Raids on the government which are made to appear to come from a certain village are another technique used for recruitment. This may provoke government troops to retaliate against the village, which then becomes alienated from the authorities (since it has been attacked for no fault of its own) and this is fertile ground for the terrorists to solicit help in. In all cases, says Molnar, the insurgents make it seem flattering to an individual that they want him. It is him they want, not his friends. But of course they tell his friends the same thing, only vice versa.

The organization of undergrounds calls for clandestine and covert behaviour. This is a dangerous aspect of their affairs, says Molnar, because of the inadequacy of the individuals who compose the underground. People have a tendency to be curious and to brag about their accomplishments. Individuals with inadequate personalities – who are prone to show off, say – are therefore excluded from the early stages. Insurgents hardly use tests for this screening, but a lot can be learned from a man's lifestyle and discreet questioning of his friends.

Molnar's document also surveys briefly the schools around the world where rebel behaviour is taught. For example, at the Lenin School near Moscow the students 'follow the rules of conspiratorial behavior in their day-to-day activities, with assumed names and false biographies'. Molnar describes how methods are adapted to surroundings – such as the use of witchcraft in Africa to control the local populations. (One example quoted is of an African who had been taught how to produce speech or noises from a skull or skeleton by the use of hidden microphones, and how to produce a phantom from a cloud of smoke.) But there are now, he says, many other schools – in Cuba, China and Latin America, for example.

In general, guerrilla training has received much attention. Perhaps only three other 'psychological' aspects need to be mentioned here because they have been overlooked elsewhere. First, is the fact that the further a guerrilla unit is from headquarters, the stronger its political wing is and its self-indoctrination sessions. This is a good illustration of a point we return to later – that the more remote active service units are those prone to defection. And a second point to arise is that the guerrillas are always taught not only to know the facts about an issue but to have an attitude towards it. This helps a person to react correctly to every political development; but at the same time it is very useful in propaganda sessions if the man who is selling a line seems to be consistent in everything he says (and it is something which government spokesmen are often very bad at). The third point is that the guerrilla schools spend a lot of time teaching the men how to remember things. Much attention is given to mnemonic systems of memory and other devices, to avoid the need to record incriminating details on paper. This is useful knowledge for government forces as it means that a guerrilla officer, if captured, can usually be relied upon to have more in his head than an officer of equivalent rank in the government forces.

THE MIND OF THE TERRORIST

'As an insurgency escalates in size and scope the kind of individual it attracts and the nature of motivation for joining change.' Perhaps the most controversial aspect of Molnar's study is the section on why substantial numbers of people become guerrillas and why they stay. For he seeks to show, not just that guerrillas join for personal reasons but that political motives are in a very tiny minority indeed.

Ideological considerations, he argues, are more important in the early stages. He concedes that most studies of guerrillas are based on reports of captured personnel who, because they have defected or allowed themselves to be captured, may be the least politically motivated of the guerrillas and who, under interrogation, may be only too happy to tell their captors what they think they want to hear. But he does quote a study of the Philippine campaign which had found that a combination of factors seemed responsible for guerrillas joining up and that a chance to obtain personal advantage was often cited, be it land, leadership or a position of authority. Family discord and the violation of minor laws were sometimes the precipitating event, since in a surprising number of cases the first contact with the guerrillas was accidental – and occurred just after one of these precipitating factors.

He is able to muster statistics to support his argument. For example, a survey of the Vietminh showed that barely 38 per cent of captured prisoners 'expressed belief' in the Vietminh cause, and that compares with only 17 per cent of a Huk sample of 400 prisoners who said they were in sympathy with the aims of the Communist Party. 'Promises and propaganda appear to have been involved in a number of cases,' says Molnar, referring still to why people join guerrilla units, though he emphasizes that the exact extent of this influence is hard to gauge. In the Huk study, for example, less than 15 per cent said they had been mainly influenced to join by propaganda, though a further 27 per cent said that promises had been a contributory factor in their joining. (Two in every five Huks became involved through their personal friends.) A number of prisoners claimed – under interrogation – that they had been forced to join. About 5 per cent of the Vietminh said they had been forced against their will, plus a further 23 per cent who, according to Molnar, 'did not appear to resent the fact'. Molnar seems to conclude that one of the largest incentives to joining is coercion combined with promises – of land, authority and so on. When the proportion of guerrillas joining for these two main motives is added up, he says, then they usually account for between a third and a half of recruits, whatever and wherever the guerrilla campaign. (It is also worth remembering that approximately one-fifth join because of the bloodthirsty reprisals carried out by government troops on innocent civilians after guerrilla raids.)

COMMITMENT AND CONFORMITY

Insurgencies always have a sustaining ideology. Molnar isolates two psychological principles that insurgents exploit to secure ideological uniformity in their members. One is that attitudes do not shape behaviour but that the process works the other way round – behaviour shapes attitudes. When you do something you have to rationalize why you have done it, so you change your attitudes in line with your behaviour. This is why insurgencies have elaborate initiation ceremonies: they are essentially ceremonies of *behaviour* where the new recruit *does* something rather than is merely told that he is now a member of the organization. This is also why symbolic acts of public commitment, at group meetings, or minor misdemeanours on behalf of the movement, have to be carried out. Not only do these provide the opportunity for blackmail but they also help persuade the recruit that he is a member of something that matters, that is worth breaking the law for.

The second psychological principle is that concerning conformity. Insurgencies are usually based on the principle that closer control of individuals brings about greater conformity. Hence the self-criticism sessions and mass meetings where people have to show their conformity by declaring their allegiance in public. Added to this, says Molnar, discussion of ambiguous topics helps because attitudes towards these are usually less fixed than towards other things. Discussion of facts is thus avoided: they cannot readily be changed. Discussions go on for a long time because exhaustion makes people more susceptible to change;

they also often take place in isolated circumstances because isolation too makes change more difficult to resist.

Three factors, in addition to the cell structure and self-criticism sessions, provide insurgencies with the organizational glue that sticks them together, according to Molnar:

(a) Endless indoctrination: this is not so much a way of teaching a particular set of values as of keeping before the members at all times what is expected of them and where any change becomes apparent as 'deviationist' or indicts an individual as being over-concerned with his 'personality cult'.

(b) Democratic centralism: to Molnar this is the system whereby everyone is allowed his say on how things should be done – provided that things are *then* done the leadership's way. 'Individuals tend to abide by a decision so long as they are permitted to voice their opinions, notwithstanding the outcome,' he says.

(c) The committee system: mainly, says Molnar, this is by no means always used to increase efficiency directly; instead it is designed to enable everybody a sort of 'proxy' leadership and in this way enables the Indians to feel, from time to time, that they may act as Chiefs. In so doing divisive ambitions are softened.

DEFECTION

Patterns of defection can also give insight into the motivations of guerrillas. It has been found, says Molnar, that defectors more often come from the rural areas than from the cities where it would appear easier to defect. He puts this down to the fact that life is not so hard for guerrillas in a city, whereas the active units often experience the harsh conditions of bandits in the hills, say; and that members of active units are more likely to belong to the underground for personal reasons – that is, for the excitement rather than for the wish to effect political change – whereas members of political units are more committed ideologically.

Certain types of cadre members are more prone to defection: underground intelligence men often have to assume a pro-government façade, and sometimes actually come to believe in it; liaison agents have the opportunity to defect at any time and are particularly susceptible to amnesties, which is why insurgents try to put only their most trusted people in such jobs. On the other hand, cadre leaders, because they get a lot of satisfaction from their position of power, are unlikely to defect, as are agitprop and propaganda leaders who invariably come to believe what they write, even if they did not when they started.

One of the largest studies of defection, of 1369 Viet Cong, showed that the harshness of the life – food and medicine shortages particularly – were the chief reasons for defection. Three-fifths of the Huks defected for these reasons; others because, they said, they were disappointed with the failures and shortcomings of the Huk organization. Still others cited government promises – the most frequent being the promise of land, second the promise that the men would not be tortured, and third that the government would pay for their weapons. Most defectors, says Molnar, gave no thought to defection during their first year, being too involved in the movement during that time. The vulnerable period, it seems, usually came after about a year and a half; this was when the sacrifices they had made in joining up either had to be paying off in terms of personal advancement or they began to worry about their future. Very often, Molnar found, the men would develop personal difficulties, say with another member of their unit, and then rationalize this into a political disagreement with the movement in general. Another source of dissatisfaction mentioned quite frequently was sex. Although roughly 10 per cent of guerrillas are usually women, sex is not normally allowed in the jungle or other remote areas and, when it is, says Molnar, it is the leaders who get the women. Other guerrillas resent this and find it easy to see this as a departure from the 'equality' ideal of the movement. Thus this personal problem also becomes political.

Defection, says Molnar, usually occurs after several minor personal crises like this. Once defection has been decided upon, the guerrilla begins to detach himself from his unit. Molnar makes the important point that he believes guerrillas decide to defect *first* and only then become susceptible to government propaganda. If true this has clear implications for the way government propaganda should be angled.

POLITICS AND PERSONALITY

Molnar's study was an early one of its kind and though he is to be complimented on trying an empirical approach in a difficult area, it cannot be said that all his arguments are convincing. His measures of economic development, for example, or of education level, were crude in the extreme. To deny, as he did in a few pages, that many political movements around the world had any claim to political legitimacy, to claim that they were merely an expression of the personal grievances and complexes of their leaders seems too simplistic.

Since the mid-sixties, however, when his study was carried out, investigations linking personality to politics have snowballed almost beyond measure. In providing some understanding of why certain people become 'political' they are no doubt of general interest to the military. On the other hand, they are mostly investigations of Western individuals operating within a Western political framework. There have, for example, been very detailed studies of the psychological factors of nationalism in the South Tyrol, of isolationists in Australia (4), and countless studies of extremists (5), patriots (6), and radicals (7) in the United States. Most show that some sort of link between personality and politics does exist, and that the extreme right and extreme left have rather more in common with each other than with the moderate wings of their respective parties' ideologies; but it is debatable how much specific help they are in a politico-military context in the Third World. We shall not, therefore, dwell too much on these studies though the interested reader is referred to three books on the subject which themselves give many other references.[2] But a few studies have added to Molnar's and have produced a better methodology in so doing. We shall confine our attention to studies which conform to one or more of the following criteria: they have been carried out by or for military units; they concern the Third World; they are concerned not simply with the link between personality and politics but with *violent* politics.

Probably the most complete non-American study was carried out by the British expert in psychological warfare, F. H. Lakin (8). The research was completed before Molnar's but it did not become available to other military forces until rather later. It is a study of terrorists in Malaya during the period from 1952 to 1955. Lakin was head of a special research unit set up to study psychological aspects of guerrilla warfare. In the course of his research, Lakin questioned some 430 surrendered terrorists (for up to a maximum of eight days), and also

2. See, for example: *Collective Violence*, edited by James Short Jr and Marvin Wolfgang, Aldine Atherton, 1972; *Radical Man* by Charles Hampden Turner, Schenkman, 1970; and *Anger, Violence and Politics*, edited by Ivo Feierabend, Rosalind Feierabend and Ted Robert Gurr, Prentice-Hall, 1972.

interviewed some 2800 males throughout the country. He drew up a number of hypotheses about the motivation of the terrorists on the basis of the first few interrogations, and these theories were then put to other defectors as they came in.

Lakin isolated seven major reasons for why people became terrorists. They show some similarities with Molnar's reasoning but are by no means identical:

(a) The dominant reason was fear of arrest for non-terrorist Communist activities. In other words, most terrorists had been successfully trapped by the Communist organizations into joining up, usually by the threat of blackmail or some such device.

(b) Fear of what the Communists might do if they failed to co-operate.

Lakin points out that the two main reasons together meant that many became terrorists because they were in a dilemma about who would arrest them. He cautions that this is a double bind for governments and is to be avoided in future conflicts.

(c) A belief in the success of the Communist revolution in Malaya. Not quite an ideological belief, perhaps, but rather more important than is suggested by the Molnar study.

(d) Attraction of better opportunities in a Communist Malaya. Again this sounds more a personal motive than a political one – though not necessarily.

(e) The wish to belong to a group fighting for a common cause.

(f) Life was hard in the community – though Lakin adds that the poor conditions of the society were *not* a strong motivating factor causing people to turn to terrorism.

(g) The least frequent of these reasons was wanting the British to leave.

Lakin's argument, then, based on these seven reasons, was that the material advantages of a Communist victory rather than anything else were what counted. In a sense this is neither a purely psychological nor a purely ideological motivation – more, perhaps, an example of pragmatic reasoning by those concerned. We shall return to this in just a moment.

Lakin also found that three out of five of the guerrillas had worked on the rubber plantations, whereas the tin-mine workers and other agricultural occupations were noticeable by their relative absence in the terrorist groups. He felt that this was due to the fact that the rubber plantation workers were nearer the remote jungle and in fairly isolated surroundings; they were thus highly vulnerable to fleeting visits from the guerrillas.

As far as he was able to tell, 8 per cent of the guerrillas captured had been convinced Communists, 24 per cent believed in Communism as a long-term goal, and 47 per cent were attracted by the promises of material advantages they had been offered before the revolution was over: 'A few even believed that there would be cinemas and stage shows to entertain them while they were fighting in the jungle.' This 47 per cent attracted by material gain, Lakin said, should be the main object of psychological warfare, though another target group was the 21 per cent who were 'barely politically conscious' and only dimly grasped even the basic tenets of Communism.

Organizationally, Lakin found the Malayan terrorists highly authoritarian, with a strict hierarchy, though there was a stress on the democratic nature of the group: democratic centralism in Molnar's words. In Malaya, says Lakin, many terrorists were certainly very disillusioned about not being treated equally after joining up (here, as in the Molnar example, the leaders got the women). The self-criticism sessions were in general not liked, partly because many did not understand the nature of Communism and were always being shown up in the discussions, but also because, as there was no possibility of prison in the jungle, the minute they showed any doubt about the 'cause' they were fighting for, they were either shipped away or executed.

Lakin's findings therefore show that purely ideological convictions did not motivate many Malayan guerrillas, though it is also true that what the ordinary person understands as psychological motives – a need for excitement, a lust for power, the wish to be part of an organization – were not particularly strong either. Communist coercion seemed to matter most, with promises of material advantages also counting. A somewhat more precise picture – and somewhat more limited – than that painted by Molnar.

The Rand studies of Viet Cong motivation and morale are probably the most elaborate studies of revolutionaries in the field, going on for much longer than Lakin's study and using only first-hand material, unlike Molnar's. However, there are shortcomings even with these.

One is that most of the smaller studies which go to make up the whole were qualitative and not quantitative. By and large, the Rand scientists did not count heads, preferring instead the in-depth interviews of fewer Viet Cong. Though this has certain advantages, the track record of Rand on this particular subject is not good and gives rise to doubts – doubts with the benefits of hindsight admittedly – about their methodology (9).

We have already seen in chapter 1 the mistakes made in earlier studies of the Viet Cong. That was in 1966. Studies in 1967 and 1969, however, reversed part of this picture. For example, by 1969 Konrad Kellen was saying quite different things from what Leon Goure had said in 1966 – that the VC were near to breaking-point. Kellen found VC morale very high, 'showed no signs of cracking', and that the VCs' resistance to psychological warfare was fairly total (10). Yet other studies have concluded that the period 1966–71 was the most effective use of psychological warfare ever – and we shall discuss the evidence in detail in chapter 24. Enough for the moment to note the changes of stance at Rand and that it was Kellen's study, based on twenty-two men, which was responsible. The size of the sample probably means also that we have to treat with some suspicion Kellen's conclusion, that *all* the twenty-two Viet Cong (who were captured after being stunned by artillery fire or wounded and did not defect) joined up for political reasons. But that is what he says. Most if not all, he says, joined because they believed in the struggle. Most believed that there were no hawks in America outside the US government (which they saw as operating against the wishes of the people). It is interesting to compare this part of Kellen's study with yet another from Rand, this time by J. J. Zasloff, written in August 1966 but only declassified in May 1968 (11). Zasloff claims, of the seventy-one intensive interviews of VC that he carried out: 'Even professed and ardent Communists were weak on ideology and largely unread in the traditional Communist literature.' W. P. Davison, in a fourth Rand study of 1967, found that the reasons for joining definitely *were* personal – ranging from promises of land to the escape the war offered from personal problems (12).

This last study was based on 200 VC – more than most of the Rand studies – but it still seems less reliable than Lakin's. The Rand studies attract further scepticism when two other reports are considered, one by E. J. Mitchell, 'Inequality and insurgency: a statistical study of South Vietnam' (13), published in 1968, and 'Economic and social correlates of government control in South Vietnam' by Anthony

Russo, published in 1972 (14). Rand chose not to publish the latter for four years, even though it was highly critical of Mitchell's earlier paper which was circulated widely in both classified and unclassified form, and was believed to be influential in policy circles. Russo's paper goes quite against the conventional wisdom of its time. Unlike the others we have considered, it is a statistical study and attempts to be objective. Russo took the twenty-six provinces in South Vietnam and examined various social and economic indicators and then looked at government intervention in those provinces and support there for the VC. He found quite clearly, that these factors *were* crucial in explaining support for the Saigon government and the control it was able to exert in the provinces. In the poor regions, Saigon was poorly represented, not popular and did not have much control; in the richer areas the opposite was true. This conclusion went quite against one of Molnar's central conclusions as well as against that of Mitchell's earlier, and more influential, Rand paper.

The combined Rand studies offer little help for anyone trying to find out what makes the revolutionary tick. It appears on the face of it that Molnar was correct in some of the things he said: ideological fervour is not as strong among guerrillas as might be expected. On the other hand traditional psychological motives – power, the need for excitement – do not fit the bill either.

At the risk of seeming unduly chauvinistic, it seems to me that Lakin's is in fact the most plausible account of what motivates guerrillas to join up. Russo is no doubt on the right lines as well. But that means that the psychology of the guerrilla is a social psychology, which looks at him as one of a group, rather than as a 'personality type' whose political violence is almost a sort of sickness and where psychological warfare should be a kind of 'cure'. We shall see in chapter 21 how this view affects counter-terrorist techniques.

SUBVERSIVE PSYWAR: AGITPROP, PASSIVE RESISTANCE AND COERCING CROWDS

The subversive use of propaganda usually means the exploitation of the darker side of human nature – grievances, greed, tribal differences, envy, and so on. The techniques are similar to those used by government forces (and which are discussed in the following chapters) but the organization differs. This occasionally means that in some countries subversive psywar is particularly effective in certain fields. One analysis of

Chinese propaganda techniques,[3] for example, noted the reliance on radio and the exchange of delegations and trade fairs – each member who visited Peking usually representing a different sector of the population. This reflected – for a while anyway – the Chinese approach of not attempting to propagandize everyone, but instead to aim only at leaders. The Chinese Defence Ministry has a General Political Department which manages all psychological warfare activities (15).

Soviet psychological warfare apparatus is far more extensive than China's. Besides terrorists, the Russians train Third World journalists in how to write propaganda, have a sophisticated way of running foreign 'front groups', all this buttressed by the insistent argument that the spread of Communism is inevitable.

In Vietnam and elsewhere, the insurgents used armed propaganda units before the Americans (16). Other techniques involved the selective use of terror against people who did not cooperate, and grievance sessions in villages, which often turned into self-criticism groups.

The main difference between the guerrilla use of psychological warfare compared with methods used by government forces is that it is usually combined with other techniques such as agitation, *agent provocateur* activities, passive resistance and civil disobedience.

Passive resistance and civil disobedience have in recent years varied in their use as effective psychological warfare, but it is in general true to say that regular troops are ill-equipped to deal with these techniques. A particular advantage of these campaigns is that they offer so much variety: ostracism, accusations, whispering campaigns, sit downs, refusal to work, and so forth. And there is little that can be done against them if everyone is taking part: not everyone can be put in jail. In many cases, too, they can be performed anonymously (such as wrongly addressing goods which a company is to send abroad). Then there is the simultaneous occurrence of demonstrations in different parts of the country; this certainly makes many movements appear more coherent and larger than they might otherwise seem. Molnar outlines a general chronology which, he says, insurgents have discovered to be the most

3. This information is taken from *Psyop Intelligence, Special Report*, 7 December 1971, headed: 'A selected bibliography of People's Republic of China foreign propaganda, information, and culture programs with explanation on the agencies and individuals responsible for such programs', 7th Psychological Operations Group, APO, San Francisco. The report includes the names and standard telegraph codes of forty Chinese personnel involved in propaganda.

effective. This begins with the anonymous, individual acts of sabotage; the less anonymous 'cold shoulder' follows; next, the work force will all stay at home at an appointed hour to demonstrate solidarity; then simultaneous demonstrations – but small ones; finally, the mass demonstration which can often turn into a direct confrontation. Each stage is difficult for the government to break.

Terror is very often misunderstood in the West. Normally it is thought of simply as isolated outbursts of brutality. But in fact terror, says Molnar, is the sharp end of threats – and the way threats are delivered is perhaps one of the most psychological of all guerrilla warfare activities.

Molnar distinguishes three kinds of 'threat demand'. First, where a threatener chooses to demand actions towards which the populace is already disposed. This is the easiest kind. Second is the demand which seeks to induce an individual or group to change specific behaviour by demanding alternative actions. The third and most difficult kind is the demand which orders an individual to refrain from a course of action he is already pursuing. The whole point of threats, however, according to Molnar, is to make them general so as to take in as many people as possible. Threats operate on the principle of irregular reinforcement: the more irregular the punishment the more resistant, experiments show, it is to extinction. Operation Black Eye in Vietnam is a good example of such psychological threat manipulation. Selected South Vietnamese troops were organized into terror squads, operating underground, trained to infiltrate villages. As a result, within a short time VC leaders started dying in their beds. On each of the bodies was a piece of paper printed with a grotesque human eye. The appearance of the eye came to represent a threat. Next, 50 000 of the eyes were printed by the Americans and started turning up pinned to the doorway of many people's homes, coming to symbolize the statement: 'You are being watched – it could be you next.' Now the eye took on a more potent form because it was an uncertain threat. Eventually this sort of thing wears off, but while it lasts it can be a powerful source of disruption, whichever side uses it.

The subversive use of crowds is fairly crude: it involves the use of cheer leaders, as it were, people trained to hold groups of people together; bodyguards; the planned and systematic use of chants and songs; the planned *agent provacateur*, who plants the appropriate rumour and then melts away. There have even been examples where people have been paid to turn up as crowds – the London *Daily Telegraph* idea of Rent-a-Crowd is not quite so fanciful as it sometimes seems. But crowds are manipulated from both sides and in terms of techniques at their disposal, government forces in general have more tricks up their sleeve than the rebels (see chapter 23).

AMBUSH PSYCHOLOGY

A final aspect of guerrilla tactics to consider is raids and ambushes. Of course these have to be carried out in the proper military fashion but in an insurgency they also have a psychological dimension. For example, attacks at weekends or paydays, when the government troops are hoping to relax, have been found to be particularly effective. In Korea, Malaya and the Philippines respectively, 55 per cent, 75 per cent and 60 per cent of government casualties are believed to have been caused during ambushes. In those campaigns, roughly 75 per cent of ambushes occurred along main roads, about 17 per cent in the hills, and 8 per cent in small villages, carried out mainly, it was found, by groups of three to five men.

Australian army research into ambushes has shown that only 15 to 25 per cent of soldiers, faced with sudden danger, respond immediately with 'fixed purpose' or 'effective activity'. The majority are stunned or bewildered. The Australians, therefore, developed a counter-ambush drill: the soldiers had to force the ambush – that is, rush through the ambushers and envelop them from the rear. It was at first difficult to persuade the soldiers to advance against concealed positions, even when helped by instructors who had used the technique in battle. But the method was found to work. The ambushed soldier immediately concentrates his fire on one point of the ambush line and runs for it. The Australians eventually became noted for their fearlessness in facing ambush in places like Vietnam (17).

The verdict on the researches that have looked into the psychological aspects of guerrilla warfare must be that so far they are highly inconclusive. There has been a tendency in the past to think of terrorists as either politically motivated or psychologically fired when probably what counts as much are the immediate groups of which people are members: students, the out-of-work, peasants, rubber plantation workers. It is possible that the way people see themselves as belonging to oppressed groups, rather than their family background or ideological beliefs, is the main factor in mobilizing social movements.[4] This has not, I fear, been at the forefront of the research by military psychologists into insurgencies.

However, that has not stopped them thinking up innumerable techniques of counter-insurgency with this or that psychological ploy which are explained in the following chapters. No doubt this inconclusive and inconsistent analysis of the psychology of guerrillas may have something to do with the controversial effectiveness of much of these psychological techniques.

4. My analysis of the psychological component in guerrilla war must inevitably seem arbitrarily divided into separate elements. Readers wishing a case study which describes these results collected in one tale are referred to Lt Col Francis Kane's study, carried out for the US Army War College, entitled *Revolutionary Psywar in Castro's Cuba* (18).

18 The Growth of Special Operations

SPECIAL INSTITUTES – AND A NEW MILITARY SPECIALITY

Whereas many theories of insurgent warfare stem from the writings of former guerrilla leaders, most counter-insurgency expertise – especially that drawing on sociology, psychology and anthropology – has been worked out by colonial powers with well-established academic populations, in such wars as Malaya, Korea, Cyprus, Aden and, of course, Vietnam. (As yet not much appears to have surfaced from the continuing struggles in Ulster.) That is not to say that other countries have no interest in this approach. Israel, for obvious reasons, is fast catching up. Australia has a few psychologists interested in the problems. And the psychological research housed in the library of the International Police Academy in Washington included, when Congress closed it down on 1 January 1975, many studies carried out by police officers and soldiers from as far afield as the Lebanon and Jordan, Venezuela and Tibet (1). Both Russia and Hungary have psychologists working on the links between personality and ideology (2), the effects of isolation and other topics. Even Rhodesia, as recently as 1977, has used loudspeakers and films to try to frighten guerrillas into defection.

In this sense, then, the social science approach to counter-insurgency is now widely pursued. Its growth has been rapid. The earliest systematic studies appear to have been carried out by the Americans and the British during the Korean and Malayan campaigns; but the main thrust started in the US at about the turn of the 1960s and the various social science agencies attached to the military grew rapidly under the Kennedy administration. Later, the agencies had their budgets pruned somewhat, but their influence for good or ill in the training of special forces officers remains strong.

These institutes (most are referred to earlier, but for details see Appendix II) began their analysis of insurgencies along the lines we considered in the last chapter and by falling back on academic studies in areas which were considered to be germane to the matter at hand. To begin with such things as the nature of conflict, the concept of deterrence, the psychological meaning of the 'enemy', the effects of personality on persuasability (the more intelligent are less easily persuaded, etc.) and similar matters were investigated (3). A series of secret studies in 1965, carried out for the US Air Force, explored the attitudes of governmental leaders towards the use of certain weapons, such as incapacitating agents and nuclear devices, in insurgencies, and a similar exercise into public attitudes was contemplated but never done (4).

These piecemeal efforts soon gave way, however, to more prosaic, more organized and probably more effective studies to see what could be learned from earlier limited wars.

One of these, a look at five low-level conflicts in which the USA had been engaged – in the Philippines, Samoa, Haiti, Nicaragua and Vietnam – was carried out by Dean Havron from Human Sciences Research Inc. (5). He concluded that two types of policy were needed: institutional preparedness and operational guides. The former covered doctrine development, specialized training and the development of personnel; the latter included how to organize psyops, amnesty programmes, grievance systems, relations with the local sources of power, etc. These were to become general guidelines for the activities of the institutes in the future.

This approach was further supported by other studies such as that by Lt Col Neil Leva, at the US Army War College, in a paper comparing the efficiency of insurgent and counter-insurgent forces (and which made use of a very valuable document produced by SORO, entitled 'Peak organized strength of guerrilla and government forces in Algeria, Nagaland, Ireland, Indochina, South Vietnam, Malaya, Philippines and Greece', Washington, The American University, June 1965) (6).

Leva, like Havron, concluded that the insurgent has several in-built advantages over the counter-insurgency forces and that to redress this greater indoctrination of the troops and the population is needed, as is greater training for the troops in social science matters and the development of new, military specialisms. An official five-year plan drawn up in February 1966, gave this reasoning the official seal of approval (7).[1] The institutes thus found themselves steered in certain directions in their efforts to help in the counter-insurgency war effort.

So much so, in fact, that by 1970 a new military speciality had come into existence in the US Army – Overseas Security Operations (OSO), a special career programme involving 5000–6000 men trained for 'assignments primarily concerned with civil affairs and psychological operations in an unaccustomed cultural environment' (8). Its goals were defined as: 'fostering a sense of nationhood among the indigenous population' and 'the development or modification of attitudes and values related to national unity' (9).

The mass of research throughout the 1960s has therefore secured a base for itself. The setting up of OSO has clearly institutionalized the whole field of special operations. In what follows we examine the specific socio-psychological techniques which the institutes evolved in the counter-insurgency war effort.

STABILITY OPERATIONS

Stability operations refer to the involvement of the soldier in non-military activities – in preserving civilian populations in cooperative attitudes. The special forces officer's 'duties and responsibilities cover a wide spectrum ranging from association with heads of state and ambassadorial representatives to instructing villagers how to drill a well or a drainage ditch' (10). Fundamental to stability operations is the nature of cross-cultural communication itself, and CRESS (formerly SORO) has spent much time looking at this area. In one study, for example, which analysed 345 incidents of cross-cultural communication – military, diplomatic, educational, political, and so on – it was found that the type of communication made no difference to its outcome, but that where the objective was definite rather than vague, and where the recipients were involved rather than just given aid, the communications were far more effective (11). Other research, in Egypt for example, showed that certain cultural types have very

1. Leva also discusses the use of social science techniques for estimating the peak strengths of insurgencies. For example, by comparing rice consumption in a village with the known population one can calculate the number of insurgents in the area. The amount can also be a guide to how sympathetic the village is to the aims of the insurgents. For other details see also 'Methods for estimating insurgent strength', E. F. Sullivan, 1969, US Naval Post-graduate School Thesis.

different psychological characteristics that have to be taken into account in communication.

The characteristics of individual societies have also been examined. Everything from tribal oaths and general levels of hostility to variations in smell and attitudes towards religion have been studied and put on file in the Human Relations Resource File at Yale (12). These, in turn, gave rise to such papers as that by Lt Col Howard Johnston, at one time US Naval Attaché to Thailand and Laos. His research, 'The tribal soldier: a study of the manipulation of ethnic minorities', looked at several tribes in South East Asia and presented detailed case histories of the Meo in Laos and the Rhade in Vietnam (13). He concluded that to be an effective ally a tribe must have its own safe area which it knows well and controls; it must be strongly organized with a natural hierarchy that can be converted to military use; it must have a history of fighting which has made it a cohesive group; and it must have a grievance. If a grievance does not exist, it must be manufactured. From his survey Johnston found that ideology is far less important as a motivating factor than a good grievance.

Perhaps the most wide-ranging piece of research in this area is the now notorious Project Camelot (see chapter 1), a secret study of the social system in Chile carried out by SORO (14). Camelot was to have been the social research project par excellence, showing how all aspects of a society are linked together to produce change or stability, thus offering opportunities for control never before dreamed of. Its demise held up the development of this side of military activity, and since then many of the institutes have concentrated on smaller pieces of more immediately practical research.

TRAINING FOR STABILITY OPERATIONS

In this section we look at the general issue of how to train officers and men to be *aware* that they are rooted in their own culture without even thinking about it. Only in this way is an army likely to make its soldiers more sensitive to the country in which they are at work (and maybe at war). Three projects will help illustrate the range of the inquiries that have been undertaken.

Early on, for example, HumRRO prepared a handbook of examples of the kinds of problems which Americans working overseas experience (15). This is, in effect, a short collection of readings of actual problems organized into a coherent way, which helps to prepare the soldier for what he is likely to encounter. An early section illustrates the tendency for Europeans abroad to think that 'Our way is the natural way'. The handbook provides a checklist of things westerners should be wary of and never take for granted. For example: control over the environment – do not assume that foreigners always think that man is the master, as say Americans tend to; progress – do not always assume that others think change is an inevitable part of life; puritanism – foreigners do not always have the same attitude about their responsibility for others as do westerners; moralistic orientation – the missionary spirit is not always welcomed; interpersonal behaviour – manners and gestures differ.

A couple of specific examples illustrate how the handbook works. The first comes from a section in which flexibility in the way an innovation is 'sold' in a developing country is being stressed. It concerns the introduction of the telephone into Saudi Arabia:

According to reports, the king wished to connect the capital with certain other cities, but the tribal chiefs would have none of it. The Koran contained no mention of the telephone, and it must therefore be a work of the devil. The king then pointed out that the devil would be unwilling to transmit the words of the Prophet and see, the telephone lines carried the sacred words of the Koran perfectly. The chiefs were thus convinced that the telephone was all right for them to use.

The next one emphasizes the differences in gestures from one country to another:

The American acted 'informally' in the presence of the indigenous people, joking and teasing in the American style, shadow boxing, fondling children, being generous to natives in need. One of his ingratiating gestures to indicate he was a 'Good Joe' was to put his arm around the shoulders of ranking native men while laughingly tousling the hair on their heads. This is the equivalent in western society of opening a man's fly buttons in public as a joke, for in Palau the head is a sensitive zone.

Not to say erogenous.

The second type of project tackles specific issues covered in the handbook in more detail to see how they affect the introduction of new ideas. Two which have been extensively explored, and which give a good idea of the military approach, are peasant fatalism and the negativism it can give rise to, and rumour.

Arthur Niehoff and Charnel Anderson surveyed the problems of 'peasant fatalism as a barrier to change' in planning 'civic action' programmes for the US army (16). Their study was a survey of 171 case histories of the introduction of new ideas or techniques into developing nations. Of these, fifty-seven were found in 'which traditional belief systems were in conflict with the innovations'. Analysing these fifty-seven, Niehoff and Anderson arrived at three types of fatalism responsible for resistance.

Supernatural fatalism: (a) theological – for example, water supplies are generally difficult to tamper with because they are often bound up with a local deity; (b) magical – these are patterns of supernatural belief which interfere with innovation but are not derived from a dominant religion. For example, in one case the local population believed that illness was caused by the evil machinations of envious neighbours. This type of fatalism may be resistant to western solutions but more susceptible to local arguments, if they can be found.

Situational fatalism: apathy and therefore resistance may be based on a real understanding of the limited possibilities for improvement. This is usually for economic reasons.

Project negativism: in this case apathy is induced by the experience of previous project failures. Improvements to agricultural practices appear to be the most amenable to change – largely, it seems, because the link between cause and effect is easier to demonstrate.

Ending their survey on an optimistic note, Niehoff and Anderson point to the fact that although there were difficulties with fifty-seven of their case histories, the remaining 117 projects, or two-thirds of the total, were successful. Peasant fatalism, according to their survey at least, is best overcome by manipulating leadership patterns, social structure and economic patterns rather than more direct intervention with a particular technique that by-passes or ignores the social fabric.

Niehoff also looked at the positive uses of gossip and the negative effects of rumour (as part of a HumRRO project for the Army) (17). He concluded that it is not easy for a foreigner to tune into gossip, but it is one of the first things which the military should encourage its informers or native coordinators to explore. The important gossip places are where rumours are picked up and where the military can spread most quickly many of the messages it wants to get across. He quotes an example where the real health problem in a village was that of intestinal parasitism. However, the gossip quickly showed that the villagers were more worried about a rat infestation. The soldiers were thus given access to the villagers' houses by promising to do something about the rats – and under this pretext they were able to clean up the public health system.

The negative rumours, by and large, have been found to fall into two areas: that information collected for innovation purposes is generally regarded as the efforts of the tax-man in disguise; or, as happened in Guatemala, Peru and Northern Nigeria, according to Niehoff, new foods may be seen as being brought in to fatten the tribe's children so they can later be carted off for some ulterior, and definitely sinister, motive. Niehoff advises the officer to have his aide, who should be familiar with the dialect and acceptable to the locals, taste the gossip in the market place or the wash place every morning.

Niehoff's latest work has been an attempt to short-circuit the civic-action field even more. He has been trying to see whether analysis of earlier programmes enables the military to predict, in advance and without any substantial expenditure of time, money or men, whether an innovation will be successful (on the understanding, of course, that successful innovations are what really change attitudes in favour of US interests). On each of his 171 case histories he has collected basic details such as location, project type, whether it was successful or not. He then constructs an equation for each element – either positive or negative. Entries, for example, may look like this:

– *Kinship:* the males in general felt it was not time for women to learn to read. They ridiculed their wives for studying.
+*Demonstration:* women in one village were sent to another to observe the success of the (literacy) programme there.

The project is somewhat controversial since, although in later phases the plus and minus signs are weighted in complicated ways and have to be run through a computer in order to get an answer (i.e. a prediction), many still regard the approach as too simple. The actual results are not readily available and I am not sure that they have ever been used in practice (18).

SOLDIER TO SOLDIER

A special effort of the army's researchers has also been directed at the kinds of contact between American soldiers and the military of the foreign country where the stability operations, or counter-insurgency fighting, were actually going on. Clearly an important yet possibly very touchy area. Dean Froehlich has been HumRRO's expert in this field, and throughout the late 1960s and early seventies he made a series of surveys of military advisors and their counterparts in the countries of the Far East, notably Korea and Taiwan. Froehlich found from his studies that it was important for the advisers to have four personal characteristics which he judged as effective (19). These four were trustworthiness, enthusiasm, competence and thoughtfulness. The more the westerners could trade on these aspects of their own personalities, Froehlich found, the more they could personalize their relationships with the host army and offer advice which would be accepted.

These four characteristics are effective, concludes Froehlich, because his studies showed that the military advisers' counterparts in the host army see the adviser's role in the programme somewhat differently to the adviser himself. He is seen merely as a middle man who can provide equipment when it is wanted. The adviser must be able to carry out this essentially subservient role well before any more enduring assistance he can offer will be accepted. Partly this is because the advisers are often junior to, in experience if not in rank, the people to whom they act as advisers. Froehlich also found that the advisers should *never* suggest that their counterparts' information was less good than their own; and that the adviser *must* make an effort to learn the customs and traditions of the counterpart's army as well as of his country, and the different ways in which things are done there.

Even so, differences will remain. For example, according to the advisers, one of the things that most got in the way of the successful resolution of problems was the difference in values between western and eastern cultures. To the counterparts, however, this was not the view at all: *they* felt that lack of equipment and materials were usually the most serious obstacle to progress. Another difference arose from the identification of new problems to work on: the advisers tended to use their own judgement on this, stemming from their own observation.

The counterparts, on the other hand, were only likely to tackle new things on specific instruction from higher authority. In spite of all this, Froehlich found that about 15 per cent more of the counterparts were favourable in their evaluation of the advisory programmes compared with the advisers' feelings towards them. So these programmes may sometimes be more successful than they seem.

These, then, are the nuts and bolts of the cultural differences that a military officer is likely to come across during an insurgency.

FAKING FOREIGNERS

Two other areas of cross-cultural training, which have been the subject of continuing psychological research, remain to be discussed. First, the use of programmed instruction for the fast acquisition of a tactical foreign language and, second, the use of various 'sensitivity' techniques to show up the differences between cultures.

We have already seen (in chapter 14) the way that HumRRO designed a special 'tactical' skeleton language of roughly 450 words of a foreign tongue so that a soldier in the front line could interrogate – after a fashion – any enemy soldier he came across. But this technique has also been used to give all men (going to Vietnam, say) an elementary grounding in the native language.

Each man was given a dual-track tape with natively spoken Vietnamese and blank periods on one. The soldier would then speak into the other tape during the blank periods and he could then compare his diction with the native speaker's by playing back both tapes in quick succession (20). This procedure also enabled men to carry the tapes with them and in effect eliminated the need for many teachers.

But perhaps most effort has been devoted to 'simulation training'. Trainees are given a role-playing exercise in which they have to interact with someone who deliberately behaves in a 'non-western' way. The problem for the trainee is to learn how to get across to this 'non-western' person what it is that he wants to say. Edward Stewart, again from HumRRO, has analysed the various forms of interaction between people of different cultures (partly by filming such encounters and then examining individual frames) and as a result he has developed a variety of happenings in which a westerner – the military trainee – comes into contact with what Stewart calls a 'contrast American' (21). His early experiments showed that cultures differ in the way they treat such things as time, the use of cause and effect (some make much more use of metaphor), the extent to which someone sees himself as an individual or as a member of certain groups, the extent to which he uses inductive or deductive patterns of thought and so on.

Later experiments explored the different emotional expressions of various cultures (to Latins and Arabs, Europeans are cold, but to Chinese and Thais they are volatile); and modes of thought – for example, Arabs habitually conceive the world as abstract first, then work towards the concrete, whereas the Chinese may attach almost cosmic significance to a single event. Stewart found it necessary to portray two types of contrast American – the abstract (or Arab) type and the concrete (or Chinese) type. His idea was that the events in his exercises would be so well chosen, on the basis of research, that only a few would be needed to train the average officer in the way other people's behaviour, attitudes and thinking patterns might differ from his own.

He devised a series of sketches (ranging in length from ten to sixty minutes); two will be described here to illustrate the sort of thing he included as significant:

In one, the American is given a description of the poor health conditions that exist in a particular locality. He is asked to speak to his counterpart, the contrast American, about developing a health project to improve the welfare of the people. The approach adopted by most Americans is usually to mention a number of observations made about polluted drinking water, the disposal of human waste and so on. In enumerating all these pointers (to western eyes) for the necessity of health measures, the soldier usually assumes that the counterpart is influenced by the enumeration of facts, and would consider that public health was one of his responsibilities. None of these assumptions was, however, shared by the counterpart (played in this case by a Turk, to whom some of the different assumptions came naturally but who also received specific training in 'contrast American' behaviour). As a matter of fact, the counterpart acted puzzled by the American's concern with living conditions that had always existed and were a natural part of people's lives.

About halfway through the encounter, the American senses his mistake and changes his approach. Now he starts raising the subject of sickness in the abstract and then carries on to deduce the need for a medical officer, a man of great prestige in the eyes of the contrast American. This was a much more successful approach using the mode of thought more common to the contrast American to achieve the aims of the US army.

In another case, there is disagreement over the use of publicity for a project:

The American wants to take some photographs of a particular field before and after the villagers have produced certain crops. The counterpart, however, is against the idea. To photograph barren fields is not seen as good and to photograph full fields for publicity is seen as showing the area as greedy and boastful. The point here is that the local culture may react strongly to the contrasts, to the fact that they are being shown to have converted barren fields into full ones – this is seen in the culture as greedy, to show this off. On the other hand, *simply* to show the full fields enables others to judge the contrast for themselves and is far more acceptable.

Having seen a couple of these encounters, the trainee has to play the role himself and actually learn to feel his way through a problem and convince the contrast American of his case.

In this fashion the soldier can gently but economically be made sensitive to the snares in other cultures and become a more effective communicator in the process, better able to fulfil his military mission of winning over the civilian population to the side of the counter-insurgents and to the detriment of the guerrilla groups.

A final way to emphasize to Western soldiers the difference between Eastern and Western cultures was tried out by the Department of Defense's Advanced Research Projects Agency (ARPA) (22). This consisted of sentence completion tests in which the same sentences had been completed by both Vietnamese and Americans so as to show the fundamental psychological differences between them. The results were circulated to psywar officers in the field in a newsletter used to keep them up to date on information they might need to attack the enemy psychologically (see chapter 21). Some of the sentences are reproduced below, each with two possible replies and the percentage of each nationality agreeing with that ending:

Social behaviour is best corrected by . . .'	'self-realization' US 71% VN 9%	'parents, teachers, leaders' US 24% VN 71%
'The changes that most hurt society are . . .'	'those that mean we lose our rights to make our own decisions' US 57% VN 16%	'those that destroy respect for the old way of life' US 4% VN 49%
'You should vote for the candidate with . . .'	'the most similar thinking to yourself' US 75% VN 34%	'the longest experience in office holding' US 2% VN 50%
'We should work together for mutual profit because . . .'	'combined effort gives more control' US 83% VN 29%	'the group bears bad luck more easily' US 3% VN 52%

All the techniques discussed so far are essentially ways of helping the military get on well with the soldiers and civilians in the foreign country wherever they happen to be. Just as important from a military point of view is the rather more aggressive persuasion that has to be directed at the active fighting units of the guerrilla armies. This is what chiefly is meant by 'psychological warfare' and it is to this that the following chapters are devoted.

19 Selection and Training of Personnel for Counter-Insurgency

The establishment of the special forces school at Fort Bragg was a direct result of President Kennedy's belief that insurgencies were a new type of war and that a new type of soldier was needed to fight in them. General William P. Yarborough, a commander of the school, has claimed that, had Lieutenant William Cally, the officer nominally in charge of the My Lai massacre, gone to Fort Bragg, he would 'never have gotten into combat'. Whether such a claim can be justified or not, a major thrust of the military interest in special operations has been to develop selection procedures for soldiers who will have to serve overseas in an insurgency (1).

In this chapter we look at the various methods used to select and train men in aspects of counter-insurgency techniques – counter-intelligence, behind enemy lines, and psychological warfare.

COUNTER-INTELLIGENCE

The counter-insurgency research which we examine later in this chapter appears to have grown out of the work on the selection of intelligence personnel which was carried out in the early fifties. At that time it was noted that people available for counter-intelligence work were of markedly lower calibre than during the Second World War. As a result the US Adjutant General's office was authorized to start a selection programme for these specialists.

The experimental design was fairly simple to begin with. Full biographical details were collected on all men and officers in the Counter-Intelligence Corps (CIC) at Fort Holabird, as well as on men then serving as agents either in the Zone of the Interior (ZI) or overseas. Over 600 men took part, including men assigned to the Far East, Germany and Austria. The test used were as follows: word-fluency – the soldier has to list as many words as possible that start with a certain letter in a given amount of time; knowledge of slang (example: 'B**F' for 'complain'); related forms – the applicant has to judge whether various geometrical forms are related or not; inspection speed – the examinee has to indicate whether forty-four pairs of forms and designs are the same or different; army clerical speed – lists of digits and reversals, the examinee has to say whether reversals are correct; broken words – twenty-six broken words, to be filled in; broken pictures – thirty broken pictures, the examinee identifies complete picture from a range.

Other tests were more directly related to intelligence work. There were, for instance, nine observation tests all presented on a continuous strip of 16-mm film. Each test is separated by a piece of blank film. The first three are visual only, the fourth is visual and auditory, the last five auditory only. After each test the examinee answers questions to test his powers of observation and recall: recall for some tests is delayed until after the next test and questions on it. This is devised to see which examinees are most affected by 'retro-active inhibition' (interference).

Although some of the tests helped to predict which men would do well in counter-intelligence school, it turned out that these men were not necessarily the ones who would do well in the field (they were all rated on at least one investigative assignment) (2). So Fort Holabird went ahead with further research to try to improve its intake. By 1957, when the next study was made, admission criteria had changed (3). An applicant had to be a US citizen, of above average intelligence on a general technical test, at least twenty-one-and-a-half and have a score of 1 (the lowest) on a test of psychiatric disorder. At the time the CIC was getting people who did not last the training course. A number of bases had had to supply the CIC with a certain quota of men. The HumRRO research showed that the requirements were so high that not all bases could meet their quota, and so the office recommended a change in the quota policy and recruitment now depends on scores on the tests listed above. Since then the recruitment of CIC men has become much more efficient and has apparently produced a better service.

Currently, the Army Research Institute is trying to understand the various psychological functions in intelligence gathering so as to be better able to teach recruits to this service.

In Britain the intelligence corps has found that it often attracts the 'James Bond' type and the psychologists now grade all applicants on a scale of 1–5 according to how much of a 'cowboy' they are. Only grades 1 and 2 get in. Intelligence work, the British psychologists say, involves long periods of boredom and tests of the ability to withstand this are now part of the induction procedure.

BEHIND ENEMY LINES: COUNTER-INSURGENCY AND SPECIAL FORCES

One of the earliest attempts to develop psychological selection methods for dangerous counter-insurgency missions was carried out in the second half of 1961, when the US Air Force called for volunteers – officers and airmen – to take part in counter-insurgency training (4). The Air Force at that stage had relatively little experience in this area and took a fairly arbitrary decision that the men selected would have to meet three criteria: be proficient in their particular military speciality; meet the necessary security requirements; and have an exceptionally 'stable and well-integrated personality'. It was this third criterion which created most bother and a research project was therefore commissioned to explore what personality would be most suited for these hazardous and isolated missions.

The psychologists had no real criteria: their candidates had not had the opportunity, at that stage, to fight in a guerrilla war. In the end, therefore, all they could do was to interview the men, assess how well they thought each would do in a counter-insurgency conflict and then compare these ideas with the men's results on the many personality tests which they were given.

The psychologists tried to measure the men's feelings of responsibility; their ability to maintain good relations with others; how original their thought was; whether they were cautious or not; how energetic they were; how willing to take risks; they even gave them a 'test' of 'success motivation' which consisted merely of the instruction to the men to hold out their arms for as long as they could, on the reasoning that the longer a man did this the greater his desire to succeed. Finally, the men were also given the more familiar clinical personality tests such as the Minnesota Multiphasic Personality Inventory (MMPI) and the Holtzman Inkblot test.

And just as well, too. For the MMPI proved one of the more useful tests. Given that this project was somewhat artificial anyway, the psychologists could not be dogmatic about anything but found that the interview performance of the men compared best with six of their tests. What appeared to count was the man's level of responsibility, his leadership abilities, his willingness to take risks, and the fact that he never got very depressed, rarely complained of minor aches and pains and finally was not too bothered about right and wrong, being in general more pragmatic than highly moral. It was these last three aspects which were all measured by sub-scales of the MMPI.

That was a beginning. The US Army, though somewhat behind at this stage, soon followed up with more – and better – studies (5).

The experience of Fort Holabird in attrition among counter-intelligence personnel had also been felt by Fort Bragg in the selection and training of special warfare men – soldiers who fight guerrillas or are dropped in behind enemy lines. The Fort Bragg people called in the US Army Personnel Research Office (USAPRO). The USAPRO staff looked at a whole series of studies already completed, into target training, Arctic exercises, jungle manoeuvres in Korea. Between these, the researchers felt, all situations encountered by special forces were covered. And between them they identified eight basic conditions which challenge the special forces officer: extreme fatigue; the necessity of performing in an unstructured situation; conflict situations; need to accept training as real combat; performance with no knowledge of personal progress; inability to leave the course voluntarily; necessity for teamwork; need to observe and retain certain military information. The research also isolated ten 'constellations of reactions' which 'appeared to have potential for differentiating the poor from the adequate fighter': malingering; lack of social responsibility; an attitude of martyrdom; unauthorized withdrawal; hostility; fear of injury; uneasiness over the unknown; psychotic-like reactions; failure to assume the combat role; inability to follow instructions.

In early 1961 a selection battery was developed and validated against success in special forces training (it was officially implemented the same year). The increasing need for special forces in the early 1960s meant not merely that more were needed but that more were dropping out – the attrition rate at times was 70 per cent. The selection battery consisted of four measures which in combination provided a means of effectively screening out men who were not likely to succeed in special forces training (special forces officers were all volunteers). The components of the battery were validated against performance in field exercises during training. Scores on the following nine performance tests constituted the criterion measure:

Weapons: assembly, disassembly and use of infantry weapons.

Communications: using a transceiver to open a communication net, transmit a message, receive one, and close down the net.

First-aid: suitable application in conditions such as gunshot wounds, drownings and other 'accidents'.

Survival: demonstration of proper methods of living off the land, etc.

Land navigation: plotting a course on a map to include four points over a distance of five miles and to traverse the course on foot, touching each of the four points, within a given amount of time.

Demolition: identification, selection, placement, priming and detonation of various explosives.

Organization and development of guerrilla forces: use of a sand table to select proper locations and lay out a complex guerrilla situation.

Aerial re-supply: behind enemy lines, using sand table.

Guerrilla tactics: demonstration of how to plan and execute raids and ambushes.

Three of the experimental tests were found to be good predictors of performance on these nine criteria at the end of training (they are also relatively unique in the psychological dimensions they measure). Used as a battery with the infantry aptitude test, they give a multiple correlation coefficient of 0·63 – very high. The three were: the special forces suitability inventory, designed to tap aspects of the personality thought to make one suitable for special forces work; the critical decisions test, which is a measure of risk-taking or chance-taking tendency – few facts and limited time for deliberation characterize the test which is presented by tape recorder and test booklet; the locations test, a measure of ability to perceive space when actual terrain features are used as the visual stimuli. The examinee must orient himself in photographs.

This technique was effective in that 46 per cent more men on the special forces training course stayed to the end and passed out.

The next step was the decision by General Yarborough at the special forces school in Fort Bragg to undertake more research on the selection of special forces *officers*. As with its research into special forces ratings, USAPRO first tried to outline the psychological requirements of the job. Previous research was examined, as were the eligibility requirements, the training programme at Fort Bragg and reports of special missions. Conferences were held with special forces personnel who had been serving in Vietnam. Officers' academic averages were looked at, their performance on a two-week field exercise, their peer ratings at the end of their course. They were then evaluated twice after graduation – at nine and eighteen months – actually in an operational environment. Each man was rated by two immediately superior officers on thirteen scales designed to measure his competence. Tests used included: the officer qualifying exam, which was mainly academic; the army language aptitude test, which measures the ability to learn the vocabulary and grammar of an artificial language; preference for army duties test, which measures interest in five broad aspects – medical, mechanical electronics, clerical and hazardous duty. Three personality tests were used: a personal profile; a survey of interpersonal values (measuring conformity, leadership, independence, cautiousness and emotional stability); and the orientation inventory, which looks at the attitude an individual has towards work – getting the job done, personal rewards and satisfaction, being concerned with getting on with others on the job. Finally, the three tests from the special forces selection battery were also used – the suitability inventory, the critical decisions tests, and the locations test (see p. 372).

The battery was given to two special forces officer classes during the first week of their training – 238 officers in all – in 1964. Against classroom grades the officer qualifying exam and the language aptitude test gave a correlation overall of 0·59, fairly good. Three other tests, the suitability inventory, the emotional stability scale and the interaction orientation scale all correlated significantly but more modestly. But after nine months in the field it was found that the army qualifying test and the language aptitude test, the best predictors of classroom success, had no validity at all in predicting field performance. On the other hand, certain personality measures *unrelated* to academic success were now significantly related to later field success. The personal relations, support and cautiousness scales all predicted performance well (a negative relationship in all cases – the more cautious the worse the performance). The special forces selection battery also yielded a significant correlation, and for the most part attributable to its personality section.

Broadly speaking then the results of the tests show that the successful special forces officer tends to be 'psychologically self-sufficient, and not overly trusting of other people, at least somewhat willing to take risks, and generally well adjusted'. Other tests showed that he tends to be assertive and vigorous as well. Further useful predictors were found to be the man's general training score and his peer rating.

Francis Medland, of USAPRO, who headed this study, concluded that it is possible to predict the efficiency of special forces officers as they begin their training. But the research is now looking at a wider cross-section of follow-ups in various countries at different points in time. There is also a special study of the men who are (or were) successful in Vietnam, but results of that study have not yet been made public.

Britain's nearest equivalent is probably the SAS, the Special Air Services. Nineteen out of every 100 applicants to join the SAS actually make it. They are given computational tests, the 16 PF (a well-known personality test), and a psycho-dynamic test. The psychologists look for those who, on the tests, are: above average in intelligence; assertive; happy-go-lucky; self-sufficient; not extremely intro- or extraverted. They do *not* want people who are emotionally stable; instead they want forthright individuals, who are hard to fool and not dependent on others. The psychologists do acknowledge that occasionally, with the SAS, there are problems of too many chiefs and not enough Indians. The Royal Navy also use psychologists in choosing men for hazardous missions. Their approach is similar, but they also select for 'Machiavellianism' as they regard this as a possible help in some of the situations in which the men are likely to find themselves (6).

SMALL INDEPENDENT ACTION FORCES (SIAF)

The most recent work on selecting and training for counter-insurgency has been concerned with the role of the Small Independent Action Force (SIAF) in which the conventional soldier is faced with fighting in guerrilla action. (The teams studied by Dr Peter Bourne, and referred to earlier in chapter 9, are an example of SIAFs in action.) The definitive available work was carried out by HumRRO at Fort Benning, and was completed in 1972 (7). The main focus of the study was to examine the physical and psychological stresses under which SIAFs operate. In the first instance the research unit collected details on missions already accomplished – mainly in Vietnam, but also from allied forces such as Thais in Thailand, and the Australians and British in Malaya. Analysis of these missions led the research team to the view that there were twenty-five types of skill needed in a SIAF unit (see Table 30). Further analysis of the comments of commanding officers returning from these units stressed that in some of them training was inadequate for the job. The critical areas identified in these ways were: land navigation; delivery of indirect and aerial supporting fire; use of camouflage, cover, concealment, and stealth; human maintenance; tracking; and communications. Research, therefore, concentrated in these areas. (A final move before beginning the experiment was that a number of crack squads were checked for their entrance requirements.

Table 30 The twenty-five basic skills needed in SIAFs

Land navigation	Civic action, language and training of indigenous forces
Delivery of indirect fire and aerial support	Use and detection of mines and boobytraps
Use of camouflage, cover, concealment and stealth	Combat first aid
Human maintenance	Use of image intensification devices
Tracking	Leadership
Communications	Intelligence
Use of aerial pictures	Mission organization
Physical conditioning and combativeness	Airmobile procedures
Use of individual weapons	Use of small boats and stream crossing expedients
Use of machine guns	Mountaineering
Demolitions	Use of sensors
Use of hand grenades	Patrolling
	Survival, evasion and escape

These were the US Army Airborne Course, the US Army Special Forces, the US Army Ranger Course, the US Marine Force Reconnaissance Company, Special Air Services Regiment (Australian) and 16 Paratroop Brigade (British). The similarities within them served as the basis for the tests used in this experiment).

The army wanted to know if psychological tests could predict performance on these tasks. So tests were given to several groups of potential counter-insurgency soldiers and compared with their performance on three types of mission: a short-range patrol; a long-range patrol; and as an element operating with other indigenous troops primarily in a civic action 'hearts and mind' programme. The two patrol phases of the exercise were divided up as follows: planning and preparation; insertion into the area; deployment; debriefing and critique. The civic action phase was structured in this way: planning and preparation; entrance into village; securing village; training indigenous personnel; defence of village; civic action.

The exercises included the following specific types of manoeuvre:

Patrolling: ability to inspect and detect discrepancies in a patrol member's camouflage; ability to assume proper battlefield positions (behind trees or low bush); ability to handle prisoners of war according to the Geneva Convention; ability to select patrol routes from a map.

Target detection techniques: ability to detect enemy's personnel on the basis of movement, noise, reflection, and poor camouflage; ability to locate stationary and moving targets (personnel) at ranges from 50 to 150 metres; ability to estimate the range to a target.

M16A1 rapid reaction range firing: ability to successfully engage at close quarters (35–100 metres) surprise targets while negotiating difficult terrain and man-made obstacles.

Against these criteria (there were many more than those listed above) the performance of the soldiers on the various psychological tests was compared. Seven tests were used: first was the interest opinion questionnaire which sampled the soldier's general intentions in life, his personal history, his sense of humour (by his reactions to several jokes), and his 'self-concept'. (In several studies this questionnaire has been found to distinguish effectively the good fighter from the poor.) Next used was the life history inventory, mainly inquiring into the soldier's family and religious background, his social and educational experiences, hobbies and childhood behaviour. This, too, has been found to differentiate fighters from non-fighters. Then came the military interest blank, mainly aimed to find which aspect of military life most suited a man but which also explored the soldier's attitudes to various military situations including garrison life. The fourth test was a specially constructed one, the SIAF activities inventory, designed to measure two contrasting attitudes – confidence and despair in situations where there is the possibility of danger. The questions were so designed as to measure the two types of danger – specific dangers and background danger – which a soldier is liable to face while on active duty.

These and other tests were given to 100 soldiers at the special warfare school in Fort Bragg and they were compared with 100 control men from the 82nd Airborne Division, who were not special forces personnel. Before men were allowed into the experiment they were seen by a psychiatrist to eliminate all those who suffered from any mild form of psychological illness such as sleeplessness, claustrophobia, severe night fears, fear of the dark.

The study showed that the special forces performed much better on all the relevant categories (like navigation and throwing bombs) although they were not too good at the test of climbing ropes. The tests were therefore judged as good – they distinguished special forces from ordinary ones. The next step was to reduce the number of tests to a manageable size. At the latest stage twenty-two sub-tests had been reduced to thirteen:

Auditory number span
Embedded figures
Verbal classification
Word grouping
Word-number
Life history fighter score
Physical endurance
Military interest fighter score
Background despair score
Team task motivation
ACB arithmetic*
ACB army clerical
ACB automotive information

(* ACB = Army Classification Battery.)

Because of the success of this programme, research is now going on to develop the same capacities for special forces acting in urban, Arctic and desert environments.

FORT BRAGG FIELD TESTS

The US Army's psychological warfare school was set up in 1950 at Fort Riley, transferring two years later to Fort Bragg near Fayetteville in North Carolina, which has been its home ever since. It is now the most sophisticated institution of its kind in the world. To begin with, the school mainly considered its role in total war, but as time went by it became more concerned with unconventional warfare and in 1961–2 the counter-insurgency function was also added. The school has three official missions: to provide resident and non-resident instruction (involving hundreds of people annually); to contribute to psychological warfare doctrine, tactics and techniques; to prepare psychological warfare training materials such as field manuals.

The school offers extended courses for officers (both from the USA and allied armies – particularly Britain, Australia and Thailand), short courses for enlisted men specializing in printing (how to forge banknotes) and photography (night techniques, for example), and supplementary courses for other institutes, including training for police officers from the Third World.

The officer psychological warfare course, which in some ways is the most important thing Fort Bragg does, is divided into five sections. First, the plans and operations department outlines the way persuasion campaigns should be coordinated – when leaflets should be used, or forged money or armed patrols; the social science section explains how psywar material should be tailored to an area, how audiences differ in their social and psychological make-up, what symbols or allusions apply in what circumstances; the command staff section deals with how best psywar units are to be organized, what the optimum size is, how near they should be to one another, how long it takes to produce a magazine and how many men are needed; and the propaganda department explains how propaganda is written and tailored for different audiences.

Most of what is taught in these sections of the course is covered in the following chapters, in some cases in as much detail as the officers themselves get. But the final part of the course is worth dwelling on now because this is a practical field exercise in which the men actually get the chance to try out what they have learned on an ordinary population of civilians.

To the north of Fort Bragg, at the top end of North Carolina, the psywar school has invented a mythical country – 'Pineland' – which takes in five counties of the state. In the neighbouring country, 'Satilla', a number of soldiers pose as 'guerrillas' camping out and from time to time leaving their mark on the Pineland community by some such stunt as leaving a 'revolutionary' poster on a wall. The job of the trainees is to alert the locals, using psywar techniques, to the presence of the rebels and then to persuade the civilians to look out for them and disclose their whereabouts to the trainees, thus leading to the rebels' capture. The exercise lasts ten days and begins with the officers parachuting into the area and then using all their skills to persuade the local population of their case. The men must organize rallies, write leaflets, make broadcasts from helicopters, and so on. Although the test is not real in the sense that the conflict between the trainees and the 'guerrillas' is not vicious or ideological, the communication problems the trainees face in persuading the locals to take part are real enough and give the men, according to Colonel John Howard, director of the school at the time I visited it, a good go at learning their future trade.

At the end the number of guerrillas caught and the survey results showing how many of the local population have heard about or been moved by the trainee officers' techniques and campaigns determines whether the men pass their course or not. Enterprising as this may appear in comparison with other psywar training courses, an army team which examined the impact the training had on the men's performance in Vietnam found that it was too academic (8) (see pp. 385 and 428). It can be expected, therefore, that the course will be changed somewhat in the future. For comparison, the British course at Old Sarum, near Salisbury in Wiltshire, is modelled on the Fort Bragg course, but lasts less than two weeks.[1]

1. Specialists may like to know also that, despite being modelled on Fort Bragg, the Old Sarum course is highly secret – the Ministry of Defence will never even concede that psychological warfare is so much as considered by British forces. Yet one restricted British document I have seen shows that an average course consists of some fifteen or sixteen members, including one from the Green Jackets, two from the Ministry of Defence, one from the Foreign and Commonwealth Office, a Squadron Leader from RAF Abingdon, and captains from the Royal Marines and from the Royal Artillery. This particular course was addressed by Mr Keith Belbin, of Coleman, Prentice and Varley, the advertising agency, who spoke about recruitment, Peter Bartlett on target analysis with special reference to the use of Hong Kong by China. In addition, R. M. Farr spoke on attitude change – Mr Farr being a psychologist and for a time an official of the British Psychological Society. But perhaps the star was Lt Col B. R. Johnston, described as the 'foremost British authority on military psyops'. He spoke on 'military information policy in low intensity operations' – mainly on Northern Ireland (9).

20 Psychological Warfare: Operational Organization

Just as America's psychological warfare training is the most elaborate, so is the operational deployment of its men. Other countries, France, Britain and Israel, for example, cannot compete in terms of numbers. They also tend to be more secretive about military affairs in general, so what follows is a rather one-sided account.

THE FRENCH IN ALGERIA

French psychological warfare was most visible during the Algerian campaign. The French had a five-pronged policy. Top priority was protecting their own forces against the insurgents' propaganda (which shows how worried they were about the logic of the insurgents' claims). Second came attempts to destroy the guerrillas' political network and that of sympathetic organizations. Next came the re-education of captured rebels or the 'disinfection' of prisoners as it was then called. The organization and education of the local population came at the bottom of the French list, possibly a mistake in the light of subsequent history.

The Algerian insurgents themselves had a rather amorphous body known as the *Service Psychologique du FLN* which was organized into three sections. One sought to conquer the civilian population, mainly by forcing or tricking them into attending illegal meetings, engaging in minor acts of sabotage and so forth and then blackmailing them into other, more serious acts (1). A second section attacked the morale of the French army, mainly by making acts of sabotage so obviously against the army rather than against any other groups, that the division between the army and the civilians was widened. A third technique was to try to gain foreign support for their cause.

BRITAIN

Three documents I have been given access to convey the range of British activities and give some flavour of the British attitude to this end of the military spectrum. The first is the paper prepared by F. H. Lakin of the Army Operational Research Establishment which describes the British psychological warfare research in Malaya between 1952 and 1955 (2). This is a very useful and detailed document (see chapter 17) and we shall be returning to it again later. For the present, however, we are simply concerned with the light it throws on organizational matters.

Lakin was in charge of a nine-man research team, jointly responsible to the Army Operational Research Establishment in Britain, and the Research Division of the Director-General of the Information Services of what was then the Federation of Malaya. Second-in-Command to Lakin was another psychologist. In addition there was one Chinese-speaking British officer, two British scientific assistants and four Chinese information officers/data collectors. For six months two men from the Operational Research Office of Johns Hopkins University, Maryland, were attached to the team, plus an Australian army psychologist. (ORO carried out the mammoth American study of Korean insurgents and US soldiers fighting there referred to earlier in chapter 5.) As early as this, then, the collaboration between allies was well advanced. Although Lakin's team was only nine strong, its research was extensive and highly useful.

The second document concerns the operational organization and deployment of British psywar units. Here, too, though combat psywar units may be small, in terms of numbers, the British army expects their influence to be quite extensive (3). The basic unit is designed to be independent, with one officer (a major or a captain) and twelve other ranks, plus any necessary civilians. These units are equipped with vans and landrovers to carry loudspeakers, tape recorders and cinema projectors, but organized in such a way that they can also split up and act in three smaller independent teams. A major difference between the British and the Americans appears to be that a main task of the British units is to produce material favourable to the army for use with the local *civilian* press rather than simply through the army's own leaflets. This, no doubt, has something to do with British ideas about credibility (and it should be said that, in spite of Vietnam and the huge amount of experience that war offered, all American officers I have spoken to have the highest regard for British experience in this field – Malaya, Aden, Kenya and Cyprus being an impressive list of insurgencies). The British seem to be stricter than the Americans about the use of civilians: the British only allow civilians in their psywar units as either interpreters or artists (for the design of leaflet symbols and so on), though the documents do advocate bringing in ambassadors, police and special branch representatives or their equivalents (4).

The British role for these units is also designed somewhat differently from the American version of psyop. The British prefer to send in their units after a *successful* military strike, whereas the Americans make more continuous use of theirs. But the British units also have a consolidation role of rallying support when there is no military action. Here again, however, it appears that the British rely more on face-to-face contact with the locals than the Americans do, and more on feeding material to the local press than publishing it themselves. (Though, according to a *Guardian* report in February 1977, the British Ministry of Defence did, by implication, admit to a 'black propaganda' role in Ulster: 'All statements issued by the army press desk at Lisburn [Ulster HQ]', ran the report, ' . . . are being vetted by the Secretary for Ulster, Mr Roy Mason, to stop the use of "black propaganda" by soldiers.' It seemed, from the report, that the general approach at a quick propaganda response by the army was not working (5).)

Lt Col B. R. Johnston, Britain's foremost psywar expert, distinguishes three phases of an insurgency, according to a classified talk which he gave at the Joint Warfare Establishment at Old Sarum (6). The first of these is the identification and isolation of hostile elements; the second, he says, is the elimination of the rebel elements and the projection of a favourable image of the British forces; and the third is the consolidation of these gains, negotiation towards amnesty, and the winding down of hostilities. Active psywar comes in at the end of phase one. At the time of this talk (1972), the war in Northern Ireland was in phase two, according to Johnston. And at this time, according to Kitson, Britain had an active psyop unit in service – almost certainly in Northern Ireland.

In bemoaning Britain's backwardness in psyop generally, Kitson laments that there is only a staff of eighteen to teach it. But even so, two hundred and sixty-two civil servants and 1858 army officers had been through Britain's psyops course at Old Sarum by October 1976, and visits are also made to several Commonwealth countries. Still, the British government is very cagey about its psyops work. For instance, according to army minister, Robert Brown, the Old Sarum course only started in 1973–4, but two of the documents referred to above were seen by me *before* that time.

But the British documents do talk about exploiting the 'psychological vulnerabilities' of the enemy – the American approach is shared, even if the financial and organizational commitment isn't.

OTHER COUNTRIES

Little is known of the organization of psywar teams in other countries. They are unlikely to be as imaginative as the Americans but, according to one military authority who should know, Frank Kitson, several countries have a rather more elaborate psywar paraphernalia than the British. The West Germans, as an example, maintain psyop units totalling 3000 men in the regular army alone – plus reservists. The Italians have one full-strength psychological warfare company in their order of battle and even the Greeks and Turks, according to Kitson, employ around 300 men (7).

Outside Europe the Brazilian armed forces have a guide, *Guerra Psicologia*, which includes three annexes on psywar plans, leaflet bombs and various tables and charts for leaflet dropping. This guide also outlines the Brazilian requirements for a psywar specialist and how to defend Brazil against psywar from other countries. It appears that the Brazilian army, unlike many other countries, can direct psywar at its own population as well as foreigners (8).

Rhodesia and South Africa, of course, have fairly sophisticated international propaganda machines but they have military outfits, too. In Rhodesia there are several small 'psychac' teams, highly mobile and seemingly very imaginative (see p. 414 below). In March 1977, a colleague on the *Sunday Times* revealed that the South African army had been issued with a 'psyop guide'. Compared to much of the American material this is fairly rudimentary. It cautions soldiers to behave well and avoid accidents – lest these alienate the blacks. It also offers some ideological indoctrination for the enlisted man to help sustain his sense of purpose. But the fact that this guide was widely issued shows that even in South Africa there is more commitment to psyop as a military activity than in Britain (9).

Many Third World countries now have a good idea of the scope of psywar. Mainly, they have gained this through officer-exchange programmes. The International Police Academy, housed in an old tramshed in Washington, DC, was probably visited by more foreign security personnel than any other single agency before Congress closed it down in 1975. Most of its visitors were policemen; but they all spent several days at the psyop school at Fort Bragg as part of their course, so many senior soldiers from these countries will have done the same. The international scholars came frequently from Thailand, Iran, Venezuela, Vietnam and Lebanon, plus a few from Egypt, Saudi Arabia and Jordan (10). Their theses – on interrogation, the vulnerabilities of guerrillas, the defection of security risks – almost invariably showed an interest in and concern for the psychological side of things. This interest is bound to go on growing.

USA

In comparison with the preceding limited psywar forces, the American edifice is huge and exceptionally elaborate. Besides housing the psyop school, Fort Bragg is also the home of the 4th Psyop Group which consists of three battalions – two strategic and one tactical. In Vietnam it played a major role in support of both American and Vietnamese forces. Its various units are ready to go anywhere in the world, at short notice, to support US or allied forces in any type of military operation, conventional or unconventional. To ensure this capability, the Group maintains detailed *Psyop Estimates* and contingency plans on the world's potential trouble spots. All personnel are trained to enter operational areas by conventional and unconventional means, such as parachute, underwater (using scuba gear) and even cross country on skis (11).

The 4th Group specializes in the Middle East, Africa and South America. There are other Psyop Groups in Germany, covering East and West Europe and Russia, and in Okinawa, covering the Far East. Capabilities also exist in Thailand to cover the Indian subcontinent.

The 4th Psyop Group is organized and equipped to conduct psychological warfare in support of an entire theatre of operations. The resources of a strategic psyop battalion include such things as a transportable 50 000 watt AM radio transmitter (for long-range broadcasting) and a mobile radio-receiving and monitoring station capable of listening to almost any broadcast in the world. The battalion also has a transportable printing plant which can produce no fewer than 800 000 leaflets per *day* in colour, professional quality magazines, brochures, pamphlets, posters, banners and even books. There is a graphics section and a photography department. Its propaganda development teams are trained to develop written material on any theme or message. Its intelligence section is able to analyse the potential audience for messages and arrive at a conclusion as to what approach and method is best suited to that audience. There are enough linguists in the battalion to translate the material into any language or dialect.

The tactical battalion provides shorter range psywar at corps or division level (roughly 15 000 men). Its equipment, however, is smaller than that for the strategic battalions, and can be transported by air. It has van-mounted printing plants, mobile graphic and photographic laboratories; it also has van-mounted units which can run films, slides, and tape, and has loudspeaker equipment, including some which can fit onto helicopters and aircraft. The battalions are divided into companies and smaller sections organized along cellular lines (that is, specializing in, say, photography or intelligence gathering) so that units can be tailored to the specific needs of particular operations.

The psychological warfare techniques used in Vietnam were among the most sophisticated ever developed, and we shall be looking at them in more detail a little later on. But towards the end of 1968 the Army Concept Team in Vietnam conducted an evaluation of the psywar units there. This, thus, makes it the most realistic picture of the way psychological warfare was actually waged in that country.

According to the Concept Team's report (12), the expansion of the 'US and Free World Military Assistance Forces' in Vietnam occurred in the mid-to-late 1960s. The team found that there was quite a bit of confusion. The 4th Psyop Group was based at Saigon (see Figure 51). In addition, there were psyop units attached to many other forces. For example, the special operations squadrons had their own units, as did at least three infantry divisions, one airborne division and two Marine divisions. The 7th Psyop Group, headquartered in Okinawa, provided many back-up printing facilities, and many smaller units had their own psychological warfare capabilities. To make matters worse, one office – the Joint United States Public Affairs Office (JUSPAO) – had authority over psywar *policy* in Vietnam, whereas another – the Military Assistance Command in Vietnam (which came directly under the Ambassador) – had responsibility for the way it was carried out.

Not surprisingly, in view of this, the Concept Team found that there was quite a bit of duplication of effort in the psychological warfare field. It also appeared that although the cellular concept was theoretically valid, in practice it often acted as a block to communication between units. For example, the concept team found that the propa-

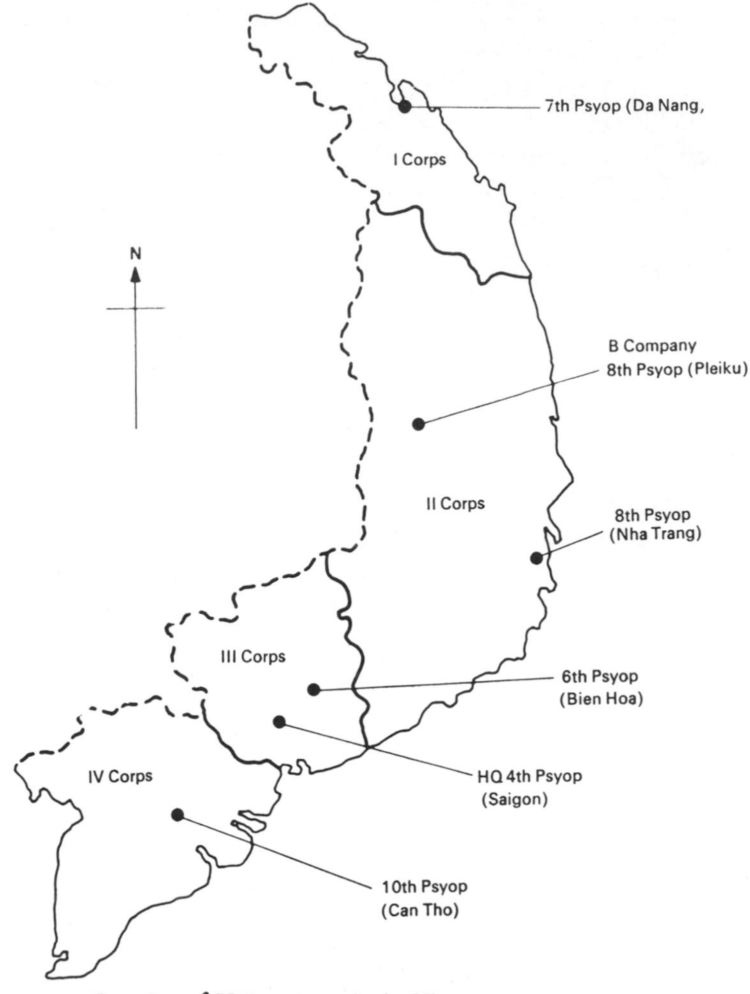

Figure 51 Location of US psyop units in Vietnam.

ganda research and analysis units, though capable of producing what was wanted, were rarely told by other units how much to produce or the geographical limits of the target population. Other worrying findings were that although the proportion of officers with Fort Bragg psyop training had risen from 50 per cent to 80 per cent in a year, the training was considered still too academic and not practical enough in such aspects as actually disseminating propaganda or even in language training.

The attitude of many soldiers to psychological warfare also left a lot to be desired. The team considered that many commanders of the small tactical units lacked understanding and were seen to be more interested in body counts than anything else. One psyop commander complained that his unit was only 70 per cent effective because the men were used for other things. Many units used the film and loudspeaker equipment to show films from home or to broadcast messages to their own troops. The attitude of many was summed up, says the report, by the comment of one unit commander who said that his 'Chieu Hoi' (open arms) programme – a defection programme developed by the psychological warfare specialists – consisted of two 105 mm howitzers, one called Chieu and the other called Hoi. Matters were not helped, says the report, by the fact that many soldiers seemed not to realize the bad propaganda effect of being involved in traffic accidents or having a patronizing attitude to the Vietnamese.

In any case, many psychological warfare officers themselves were not as keen on their speciality as they might have been because it was regarded as lying outside the mainstream of career development. As soon as they were in it, therefore, they tried to get out.

However, psychological warfare is probably more effective than many of the Vietnam commanders gave it credit for. (Its use is mentioned frequently for example, in the Pentagon papers.) The more imaginative, and so presumably the most effective, techniques are described in the two following chapters.

21 General Psywar Techniques

There are several ways in which the psywar intelligence officer gets his information: by analysis of the long-term psychological strengths and weaknesses of a nation (see chapter 17); by a constant monitoring of a particular situation to ascertain vulnerable points; and by interrogation of prisoners and defectors. From all of these sources the psywar intelligence officer will glean information which he can then use in propaganda.

PINPOINTING THE VULNERABILITIES

Captured official military documents may give much information about the morale and motivational state of the guerrilla fighters themselves. Social analyses also help in understanding psychological staying power of revolutionary movements. More specific documents or analyses, however, help the psychological warfare specialist to get at what are called 'psychological vulnerabilities' in a particular movement or country which he can then exploit.

For example, a project at the Human Resources Research Institute, Maxwell Air Force Base, Alabama, initiated as long ago as 1951, analysed the letters to the editors of several newspapers in Communist China (1). At this time Chinese newspapers were required to establish 'correspondents' networks' in which the letters columns were used as a means by which a large number of people could carry on a dialogue with the paper. The columns were thus a useful source for psychological warfare information. The HRRI project showed that thirty-seven per cent of the letters were negative to Communist Party and that worries about corrupt officials, treacherous merchants and production efficiency were voiced as often as diatribes against US aggression. In general, criticism of the Communist Party and its structure was indirect suggesting that, at that time anyway, US propaganda should concentrate on such things as production inefficiency and corruption and leave the Chinese themselves to make the logical link – that the CP was to blame for this – rather than point it out directly which, it was felt, would have been counter-productive.

Another example of this type of analysis, also carried out at the Human Resources Research Institute at Maxwell Air Force Base, was an examination of the 'Family revolution in Communist China' (2). This study examined traditional marriage in the country, the new marriage laws which had outlawed arranged marriages or the bartering of brides, and sought to analyse the social-psychological effect which this had. For example, the number of wife murders and female suicides rose at this time, as a result of the tensions due to the new marriage laws. The Air Force, should it ever need to do so, could no doubt exploit this type of difficulty. No propaganda was ever produced as a result of these studies – it was never needed. But one can see from this how it would have been done, had the need arisen.

This type of study was carried out in enormous numbers throughout the fifties and sixties – many for each of several countries which were regarded by the US Army as potential trouble spots. The subjects range from analyses of crime to descriptions of local religions and taboos, to accounts of ethnic rivalries in remote areas. Over the years, they were put together to construct 'psychological profiles' of various countries, updated from time to time, which could be used as field-guides to help the psywar officer know immediately how to take advantage of the people in the country in which he is operating. Table 31 lists when these field-guides of various countries were completed; and gives some idea of the USA's developing interests over the years. All the guides in the list are classified. However, one field-guide that is available (being de-classified on 3 April 1973) was prepared by Alexander Askenasy and Richard Orth in 1970 (3). The guide gives clear instructions on how special forces officers can quickly construct questionnaires, identify proper (random) samples, conduct interviews, solve translation problems. All this is straightforward social science. What is less straightforward are the sections on the use of prestigious persons in the community to facilitate attitude change and of 'unobtrusive measures' of change. The first section clearly identifies the clergy in the country in question as the most accessible prestigious person to a special forces officer. This is in marked contrast to the politicians, headmasters, doctors and so forth, who are normally seen as the influential people in most societies. In this particular country Askenasy and Orth saw the clergyman as someone who was inherently conservative (and therefore predisposed to the major power) and someone whom the insurgents could not properly indoctrinate because he had usually already been indoctrinated. The second type of person identified by Askenasy and Orth was the technical man, the man who believes in technological advancement and spends a lot of energy devoted to popularizing these advancements. These activities, Askenasy and Orth found, gave the man a high credibility and his work brought him into contact with many other influential people. It also made him more sympathetic to the West.

They also dwelt on the use of 'unobtrusive measures' of behaviour. We have already encountered this earlier in the mention of rice consumption to assess whether insurgents are in a particular area. Others that spring to mind are water consumption, if this can be monitored, plus the simple counting of all movements in a village. If these sociometric indices do not correspond to accepted population estimates or if there is no apparent need for such traffic, investigation is in order.

This guide does not enable us to be specific about the country in question, but it does suggest that the US psywar people are interested in the following aspects of the countries listed in Table 31.

Prestigious persons
Common gifts used by people to get to know each other
Waste and disposal patterns
Attitudes to leaders
These leaders' opinions
Well-known people who are 'marginal' in some way – i.e. belong to a racial or religious minority, or who are perhaps physically deformed.

INTELLIGENCE DOCUMENTS

When he finds himself interrogating a prisoner, or about to draw up a leaflet or make some other device, the psywar intelligence officer should be thoroughly familiar with the social and psychological background from which the captured soldier comes or at which the leaflet or device is about to be aimed. However, besides the more general psychological vulnerabilities, just mentioned, psywar organizations routinely prepare other, more short-term aids, which are observations in the same vein but are designed to help the psywarrior take im-

Table 31 US Psychological Warfare Guides: years when prepared

1958	Lebanon
	Saudi Arabia
1959	Burma
	Cambodia
	China
	Egypt
	Iran
	Iraq
	Laos
	Syria
1960	Thailand
1961	Indonesia
	Pakistan
	Turkey
1962	Russia
1963	Colombia
	Congo
	Cuba
	Ghana
	Sudan
1964	Borneo
	Russia (2nd time)
1965	Brazil
	Himalayas
	Kenya
	Sumatra
	Venezuela
1966	Dominica (2 volumes)
	Russia (3rd time)
	Somali
1967	Congo (2nd time)
	Ethiopia
	Nigeria
	Thailand (2nd time)
1968	Thailand (3rd time)
1970	Philippines
1971	SE Asia

mediate advantage of some recent event or trend. In the field, the psychological warfare troops get four types of intelligence document to aid them. These four, which we shall look at in turn, are *Psyopportunities*; *Psychological Vulnerabilities of Enemy Troops*; *Propaganda Highlights and Trends Analysis*; and *Psyop Intelligence Notes* (4).

Psyopportunities

These are mainly classified, though some of those relating to Thailand are available.

Psychological Opportunities is a fairly frequent publication and consists of roughly twenty items which the psyoperators believe can be exploited. Each item is described and a comment added: the comment being usually an indirect pointer to how the episode may be exploited. Here are two items from issue number 38 of the *Thailand Psychological Opportunities* (for 14 August 1967):

Telephones Under A Cloud
The Minister of Communications recently complained that 1967 is the unluckiest year for his ministry. It has been relentlessly attacked on a variety of subjects ranging from bus transportation to telephone services. *Siam Nikorn* [a newspaper], one of its frequent critics, came out with another round of attacks on the telephone services. The paper said that . . . the telephone business and services have plummeted into a paralytical situation . . . . 'This is because the present process of laying cables is not a right one . . . . Applications [for phones] have accumulated as high as a mountain now . . . .'

Comment: Telephone service is important to the economy of any modern nation and, in the case of Thailand, it is also important in combating insurgency. The availability of communications is important, whether it is in urban areas of Thailand or rural areas. One major reason for the success of insurgents in certain areas of Thailand is a lack of communications.

In other words it is in the USA's interest for the telephone service to be improved; it will help the war against the insurgents. Psywarriors should stress how the USA can help in this.

Thai Military Aide Slain
The secretary of the Thai Supreme Command Headquarters in Bangkok was bludgeoned to death and his body dumped on a construction site. Thai police suspect that there were political implications in the slaying of Colonel Chamnong Sagkadul, 54, but declined to elaborate.

Comment: Politics have been notably 'dirty' in Thailand in the past. This is one of the problems which the present Thai government has been trying to resolve, since it gives the political scene in that nation a bad name in both national and international circles. The Thai are not always the passive, delicate people that some imagine.

In other words, show how civilized the government is in trying to improve the image of the nation in international eyes and, by comparison, how primitive and dirty the opposition – or guerrillas – are.

Psychological Vulnerabilities of Enemy Troops

The Rand Corporation, as we have already seen, conducted a study of Viet Cong motivation and morale in the mid-sixties. From time to time, the Corporation provided an update on the study to help psychological operators take advantage of the changes that were occurring.[5] The first supplement to the first guidance they gave, for example, came at a time when the Rand psychologists had studied and interviewed

450 captured personnel. The study was designed to concentrate on the weakness of the VC position, and *deliberately* neglected the strengths. Here are some of the details circulated to psywar troops during the period June–December 1965.

Our general impression is that the trends in VC morale . . . identified in earlier presentations . . . have become even more pronounced. This is also true of the effects of VC recruiting, taxation and population control policies which, as was reported earlier, have tended to alienate the population from the VC. Our recent interviews indicate that . . . at least among those interviewed, VC morale has become more brittle . . .

. . . the VC soldiers must fight more often, take greater precautions against surprise attacks and ambushes . . . increased harassment of the VC units by air, artillery and ground forces is having an effect on morale. Many complained of being exhausted and discouraged by the frequent moves, the disruptions of their rest and cooking . . . they indicated that the VC casualty rate had increased.

Strafing by helicopters continues to be generally effective, although the VC were reported to try to cover their trenches and foxholes for better protection . . . .

The intensification of military activities appears to have resulted in some decline in the VC soldier's expectation of surviving the war. VC cadres noted that in the recent months the soldiers speak more often of dying in the next battle, of never seeing their families again . . . .

The VC troop indoctrination has tended to describe American soldiers as being soft, unfamiliar with the terrain, and unable to wage effective jungle warfare. VC soldiers who have been in contact with US troops note that contrary to what they had been told, the US soldiers were aggressive and would stay to fight even in a tight situation . . .

. . . the proportion of poorly motivated draftees, whom the cadres often describe as being cowardly and unresponsive to indoctrination, is growing . . . among the defectors who have surrendered to the Government of Vietnam in the past six months, a large proportion were youths 15 to 18 years of age, suggesting that the VC have been forced to draft persons below the minimum official draft age . . . there appear to be a considerable number of instances of corruption, favouritism or mismanagement in the VC civilian administration at village levels . . . .

Although many VC and civilians believed Hanoi's claim of a large number of US aircraft shot down, they also noted the ability of the US to continue the attacks . . . many discounted Hanoi's assertion that the raids did no harm . . . in general, however, it appears that for most VC soldiers the attacks on North Vietnam seem very remote and as long as Hanoi claims to be able to hold its own the raids generate little anxiety.

Communist Propaganda Highlights and Trends Analysis

These are weekly documents prepared by the HQ of the 7th Psychological Operations Group headquartered in Okinawa.[1] Propaganda 'highlights' are listed by country, with a comment which tries to assess the propaganda value. Here are two examples (abbreviated):

North Korea: Fisheries Upsurge a Must
Pyongyang domestic (radio) said that the development of the fishing industry in NK, which is surrounded by sea on three sides, is of great significance in improving the living conditions of the NK people. . . . another upsurge in the fishing battle . . . is to bring the true superiority of our socialist system into full play. . . . fisheries must also increase the production of canned and bottled fish and cod-liver oil . . . all red fishermen are urged to rally more firmly . . . .

Comments: Fishermen, like soldiers, must follow the NK mystique and be politically motivated or they will fail to show the true superiority of socialism as applied to fishing. The domestic pep talks on fishing are hard to assess. There is a quality of desperation and urgency about them, but also NK has done many things to improve fishing and the pep talks may be an attempt to bring these efforts together and push the fishing industry over the hump. . . . transport that is co-ordinated with catch, processing and packaging – long an NK problem – seems not to be running smoothly. Local industry is supposed to do something of this, but are the local industries as well built up and as efficient as NK claims?

Not a very conclusive piece of propaganda or of analysis but a good example of what the psywar officer hears about all the time. Clearly, if the North Koreans have to propagandize their own people about fishing, there may be something there for the USA to capitalize on.

The next example concerns an event inside the USA:

Black American–Vietnamese Solidarity
On 15 September, Radio Hanoi in Vietnamese to South Vietnam broadcast a Nhan Dan editorial condemning what was termed the 'murder' of Soledad brother, George Jackson. Jackson was depicted by the Communists as another combatant against racial discrimination whose continued existence was a source of embarrassment to the Nixon administration.

1. This account is based on two of these documents in my possession: issue nos. 23 (of thirty-eight pages) and 38 (of fifty-one pages) for 1971.

Comment: . . . The desired effect of the broadcast was to arouse the South Vietnamese audience – also non-white – to dredge up the negative evaluations of Americans because of racial troubles in Vietnam. Members of the audience may have developed an unfavourable attitude through disagreeable face-to-face encounters, specifically encounters which may have revealed racial prejudice by Americans. . . . the audience relating its own experience with American racial prejudice can thereby relate to the plight of the black American, about whom very few Vietnamese are likely to know anything. More important for Communist propaganda purposes, questions are raised in the minds of the target audience about the good faith the American exhibits by allowing what may seem to Vietnamese to be disproportionate numbers of blacks to fight in what the Communist call a colonial-imperial war against a non-white race.

This episode appears to concern the writer, as if the propaganda of Radio Hanoi might be truly effective – perhaps because there was more than a grain of truth in the editorial.

These propaganda highlights number some two dozen items a week from most of the countries of the Far East – Communist China, Laos, Malaysia, Singapore, North Korea, Vietnam, Republic of Khmer, Thailand, Burma.

Now to another set of circulars for the psywar office, *Trends Analyses*. According to the 7th Psywar Group's documents,

The weekly *Trends Analysis* presents a continuous account of radio and news service propaganda released by the Communists to the nations of Asia. This section attempts to categorize Communist propaganda in major recurring themes so that trends can be noted. Both total time and frequency of each are presented. The frequency is often more significant than the total time. Analysis is based on material received by this office, and cannot represent all the programming of these stations for their intended audiences. . . .

An example of this trend analysis plus comments is shown below (6).

LAOS

Radio Pathet Lao (RPL), 11–17 Sep 71 used two major themes. Support for and success of people's revolutionary struggles, highlighting PL victories, is the center of attention this week; thus replacing negative treatment of established governments, which received slightly less coverage. As usual, US Indochina policy was criticized. Significantly, a greater emphasis was placed on anti-Thailand issues. *RPL* did not use the praise of the leader theme again this week, but as happened last week, an increasing amount of time was devoted to anti-Souvanna Phouma themes.

Radio Patriotic Neutralists Forces (RPNF), 11–17 Sep 71, as did *RPL*, used negative treatment of established governments and support for and success of people's revolutionary struggles as the main themes. This week an unusual amount of air time was devoted to anti-Royal Lao Government, anti-Souvanna Phouma themes. The US Indochina policy was also criticized under this heading. Success of people's revolutionary struggles received about half as much broadcast programing time as did the main topic. Neither the praise of the leader theme nor the peace proposals were mentioned this week.

Liberation Radio (LR) in Laos, 11–17 Sep 71, concentrated all of its programs on the theme categories of negative treatment of established governments, with programs criticizing the Republic of Vietnam and the US Indochina policy, and support for and success of people's revolutionary struggles, with the emphasis on victories in South Vietnam and Laos. Significantly, the only other topic mentioned was in programs considering the Paris peace talks.

Radio Peking (RP) in Laos, 11–17 Sep 71, as usual concentrated on topics related to the theme of international prestige; China and North Korea receiving most of the emphasis. The success of socialism was given greater coverage than normally. Anti-US themes were given a reduced amount of air time this week.

Themes	*RPL*		*RPNF*		*LR*		*RP*	
	Items	*Mins.*	*Items*	*Mins.*	*Items*	*Mins.*	*Items*	*Mins.*
Praise of the leader	0	0	0	0	0	0	3	24
Promulgation of the ideology	46	69	6	47	0	0	3	35
Present success of socialism	3	3	10	17	0	0	13	61
International prestige	2	4	1	1	0	0	23	118

Psyop Intelligence Notes

These are perhaps the most extraordinary forms of 'intelligence' which the psywar officer receives. They are a series of fact sheets, often only one or two sides of typed paper, in which some aspect of enemy society is described with the obvious implication that use be made of it for either propaganda or other psywar purposes. Of the many which I have seen (7) they can be divided as follows:

Firstly, the *psychological:* North Vietnam's war worries; can North Korea stand more indoctrination?; jokes in North Korea; attitudes of

North Korean people to indoctrination; North Korea wants food variety; interpersonal relations in North Korea – romantic exploits.

The *social:* use of civilian food coupons in North Korea; distinguishing characteristics of the North Korean underworld; daily routine of the North Korean gang member; mode of operations of North Korean gangsters; a North Korean criminal organization; tattooing in the North Korean underworld; weapons used by North Korean gangsters; telegrams in North Korea; telephone directories in North Korea; social conditions and related psyop themes; prostitution in North Korea; adoption, orphans and orphanages in North Korea.

The *anthropological:* signal words used in a North Korean urban area; use of scribes in North Korean offices; graffiti in North Korea; handling of time in North Korea.

The behaviour of the enemy soldier: parties held during the year at a North Korean general's mess; ideology study sessions at a North Korean army unit; recreational facilities at a North Korean army unit; the library at a North Korean army unit; behaviour of staff officers at a North Korean general officers' mess; the attitude of North Korean enlisted men to their military service; topics of conversation among North Korean soldiers; the 'no smoking' campaign in the North Korean army.

To give some idea of the flavour of these intelligence notes, let me quote just three examples.

The note on graffiti in North Korea says that most of it is sexual and consists of drawings and writings on lavatory walls. It was also reported – and this is clearly regarded as most important in the note – that on a few occasions the lavatory paper in North Korean toilets was newspapers with a picture of Kim Il-Sung on them. In Korea it was strictly forbidden to use paper in this way. The implication is that disaffection in North Korea could be incited by encouraging this behaviour – not necessarily with newspapers but perhaps with specially prepared toilet rolls with Kim Il-Sung's picture printed thereon many times. Lavatory graffiti, in general, was not political and again the implication was that were it to become so it might well start to signify to people that some at least were becoming disillusioned.

In the study of North Korean jokes it was noted that the North Koreans used to call a widow a *kuramang*, a word apparently derived from the name of the Grumman airplane used by the United Nations forces in the Korean War. It was considered a slow aircraft, especially compared with a *ssekssegi* – a jet, the nickname for an unmarried girl. *Kuramangs* were therefore easier to 'shoot down' than *ssekssegis*. On the basis of this, the psyop officers judged it particularly effective to have propaganda which emphasized that US aircraft were faster than anything the North Koreans could offer, and furthermore that in reality a fast aircraft is, of course, a better 'widow-maker' than a slow one because it could kill more. The prevalence of this particular joke made this piece of propaganda stick in the minds of the people.

A study of North Korean gangsters, circulated on 6 July 1970, was based largely on an interview with just one captive who had been in jail for a large part of his twenty-five years (8). He had been part of a group of fifteen young men specializing in pickpocketing, theft and violence and known as the *Haebeng-dong kalmaegi* (Haebeng-dong Sea Gulls). There was a boss (*wangcho*), three assistant bosses (*pu wangcho*), and three sub-units with the following specialities – pickpockets (*ssuri*) of four men, theft (*choldo*) of three men, and violence (*pokryok*), again of four men. After operating for the day, either in pairs or individually, they would meet in the evening near the North Korean–Soviet cinema to discuss the day and split the takings equally among them. On very successful days, part of the loot would be kept back for emergencies or 'rainy days'. In the violence sub-unit, for example, the men worked in threes. One would keep observation on and select the target; the second would hit him and knock him down; and the third was responsible for robbing the victim after he was down.

Here the object of the psywarrior is, of course, to use this knowledge of the underworld and its argot, to suggest in words that Koreans understand, that political leaders are really criminals. As a result of this intelligence, the propaganda can make use of the criminal slang and organizations to characterize the political leaders and their organization as criminal. Political leaders thus become *wangchos*, for example, the parties acquire, in propaganda, the 'Sea Gull' epithet and all that that implies, and political triumvirates are likened to the violent sub-units. The intelligence notes are thus used as a way of getting into the enemy's thought patterns and encouraging him to think by metaphor which, as we have seen, is commoner in the East than in the West.

PRISONERS AND PSYOPS INTERROGATION

In the front line most of the effective psychological warfare manoeuvres are derived from an analysis of intelligence obtained from interrogation of captive enemy soldiers. The psychological warfare intelligence officer is less interested in what armaments are where and how many soldiers man them than in more general aspects of the enemy's military life: the morale of the troops (as reflected, for instance, in the AWOL rate), specific grievances of specific parts of the army, grievances on the home front, morale of the civilian populations friendly to the enemy, whether there is a black market or a currency problem, whether the military has its own currency, and the situation regarding strikes or absenteeism. This list of subjects is taken from a psychological warfare officer's interrogation guide which I have seen (9). This also recommends that interrogation should take place within forty-eight to seventy-two hours of a man being captured to take advantage of his state of shock, disorientation and fear.

One way of eliciting the required information, the guide says, is to take the captured enemy soldier through his last few days in detail, looking especially for evidence about the physical and psychological shape of the troops. The interrogator will ask, for instance, whether there were foreigners present and the attitudes of the troops towards them; were there shortages of food; what were/are the attitudes to officers; are there any inter-service quarrels; what leaflets have they seen; what are their listening habits so far as the radio is concerned; when was their last leave; how many wounded do they have; what is the proportion of non-combatants and what are the attitudes of the fighters towards them; what has been the effect on morale of major defeats; what is the role of rumour; do the soldiers get letters from home?

The interrogator ends up with a lot of fairly innocuous questions but ones that can be crucial in future actions. For example, questions of climate (a high wind makes loudspeakers useless, snow and rain reduce the range of broadcasts), literacy levels, attitudes to colours – yellow often means good fortune in the Far East, though elsewhere it means 'be cautious'.

THE DILEMMAS OF DEFECTION

Many of the techniques and devices used in psychological warfare are designed to worry the enemy, to make life more uncomfortable and to undermine morale. Directed at the active guerrilla units, the end product of this mental wearing down is, of course, their defection. Defections are important for numerous reasons: they show that a particular military policy is working; they sap the morale of their colleagues and boost that of their captors; they provide intelligence; they can be used as further psychological warfare material. A systematic strategy to induce defection is therefore an essential aspect of counter-insurgency operations. Two papers which spell out more systematically than most the strategies used to induce defection are Lakin's paper on psychological warfare research in Malaya during 1952–5 (see chapter 17 and p. 430 below) and a paper by Emmett O'Brien, a colonel in the US Army's military intelligence, researched and written in 1971, when he was at the US Army War College. Between them they show just what details the psychologists are interested in so far as defectors are concerned and the overall strategies which are effective in bringing about defection and capitalizing on it.

Lakin found that only a small minority of defectors were politically committed, and that defectors were slightly less intelligent than the population from which they were drawn. Units with discipline problems or those exposed logistically were prone to higher than usual rates of defection.

By and large this told the British how to try to induce defection. Propaganda was aimed at the less intelligent and the arguments used were not intellectual or political ones about the British in Malaya but simple emotive ones about the hardships of jungle life. The research also showed that leaflets were the most widely seen psychological warfare technique and that promises about the way the guerrillas would be treated if they surrendered had the most influence on inducing defection. After the middle of 1954, however, Lakin also noticed that more and more of the hard-core terrorists were appearing among the captured men. He put this down to the fact that, by this time, many could see no real hope of victory.

What this research shows (and other psychological studies, carried out in Korea by the Operations Research Office of Johns Hopkins University, Maryland) is that disaffection grows slowly among the guerrillas and then, after some almighty battle or similar event, a large number give themselves up at one time. Lakin's research therefore suggests that most psychological warfare material should try to tire the men out, try to sap their energies and then, all of a sudden, capitalize on this by inducing them to surrender at the appropriate point. It has to be sudden because many guerrilla groups recognize the danger and keep a look-out for possible defectors and ship them away from the front line or, in special cases, execute them.

Colonel O'Brien's study was far more wide-reaching (10). He looked into peacetime defections as well as war-time ones, including in his survey, for example, Cubans entering the USA, and people leaving Russia since the Second World War. O'Brien concludes that the defection process has five phases: inducement, reception and interrogation, training, resettlement and employment, follow-up. The most important point, he says, is to communicate to the guerrilla the promise which will induce defection. He quotes figures that make the mind boggle. For instance, in March 1969, a typical month he says, the psychological warfare people in Vietnam dropped 713 *million* leaflets and distributed a further three million by hand, all inducing, or trying to induce, defection. In addition, 156 000 posters supporting defection were distributed, more than 2000 hours of broadcasting and 12 000 'face-to-face situations'. O'Brien's main point, however, is that the most effective technique to induce defection is to make use of the voice or stories of guerrillas who have already surrendered and the greater part of his paper is devoted to that.

Studies show, O'Brien says, that many defectors agonize silently for six months or more over defection. He concludes from this that they should be treated fairly gently – otherwise others might not follow them in the future. He recommends that, early in the procedure, the defecting insurgent should be asked whether he wants to *volunteer* knowledge or help find weapons caches or take part in some similar activity. This, he says, establishes the insurgent's *bona fides* and also confirms him in his defection. As many soldiers have noticed, once a man has decided to give himself up, he is often only too willing to lead officers to arms caches and hideouts – he may even become quite detached psychologically from his former colleagues and allow them to be killed. This, O'Brien suggests, usually happens a while after capture so the captors should be on the look-out for it, and not press too hard early on or the defecting rebel may not play along.

O'Brien also found that, in the Philippines, Malayan and Vietnam wars at least, though there was no single profile of the insurgent that could be constructed, the lack of necessary skills for earning a living often induced a man to join a terrorist organization. He implies, therefore, that the captured soldier should be trained in appropriate crafts once he has defected, to prevent him slipping back into his old habits. Such training, he says, should have political, vocational, and literacy aspects. Resettlement is of course desirable, but O'Brien found that defectors remain unpopular with most people for some time and so prefer to remain in the military, albeit on the other side. (In general they were used, in Vietnam at least, in paramilitary outfits such as armed propaganda teams.) Isolating defectors in hamlets or compounds solves some problems but raises others. In Vietnam, for example, these compounds became frequent targets for attack by the guerrillas.

In Malaya and Vietnam at least, defectors did not return to the rebels in any significant numbers (2 per cent only, according to one study). The main reason for this is that every opportunity was taken to inform a man's erstwhile colleagues of what he had done. This, above all, prevented his return to that particular fold.

O'Brien concludes his paper by saying that the US Army, like others, has shied away from defection as a systematic military programme, though he believes it can be effective. He therefore recommends that the army shed its reluctance to use the lure of defection as a tactic, and henceforth train its men, particularly its psychological warfare officers, in appropriate techniques.

To the psywarrior all we have considered so far may be grouped under the heading 'audience analysis'. The material helps the psywarrior to know just how and where to pitch his special techniques to take advantage of the enemy's psychology. At times broadcasting will be appropriate, at others leaflets will be a better bet. He will by now know the levels of literacy, any natural or man-made barriers to broadcasting reception, what time of day is best for a psywar attack, what the local taboos are, whether the colour of a leaflet is likely to raise problems. And he will have begun to have some idea as to which of the special techniques treated in the next chapter are the most suitable (11).

22 Special Psywar Techniques

So far we have considered the more systematic uses of psychology in counter-insurgency operations – selecting the right officer, inducing defection, the construction of elaborate psychological profiles of entire countries. However, some of the more imaginative and colourful uses of this science in guerrilla wars have been one-off, *ad hoc* affairs, which use psychological or sociological concepts to enhance the effects of propaganda; or even take advantage of some particular way of thinking or behaving to create fear, suspicion and hate.

DATES AND DEATH

Perhaps the most widely researched programme has concerned the extent to which the forces of the West can exploit primitive beliefs in witchcraft and sorcery. In August 1964, for example, James Price and Paul Jureidini produced a report for CINFAC, the Counter-Insurgency Information Analysis Center, entitled 'Witchcraft, sorcery, magic and other psychological phenomena and their implications on military and paramilitary operations in the Congo' (1).

The report was prepared in reply to a query from the Department of the Army which was concerned about the use of witchcraft and sorcery by 'insurgent elements' in the Congo (in Leopoldville, as it was then). At that time, magical practices were said to be effective in 'conditioning dissident elements and their followers to do battle with government troops. Rebel tribesmen seem to have been persuaded that they can be made magically impervious to Congolese army firepower. Their fear of government forces has thus been diminished and, conversely, fear of the rebels has grown within army ranks.' CINFAC was therefore asked to assess whether witchcraft could be used in a counter-insurgency campaign in the Congo (2).

The analysis of the situation made by Price and Jureidini was along social, psychological and anthropological lines; they concentrated on the significance of tribal change at that time. They noted that many younger men had, since the Second World War, migrated to the towns where they had become *évolués*, evolved, and had rejected magic. On the other hand, despite attempts for hundreds of years to stamp out magic and witchcraft, the practices still flourished in the villages of the hinterland. This, said Price and Jureidini, was partly to do with the fact that since many tribes were in danger of breaking up, so the idiosyncratic nature of magic lent them identity. It was this which gave magic at that time its military significance.

The report also analysed the long-term effects: that by perhaps changing the behaviour of tribesmen by counter-magic (as had been suggested), any military success which followed also meant that the government, or the USA, would merely be adding to the force of superstition and witchcraft in general. Price and Jureidini thus doubted whether counter-magic would have any beneficial effect in the long run. They concluded therefore that special magic potions 'concocted' by government troops could work and might with profit be used in tactical situations, but in the longer term and at a strategic level their use was inadvisable. This also would require a thorough knowledge of the potions used by the rebels so as to know what antidotes were recognized in the tribes. Another technique considered was to have a double agent posing as a witch doctor among the rebels so that he could manipulate their feelings and attitudes away from troublesome potions and to the favour of government troops. But this, too, was thought unsuitable in the long term.

Another belief taken seriously by psychological warriors concern dates relating to ceremony. In Vietnam, for instance, a long seventy-one-page document was prepared in the late 1960s examining the psychological value of, for example, New Year's Day, weddings, pregnancy and childbirth, death and funeral ceremonies, and anniversaries of death (3).

The celebration of Tet is the Vietnamese New Year. The idea of the propagandist is to make the celebration as unhappy and morale-sapping as he can. Therefore, the more he knows about the event the more control he is likely to exert over it. For example, the eve of the New Year is a time when all social and personal business should have been settled and debts and obligations honoured. Propaganda at this time then – just *before* the New Year – will stress these unfavourable aspects, will try to make people feel extra guilty about the debts they have not paid, and will remind those, immediately after Tet, who have not paid their debts that this will be a bad year for them since custom dictates that it is a bad omen to begin a New Year in debt. Another part of the New Year ceremony is to avoid meeting a female first in the year, to avoid unwholesome thinking, heated discussion and vulgar language. Propaganda will therefore try to provoke all this wherever possible, reminding people who were unfortunate enough to meet a woman first that they too are in for a bad time.

So far as deaths are concerned, it is very important for the Vietnamese to die in their homes. Leaflets are therefore likely to remind families of their relatives who were soldiers and who died far away, or to remind the soldiers themselves of the unlikelihood that they will die at home. Furthermore, the Vietnamese remember deaths after forty-nine and again after 100 days, as well as at anniversaries. Consequently, leaflets coinciding with these days after large battles where large numbers from a certain area were killed together will aim to provoke yet more misery. At these times, leaflets might ask whether the land has been split up among descendants – again according to custom – or whether some family members are profiting at others' expense.

Dates also vary in whether they are propitious or not. CRESS produced a memorandum on propitious and non-propitious dates in the Vietnam (and Cambodian) calendar (4). The authors say that some dates are 'absolute' whereas others depend on the age of the person. For example, in traditional Vietnamese culture it is unpropitious for a woman to wed if her age in years is uneven – she must wait until the next year. More important for the military is to know that there are *Nguyet ky* – forbidden days – and *Con nu'o'c* – a day of water. During specific hours of these days the times are clearly not propitious and therefore attacks – physical or psychological – should by rights have more impact. It also means that people are more likely to be at home – so bomb attacks will kill more. The same document says that in Cambodia thirty-day months are 'female' or 'full' where twenty-nine-day months are 'male' and 'lacking', the implication being that many good things like marriages happen only in the female months – thus attacks in the male months might have more effect. Monday is generally propitious apparently, but Saturday is unlucky. Some days of the month are even reserved for death – an appropriate invitation to an enemy.

Other social psychological studies carried out for military purposes include three by Rand scientists, one into folk tales which deal with swindling and dishonesty (a common theme throughout the Third World from Africa, Ceylon and Yugoslavia to China and the Yemen), and the other into graft and corruption among government officials. Nigeria, the Philippines and Brazil were three of the countries chosen for this study (5). Daniel Ellsberg, also from Rand (and of *Pentagon Papers* fame), produced a paper on blackmail (6). This was more theoretical than most but nevertheless explored this backwater of behaviour with a view – however indirect or distant – to its being useful to the military at some stage.[1]

LEAFLETS

A different kind of psychology is needed, once you have decided on your theme, and you wish to make sure that your message gets across. There is nothing particularly 'psychological' in leaflets themselves, of course, though they are possibly the most widely used form of psychological warfare. The 7th Psychological Operations Group, for example, prepared a mammoth 115-page booklet for use by psyop personnel which explained all about leaflet drift from various heights, in various winds, in wet and dry weather, at different aircraft speeds for safety's sake and the calculations that need to be done in order to achieve a particular density of leaflet distribution at any one point (7).

Two traditional ways have been used to make sure that a leaflet is, at the least, picked up. One is to print a replica of a banknote on one side. Since there is a one in two chance of it landing with its banknote surface uppermost, these will usually be picked up – just so that the individual can be certain the leaflet is *not* a real note (8). But since in many rebel movements *possession* of an enemy propaganda leaflet is punishable by very severe sentences (even death in one or two cases), the best way to ensure that these are picked up is to make them safe-conduct surrender passes – a sort of offer-while-stocks-last. It can be picked up now and will provide safe conduct at any time in the future, should the rebel choose to surrender (10). But the offer will run out – the implication clearly being that the leaflet must be picked up now

1. Also in the same vein is a paper on forgeries as a propaganda technique by Russia. E. W. Schnitzer, in another Rand document, surveys the many Russian attempts to gain advantage by leaking forged documents (or alleged forgeries) that show Americans, or some other western nation, in a bad light (9).

and stored for possible use later. Forged notes have sometimes been dropped alongside leaflets which are a note on one side only. This improves the chances of them being picked up and does not do small economies much good either.

Getting the message across is a much more elaborate business, of course, and one which can never – or only rarely – be judged in immediate terms.[2]

In practice, however, what are the real propaganda efforts like? Let us look at two examples, taken from both sides in the Vietnam War. First, an analysis made of Viet Cong propaganda leaflets that were captured between March and the end of May 1964 (11). The target audiences of the 418 different leaflets found during this period were:

Military and paramilitary	45 per cent
General public	36 per cent
Government officials	4 per cent
Americans (in English language)	4 per cent
Buddhists and other religions	1·3 per cent
Cambodians	0·3 per cent
Specifics: youth, students, women, small businessmen, etc.	7 per cent

Those that were especially psychological in their approach, that is followed emotions rather than facts or political themes, comprised 30 per cent of the leaflets. Those that attempted to be divisive and therefore stressed ethnic differences and other social and political divisions occupied 69 per cent of the leaflets.

Analysis shows that the themes used in the VC leaflets in this period were as follows, in descending order of importance:

Khanh the lackey of US imperialists – sweep away the wolf.

Coup leaders are lackeys.

Coup was US-inspired, the change has come in midstream (this particularly used for Buddhists, youth and students).

Be against the blood-thirsty US officers who put you on operations.

You'll get shot and thrown in the river.

Avoid locations where the US are; do not use military transport.

2. In 1965, SORO was asked 'urgently' for a guide to persuasion, with particular reference to Vietnam. The office produced a 153-page booklet which contained twenty-four examples of successful persuasion by military commanders in Vietnam and, for comparison purposes, one unsuccessful attempt (12).

VC policy is peaceful.

Recruitment.

Details of military victories of the VC.

Besides these there were many *ad hoc* leaflets with such messages as: the USA was secretly plotting to militarize women; the publication in leaflet form of letters from captured VC soldiers showing how badly they had been treated; and exploitation of, for example, the Kennedy assassination – the day after his death the VC apparently produced a leaflet saying that the new President, Lyndon Johnson, had also been slightly wounded in a *separate* incident.

Turning now to the American side, a few years after the study of VC propaganda just mentioned, the Army Concept Team in Vietnam examined the output by the US psyop groups there and found that their leaflet programmes were as follows (13):

Chieu Hoi Inducement and Dai Doan Ket Programmes

Fifty-five per cent of leaflets were in support of these programmes, urging members of the VC and their supporters to leave and return to the government's side. 'Tactical Chieu Hoi leaflets were designed to provide assurance of good treatment to ralliers and explain how and where to rally. The five major vulnerabilities exploited were hardship, fear, loss of faith in victory, disillusionment with the enemy cause and concern about families.'

B-52 Follow-Up Programme

About 5 per cent of the leaflet effort was directed to this programme. Within four hours after a bombing strike, leaflets were dropped into the area informing the enemy that he had just experienced (in case he hadn't noticed) a B-52 raid. The leaflet reminded him that the B-52s would soon return to strike his unit again and therefore urged him to defect ('rally to the government side' sounds better). These leaflets were also designed to engender friction between cadre and soldiers, encourage malingering and desertion so as to 'spoil' the enemy's preparation for impending operations.

North Vietnam Army in Republic of Vietnam Campaign

Roughly 10 per cent of the leaflet effort was conducted against North Vietnam soldiers and stressed such things as their poor chances of being buried in unmarked graves away from home or tried to cast doubts in their minds about their relatives in the north.

Frantic Goat Campaign

This accounted for about 20 per cent of the leaflet campaign during the period in question. It was conducted outside the Republic of Vietnam with the object of disseminating news and propaganda to North Vietnamese civilians. The idea was to counter the impressions given by the North Vietnamese government. One leaflet merely listed the names of soldiers who had been taken prisoner.

Trail Campaign

Ten per cent of the leaflets were directed at the soldiers using or maintaining the Ho Chi Minh Trail. Along this trail the men's loneliness was exploited by using leaflets with nostalgic poetry written by North Vietnamese soldiers about their life at home. The themes of hardship and probable death were constantly brought to the soldier's attention as he moved down the trail.

In general a system of priorities was used by the Americans with regard to leaflet campaigns. The main priority was for quick reaction support leaflets for tactical operations; 'priority two' was the exploitation of intelligence gathered from ralliers; and the third priority was for support of operational movements.

The Americans also printed a number of propaganda oriented newspapers, ranging from *Ban Tien* (*News Clips*), two pages, twice weekly, with a run of 72 000 copies, to *Tudo* (*Free South*), one page, every two weeks with a circulation that varied from 700 000 to 2 600 000. Plus, of course, magazines, booklets and calendars (14).

Leaflets are the most traditional and the most widely used tool, but there have been several somewhat more imaginative techniques, many of which were developed for the first time in Vietnam.

OPERATION TINTINNABULATION AND WANDERING SOULS

Operation Tintinnabulation was a technique new to the Vietnam war, which according to the Army Concept Team's report, was used by the 10th Psyop Battalion and the 5th Special Squadron (15). This exercise involved two C-47 aircraft, one known as 'Spooky', equipped with miniguns, and the other known as 'Gabby', fitted with several loudspeakers. During the initial phase the Gabby employed a frequency pulsating noisemaker designed to harass and confuse the enemy forces during night hours, while the Spooky provided air cover. During the second phase, the harassing noisemaker continued, but emphasis was given this time to the use of Chieu Hoi tapes. The first phase 'is designed to eliminate the feeling that the night provides security to the target audience', while the second phase was designed to reinforce the enemy's desire to defect.

One operation involved twenty-four missions with the aircraft spending on average two hours over the target. During this time the number of defectors more than doubled from 120 a month to 380.

A similar but perhaps even more eerie approach was known as the 'Big Red One'. This was used in Vietnam by the 1st Infantry Division, according to the *Psyop–Polwar Newsletter* for the 30 November 1969 (16). It involved 'hounding the enemy in the daytime', mainly by armed propaganda teams (see page 412) and loud broadcasts, the main aim of which was to keep awake the VC guerrillas hidden in a village so that they were too drowsy to work effectively at night. Then, after dark, the villages would be overflown with helicopters, broadcasting loudspeaker messages far into the night. For two hours between midnight and six a.m. psywar tapes would be played – the themes being either nostalgic ones directed at male Viet Cong, to make them think about their loved ones back home, or else eerie sounds intended to represent the souls of dead VC which had not yet found peace. This tactic was called 'Wandering Souls' and was based on the reasoning that the VC were very superstitious about being buried in an unmarked grave. According to the Army Concept Team, the soldiers realized that the sounds were coming from a helicopter but with their resistance lowered, having been hounded all day, the tapes still had an effect.

In the Congo, it was also reported that tapes had actually been made to simulate fearsome local gods (17). They were never used, but the apparent object was to broadcast them in the jungle at night to keep tribes in their villages as a sophisticated form of population control.

In other cases (18), recordings of gunfire were dropped in the jungle, timed to go off at irregular intervals, accompanied by flashing flares to simulate troops and thus keep villagers and rebels firmly in their place at night in the belief that they were surrounded.[3]

PSYOP AIRSTRIKES

The US 1st Cavalry Division (an air-mobile unit) employed quick-reaction helicopters to 'psychologically exploit contacts with the enemy in battle'. A helicopter, equipped with broadcast gear, was located near the operations centre of each of the three brigades. When contact with the enemy was established the helicopter would scramble to the area, the psyop effort then being 'integrated with artillery fire, tactical airstrikes and helicopter gunships' (19). Its main advantage was that, if a VC was captured, his confessions could be transmitted to the helicopter and then *re-broadcast* to the fighting zone. The captives' words were therefore heard by former colleagues within minutes of capture (20).

In the Dominican Republic in 1963 a different tactic was used (21). A group of rebels had taken to the hills. Though their general whereabouts and direction were known to the US commandos who had helped quell the incipient revolt, the rebels' specific location was a mystery. This time four aircraft overflew the area where the rebels were located in two two-hour shifts. The planes saturated the area with leaflets which said that the rebels could be clearly seen from the air (though they could not be) and that if they did not give themselves up they would be bombed. It took just a few days and the rebels came out of the hills with no further resistance.[4]

Other techniques used included mobile television, with music, news and propaganda; motorized sampans were used to reach villages not

3. Donald Tepas, of the Honeywell Military Products Group Research Laboratory at Minneapolis, conducted quite a few theoretical studies into harassment techniques. His paper, 'Some relationships between behavioral and physiological measures during a 48 hour period of harassment: a laboratory approach to psywar hardware development problems', starts with the sentence: 'Psychology can be manipulated to yield extensive deficits in the effectiveness of behavior . . .' (22).

4. The defection of pilots is of enormous propaganda value if it can be engineered. At one stage in the Korean War the Americans offered £100 000 to any Communist pilot who defected with a Mig. The gambit was successful in so far as many combat planes had to be accompanied by 'security patrols' to prevent defection and other pilots had their families kept hostage while on missions. This way people were tied up in useless jobs (23).

serviced by roads. Each sampan contained a TV set or a movie projector and a Honda generator. The sampan would move into an area by day where the crew would talk to the villagers, giving them a TV show in the evening (24).

ARMED PROPAGANDA TEAMS

This method tends to be popular among rebels themselves. Experience in Vietnam taught the US Army that armed propaganda teams should be used in Viet Cong-controlled or contested areas and that they should operate at platoon size in a highly mobile unit. 'For effective organization the platoon should be divided into three squads with one squad designated as the propaganda squad and the other two as security elements.' The security squads are extremely heavily armed while the propaganda squads should contain an ex-rebel who has defected from that area; its primary task is face-to-face communication and tries to root out the rebel infrastructure in a locality. The heavily armed units cordon off a village, including all possible ambush sites and then the propaganda squad goes in, deliberately *not* heavily armed. After security is established, the propaganda unit go from door to door talking with families, and give out propaganda urging them to rally. This was regarded in Vietnam as a very effective technique (25).

Another approach is for the team to clear a village, telling them that it is to be a target. This way lives are not lost – though homes are. The point is that this rubs in the accuracy of modern artillery and has been found to be an effective way of lowering morale and the will to resist. The technique travels quickly by rumour since the uprooted have to find homes in other villages until they can re-build their own (26).

Similar to the armed propaganda teams, in Vietnam, were the mobile motivation teams, also armed. These had only six to ten members and were, as their name suggests, highly mobile.[5] The teams always showed 'of-the-people behaviour' – that is, they lived, ate and washed with the peasants. The men – usually indigenous – were not so much trained as *selected* for their ability to get on with almost anyone. The main object of the team was tactical, face-to-face psywar, though they had a limited capacity to produce leaflets. The main quality looked for, says the handbook, was the ability to argue persuasively with villagers – to

5. Information for the following paragraph is taken from *A Handbook for Mobile Motivation Teams*, produced at the Special Warfare School, Fort Bragg, in 1968.

let them have their say but to change their views with argument delivered in discourse, during a meal, say, or on the way to work. (The only training given the teams was in discussion; this lasted two weeks.) They were told *not* to try to teach the people anything new, but to draw on their hopes, frustrations, social and tribal divisions, to discuss social problems rather than political ones. The team stayed not more than three days; this made it appear as an up-to-date information source the next time it came back. Informality, politeness and humility were the main techniques. They also need a good fund of jokes, folk tales and songs.

The armed propaganda teams were the sharp end of the face-to-face propaganda in Vietnam, the mobile motivation teams, the soft end.[6] Together they added up to more than they were separately.

PSYOP DEVICES

Besides these tactical techniques, special devices may also be used to capture attention. These included psyop soap, psyop grocery bags, flexagons, and puppets. The 7th Psyop Group, for example, managed to develop a technique of disseminating propaganda messages embedded in successive layers of soap 'thus enabling the originator to convey several messages to the user over a considerable period of time'. The soap was eventually given out in a wrapper which said that it was a gift of the government (27). The grocery bags had a propaganda message stamped on them. As there was at the time a marked shortage of grocery bags they were kept – for household purposes – for a considerable while.

The flexagons and the puppets provided much the same function for children. The propaganda messages were stamped on them or they were made in the image of a well-known personality. Because of the general shortage of toys, they, too, tended to last. A host of 'sham

6. Some special equipment was developed in Vietnam when unique psyop problems were encountered. Two are worth noting.

There were always problems in disseminating leaflets from a helicopter because of getting the leaflets clear of the rotor blades' turbulence. This led to the 'hurricane hustler' – a kind of massive vacuum cleaner which sucked in the leaflets and regurgitated them into the slip steam of the aircraft below.

The 4th Psyop Group also invented an image projector (the Mitralux) with which, using an 85-mm slide and a 1000 watt projector bulb, the projector was able to use buildings, the sides of mountains and even low banks of cloud as projection screens. This was not too successful, but gives an idea of the devotion which some psyop groups brought to their work and also some idea of what we can expect in future wars.

supplies' was also dropped on the Vietnamese – ammunition, food tins, radio equipment – real-looking but in fact sham (and so cheaper to produce). The purpose was to sow confusion as to the whereabouts of the American forces (28).

The Rhodesian PSYAC troops attempt to get to black children through prizes in school for essays on controversial terrorist incidents. Harrowing, childlike accounts can always be published later as propaganda and the essays may also provide useful information on the way a particular black community is thinking – how 'loyal' it is, for example to the white government (29).

CHORAL SINGING

Another technique, not used so much in Vietnam, is well illustrated by a psyop intelligence analysis of 'Group singing as a highly effective propaganda technique exploited by communists and other radical organizations in Japan and Okinawa' (30). The document contains many examples of actual songs used, arranged according to whether they are youth songs, old songs, folk songs, workers songs, songs of the world, songs of the struggle, songs of peace and so on, plus an analytic introduction. This argues that the Communist-led choral singing movement (known as the *utagoe undo*) began during the American occupation of Japan and appears to have grown rapidly. It grew out of the Communist Youth Central Chorus Group, and in the late sixties very successful festivals were held all around the country. Typically, the document argues, the blend of old and new folk songs, the 'togetherness' of the idea, attracted young people to the *utagoe* meetings where, unaware to begin with of the movement's Communist origins, they could then be gradually indoctrinated 'with the ideologies of the Communists or other more radical organizations'. Further, the document adds, the group singing heightened the 'clique spirit' and created a natural 'we–they psychology' which 'greatly facilitated the left's efforts to mobilize opposition to the US'.

COOKING FAT AND FEAR

I have deliberately saved until the last what, to me, is the most imaginative, yet macabre psychological device of all. It was part of Project Agile, a whole series of military science studies in Asia, and carried out by the Battelle Memorial Institute in May 1966 for the Advanced

Research Projects Agency. The aim of the study was, quite simply, to develop 'stink' bombs which would get at different races through their noses (31).

Stuart Howard and William Hitt surveyed most of the relevant anthropological and psychological research around the world with a view to defining cultural differences in the sense of smell. Particular attention was paid to the smells which various races found offensive, the search being particularly concerned with those smells which actually produced symptoms such as nausea or fright. The idea was to develop smell bombs which could either flush guerrillas out of the jungle where they could not be shot at with more conventional weapons, or to keep the tribal guerrillas away from American troops, who, being of a different race, could traipse through the undergrowth doused in smells which to them were not offensive at all. A third possibility was to pair ordinary explosive bombs with an offensive smell, especially a smell with a central place in a particular culture, so that the mere presence of the smell would, in time, come to signal fear and thus disrupt ordinary life by its continual and dangerous associations.

For example, Howard and Hitt's paper explores the fact that there is nothing the Karen people of Burma fear as much as the odour of cooking fat. The race, the document says, is impervious to some of the most awful smells, but they believe that the cooking odour can enter a person's body, especially if there is an abrasion in the skin and that this can make people sick, even kill them. There have even been cases in this tribe where accusations of murder have been levelled on the basis of someone cooking in the vicinity at the time of another's death. In a similar case, the Andanise believe that after eating pork or turtle the body emits an odour which attracts evil spirits in the jungle (this is why they cover themselves with clay). It follows that they will steer clear of any jungle area smelling of cooking pork or turtle.

Other uses are possible, however. One of Howard and Hitt's other findings was that vegetarian races often develop a keen sense of smell for people who eat meat – which they intensely dislike. It thus follows that in fighting such a race, it might make sense to flood the jungle with the concentrated body odour of the tribes' meat-eating enemies. This would not only add to the general distaste for fighting but also prevent the vegetarians from being able to smell the enemy when it was near, which can apparently happen.

Tribes with exploitable olfactory weakness may be thin on the ground, but Howard and Hitt identify two other uses of smell: to track game and as part of one race's dislike for the other (the Japanese hate the smell of the Ainu, for example, and the Chinese think Europeans smell like sheep). It is clear that the second of these, especially, lends itself to treatment by psychological warfare troops to foster discontent between races. Howard and Hitt certainly end their paper advocating more research in this area, particularly the linking of bombs to smells which are common in a society but thought of as obnoxious. Howard and Hitt's idea was that this would be a continual drain on morale. Their paper is, however, the only one of its kind that I have come across and I do not know if their ideas have been followed up.

23 Psychological Aspects of Population Control

This is a huge area of study in its own right and, apart from leafleting, crowd control is probably the most-used psychological counter-insurgency technique. Quite a lot of study has gone into the psychology of force, the way to deliver threats and warnings so that they are obeyed without delay and without panic, the effects of police formations and actions on crowds and the use of special equipment and techniques to produce desired effects. Before we come to the psychological aspects of crowd control, however, let us briefly consider some more general issues of controlling much larger populations of people.

Paul Wehr, a sociologist at the Institute of Behavioral Science at the University of Colorado, has looked at those factors which lead to non-violent resistance in a country when it is occupied by foreign troops (1). Using Norway in the early forties and Czechoslovakia in 1968 as examples, he found the most important factors were communication, psychological polarization and sub-national divisions within a country. Wehr found that resistance invariably sprang up through people with access to the various media – the press, broadcasting, even telephone operators. Secondly, the more obviously the occupying force labels collaborators as such (by giving them uniforms or privileges), the more entrenched resistance becomes.

He also found that the more divisions are used along sub-national (regional or tribal) lines the greater cohesion resistance movements have. It thus follows that to keep down a resistance movement the media system should be broken up and created on fresh lines, privileges to collaborators should be as invisible and as long delayed as possible and the differential treatment of different sectors of the community should be avoided at all costs. This probably will not prevent a resistance movement beginning, but it might limit it.[1]

CROWD CONTROL

Far more specific studies have been carried out in respect of the behaviour of crowds. One man who has devoted a great deal of energy to these questions is Colonel Rex Applegate, formerly a US infantry officer and now a prolific author of technical books on security affairs.

According to Applegate, the most frequent mistake which security forces make is not to use force early enough (2). The most significant reason for police forces not preventing the spread of mob violence is a general timidity by the political forces which normally control the police or other security agencies. This causes, he says, delays in police reaction. By the time such permission to act comes through, the situation has often deteriorated to the point where the National Guard or the military have to be called in; escalation is thus under way (which, in many instances, the insurgency forces wanted all along). Applegate's theory is that, by and large, the firm use of force at the right moment has a deterrent effect on rioting.

He quotes with disapproval the political instruction to the police during the Washington riots of 1968 to be permissive in their reactions towards looting. This was in the belief (Applegate says mistaken) that lower mob pressure and aggressiveness would result.

Basic psychological riot control measures, Applegate contends, involve the identification and apprehension of agitators and leaders, dispersal of crowds, especially small hostile groups, the arrest of looters and the prevention of mobs regrouping. He argues that the measures should preferably, but not necessarily, be used in the following order:

Show of force: the surprise appearance of a large unit of specially equipped police in full view of the mob can have a huge psychological impact. The measures must be carried through quickly and with

1. A Rand study in Germany in 1969 looked at the public's attitude towards the US troops' occupation. Clearly this was not an insurgency, but variations in public opinion are instructive. Men were more positive in their attitude to the troops than women, the over-twenty-fives more positive than the under-twenty-fives, the better educated more positive than the uneducated. Among those who were both young, uneducated, poorly off and living in the countryside, those against the troops out-weighed those in favour. This gives some clue to where occupying troops may expect to find the strongest resistance (3).

sufficient personnel between the mob and the area where the special unit is grouping to prevent rapid escalation.

Orders to disperse: these can be made by the commander or local personalities who are respected. They should be clear and fully communicated to the crowd, which usually means a powerful public address system. In large mobs, undercover men in plain clothes should be planted so as to lead the way in dispersal. Bluffing should not be used.

Use of formations: this is the point, Applegate says, where psychological force has to be replaced by physical force. But there is still a psychological element in that the force must be applied in as controlled a way as possible to keep the actual violence to a minimum.

There are a number of formations which have been worked out by police forces around the world though they vary from place to place. In general, though, they consist of small units of about twelve men which can split up into smaller groups of six, four or even three. The main point in training, however, is to instil into the riot squad the value of acting *as a group* and the psychological impact which a block of well-armed men in identical uniforms has on mobs. The men should all be trained so that they always occupy the same position in the unit and therefore know exactly where everybody else is. They never, therefore, feel exposed, as the rioters might from time to time.

In general, it has been found that as a squad of police moves towards the mob it starts to bunch, the wings especially start closing in. The commander must recognize this happening and take steps to prevent it. Once movement is under way the squad should not stop; if it does it should not have gone in the first place. Any retreat against a mob should be slow; if any single man is attacked his aides should immediately take his place, reinforcing the idea that the mob is dealing with a unit, not with individuals. The forward squads should not weaken themselves by making arrests; this is left to the back-up units in the rear of the patrol.

After dispersal, the squads should actively and 'aggressively' (in Applegate's words) patrol the area, picking up any individuals left in the vicinity to prevent the mob regrouping. It varies, of course, according to the demonstration and demonstrators. However, resesarch carried out by Richard Teeven, on behalf of the Office of Naval Research, showed that there were definite personality differences between students who (in the mid-sixties) – joined anti-Vietnam demonstrations and those who did not. The research showed that the demonstrators were more 'anxious' and 'paranoid' than the others. This result appears to mean that these students are the type who respond aggressively to equivocal demonstrations of power yet give in mutely to overwhelming displays.

The use of chemical agents and individual fire: if it is not possible to disperse the mob through the use of formations, then, says Applegate, 'chemical agents may be called for or selected fire by marksmen'. It has been found that the more timid individuals tend to cluster at the back of a crowd away from the police; consequently, it can be very effective in inducing dispersal if the first rounds of chemical gas (or whatever is fired) are actually lobbed over the heads of the mob to the back.

Fatigue is obviously a matter of some importance in prolonged riots: soldiers or police may become tired and emotional and escalation may be more likely. At all times, therefore, thought should be given to some form of duty rotation to prevent this.

Other human factors considerations for use in riot situations are to be found either in a special bibliography on the subject, 'A selected bibliography of crowd and riot behavior in civil disturbances', produced by CRESS (4) or in the proceedings of a conference on psywar for internal uses, held at Fort Bragg in 1966 (5). Measures outlined include the following recommendation:

The police should never be under too specific instructions as to what they can and can not do; the commander on the ground should have discretion – otherwise the men are apt to feel isolated and get more frightened and this jitteriness may simply make things worse.

Local criminals and professional 'fringe operators', says Applegate, will normally join the riots for personal gain. Police intelligence should aim to stop it by routine road blocks and so forth.

There will be plenty of hoaxes designed to lure the police into dangerous situations. These should be guarded against by cross-checking all information where possible.

Mobs will try to bait the men by name calling; police must be thoroughly trained to ignore this.

COMEDIANS AND COLOUR

Several elaborate 'psychological' approaches have been suggested for the control of crowds. Perhaps the most extraordinary suggestion was made by Joseph Coates, of the National Science Foundation, in a paper released in 1972 (6). In this he said that the police could diffuse many tense situations by telling jokes to the rioters. Perhaps he was not entirely serious himself, but Coates maintained that the injection of humour into a riot confrontation could keep the crowd's aggression focused on their main objective and away from the police. It could also help the police keep their own tempers from boiling over. Coates cautioned the avoidance of coarse ethnic jokes or humour exploiting physical or social differences. Instead, he said, inversions should be used to stand a particular situation on its head. So far as I know this technique has never been used – at least intentionally.

Other equally 'psychological' techniques have been successful, however. Robert Shellow gave a special civil disturbance squad some sessions in group therapy before it policed a Negro civil rights demonstration in 1963 (7). The forty-five police officers were shown films and also given the opportunity to discuss their feelings towards blacks in front of the psychologist. Particularly, they were allowed, and encouraged, to share their fantasies about the bloody battles that *might* ensue. This seems to have helped the men deal with their emotions for this group was actually commended by the black leaders after the demonstration for the restraint and sense of fair play which they had shown.

An equally unusual discovery comes from Japan. There, when the British Queen visited the country in 1974, the Tokyo police found that crowds appeared to do less damage if they were fenced in with 'soft, cool' coloured barricades, like yellow or blue, rather than 'harsh aggressive' ones such as red or brown.

More reliable perhaps are the newer police methods to break up the rhythmic chanting of mobs before this sloganeering becomes an incantation. Chanting and other rhythmic activities are seen by several people, Applegate included, as self-induced forms of mass hypnotism designed to encourage people to lose their individuality, and thus become susceptible to mass control. Leaders who use these methods or add to it with bull horns, drums or cymbals should be quickly ejected from the crowd. Very powerful sound systems can break this sort of thing up very quickly.

It has also been found, however, that noise has other, more sinister uses. Not only is very loud noise extremely painful, but when it is pulsed out at certain frequencies it can make people sick – even, in some cases, induce epilepsy. Two types of 'sound weapon' have therefore been developed. One is a 'squawk box', as used in Northern Ireland, where the piercing, shrill notes of the box were so painful to the people that they had to disperse, or at the very least stop whatever aggro they were up to and escape the noise. Second, in 1973 Allen International publicized (or had publicized for them) a new machine – the 'photic driver' – which not only pulsed out sound that could reverberate off buildings, but also pulsed out flashing lights. This too can be reflected from the walls of public buildings, compounding its effect. The noise and light together are reported to have a marked nauseous effect on crowds and the risk of epilepsy is also said to be greater with this machine.

The actual use of weapons in a crowd is quite naturally fraught with controversy, and views differ widely. Applegate insists that it is useless to carry firearms in a riot without ammunition or with instructions to the police or military not to use them. He argues that there will inevitably be a loss of morale in the police if this happens; and, second, should the crowd find out that, say, the police have no ammunition, the results will be disastrous. It is best, he says, for the firepower to be kept in the reserve element, not the front line. This gives better control and also allows the build-up of the threat. He also maintains that the practice of sending in men with unloaded but bayoneted rifles is also questionable. He has found that this, too, lowers morale.

Applegate also questions the practice of firing volleys of live ammunition into the air over the heads of rioters. He says that this can simply inflame a situation. What he recommends is for the police to fire into the ground in front of a marauding mob; this reduces the risk of fatalities, he says, and instead the ricochet bullets hit the lower parts of the body, injuring but not killing. The psychological effect of this, he maintains, far outweighs anything achieved by firing over people's heads.

Upon first confronting a crowd, he writes steps should be taken to show them that the police are armed *and* that their guns are loaded. This show of force, he believes, needs careful timing but is usually advisable. Silencers may be used – the noise of firing is what sometimes inflames people some way away. At night railway-type fuses attached to bayonets can be useful for their psychological effect as a police unit advances on the crowd. Finally, he points out that commanders should always make a distinction straight away between whether someone has been wounded or killed – they should never simply say 'shot'. The statistics can affect the reactions of many people to events.

24 Psychological Warfare: Some Final Considerations

PSYWAR IN PEACETIME

The special forces stationed at Fort Bragg have undertaken various projects in peacetime and these have often been designed as much to keep the army's image up to scratch as to offer training opportunities for special forces personnel. The projects usually centre on disadvantaged groups in the USA, and include work on Indian reservations, with Eskimo groups and with underprivileged children (1).

Ports and propaganda

However, the main peacetime propaganda effort directed at foreign countries has been carried out by the US Navy – in the form of port calls. For example, in Project Handclasp, toys, clothing and medicine were distributed to Third World countries. Various studies have been made of these port calls. One, by Dr A. Jenny for the Human Sciences Research Institute in 1968, outlined various considerations: for example, the port's place in the nation (what might go down well in a port might antagonize the nation as a whole); the size of the ship; the frequency of the visits; how quickly the effects of a visit wear off – the naval authorities need to know when to organize a second follow-up call (2).

A series of studies on port calls was made by Dean Havron and his associates who considered riverine and shallow water operations against guerrillas, and the psychological impact of harbour maintenance and other modernizing skills by which a western navy can help a country racked by insurgency (3). Havron recommends that in times of peace naval ships should follow the lines of major earthquake zones to be immediately on hand in a disaster; and that ships' commanders should study the psychology of bargaining and negotiation so that they are better equipped to deal with situations such as the Cuba blockade and the *Pueblo* and Tonkin Gulf incidents.

Along the same lines is the more general 'show of force'. A recent study by the Brookings Institute, Washington, showed that, by 1977, the USA had staged 215 shows of force abroad since the Second World War, beginning with the coup in Haiti in January 1946 and ending with the seizing of the *Mayaguez*. 'Most of them,' concluded the report, 'met with short-term success.' This was twice as often as Soviet Russia but was rarely successful in the long run, 'in an overwhelming proportion of the incidents the success rate eroded sharply' (4).

Analysing the weaknesses of major enemy figures

At a more esoteric level, a conference held at the Institute for Defense Analysis in 1961 explored whether, in international politico-military gatherings – conferences or treaty negotiations – an understanding of the personalities of adversaries and allies could be of help in influencing the outcome of the conference to the USA's advantage (5). Psychologists present included Urie Bronfenbrenner, well-known for his work on the relationship between childhood experience and adult personality; Charles Bray, an authority on the uses of psychology in the Second World War; Leon Festinger, an expert on conflict resolution; Robert Goodnow, who works on psychological assessment techniques; and Otto Klineberg, an expert on cross-cultural psychology.

A paper by Joseph Barmack was circulated before the conference and discussed the idea of indirect personality assessment of leading enemy figures by studying their handwriting, public speeches and any literary work they had produced, whether they believed in numerology, their previous occupations, their hobbies, and whether they were subject to psychosomatic disorders. He even advocated the use of telephone tapping as one method of obtaining relevant information. So far as the moral issues raised by such methods were concerned, Barmack argued that provided one was convinced of the rightness of a position, any technique to further it was justifiable.

The conference itself was a somewhat indecisive affair; they were, after all, tackling a difficult topic. But the general conclusion deviated quite a bit from Barmack's view. National and cultural factors were judged likely to be more helpful than the personality traits of any one individual. However, these were still thought worthy of study. In general, also, attempts at overt persuasion of leading personalities were seen as unrealistic; better, decided the conference, to work towards 'mutually satisfying compromises'. And rather than dwell too much on persuasion, smaller aims should not be neglected, such as controlling who attends a gathering, what the balance between formal and informal sessions should be, how to get a conference to listen, attend and understand, and how to increase or decrease tension. Among the specific research programmes suggested were: debriefing available people like ambassadors or defectors; studying critical incidents of previous crucial conferences; studying the interrelations between a person's biographic events and his personality, especially before he entered the power game (this, it was decided, should indeed include the person's choice of hobbies, selection of a mate, choice of companions at various stages of his life); small-group experiments with diplomats to examine how various status needs, anxiety, impulsiveness, authoritarianism, and so on, can affect a discussion; and a study of how individuals react to provocation. It was also felt appropriate to develop indirect measures of a person's behaviour to assess his more transient states, e.g. whether he was tense or experiencing internal conflict; such research could examine whether there is a relationship between speech disruption and anxiety, and include an analysis of a person's language to determine whether there is a link between repeated phrases, pet phrases, language redundancy on the one hand, and conflict, shift of ideas and agreement or disagreement on the other. By systematically reinforcing someone's replies – by agreeing with some of his utterances and remaining silent in relation to others – it was felt that it might be possible to influence his attitudes. Also suggested was a study of which *pairs* of personalities get on well together and which types do not. There was found to be a training problem not only for the diplomatic corps but also for interpreters who had a crucial role and could defuse the emotion in many situations by clever transmogrification of words and phrases.

The study of a person's handwriting, which Barmack had suggested, was generally not considered promising, but it was felt that an important area of research could lie in spotting whether a negotiator was genuinely negotiating or in fact acting on higher authority; some behavioural indices might be of help here. A person's dress, eating habits, jokes, his reactions to photographs, his betting and card-playing habits could all provide useful bits of information. (It was revealed incidentally that at the time the State Department had four files of information on Mr Khrushchev; these had been examined by many people, including psychologists, but no positive results were reported to this conference at least.)

The ideology of negotiators, the moral rightness or wrongness of the two sides was ignored; only the psychological manipulation of a situation was considered. Jesse Orlansky, from the IDA, in summing up, invited the participants to help 'design a research programme in this area', but I have not been able to find out whether any subsequent work took place.

However, in Britain, Alan Smith, a psychologist at Powick Hospital, Worcester, has demonstrated with the Dektor Psychological Stress Evaluator (which measures stress in a speaking voice) how possible it is to detect stress in the voice of a statesman making a speech (6). This, he feels, could be a clue to when the politician is lying and so could make an intelligence source.

HOW EFFECTIVE IS PSYCHOLOGICAL WARFARE?

The techniques of psychological warfare take up a great many pages, as do the other psychological aspects of counter-insurgency. I have, in a number of places, tried to indicate the effectiveness of some of these measures – the number of defections, for example, or the effects of this or that campaign on enemy morale. But, more generally, what does psychological warfare add up to?

Opinions vary. There have always been sceptics. The British, by and large, have always called psywar 'polwar' – political warfare – and as a result tend to think of its effects less in terms of immediate changes of attitudes and more in terms of the political advantages of the victories which derive from a psywar exercise. (Though the British invented the phrase 'hearts and minds', much of their practices in Malaya, for example, were not so much directed at changing attitudes as controlling *behaviour*.)

In the USA, no doubt, the whole counter-insurgency network which grew up in the 1960s had its share of enthusiasts who were not critical enough of what its role was or should be limited to. In the US Army, however, psywar is regarded as having finally come of age in the Vietnam War and though there are many who would be sceptical even of this, there can be little doubt that the more positive role described in the preceding pages is the one which psywarriors see themselves playing in any wars in the future. In Britain this is also slightly more likely since the appearance of General Frank Kitson's book, *Low*

Intensity Operations, a controversial but influential text which argued for more psywar in future conflicts (7).

Probably the most comprehensive field report on the effectiveness of modern psychological warfare is that published in May 1967 by the Joint United States Public Affairs Office (JUSPAO) and entitled 'Psyops in Vietnam: indications of effectiveness'. Although JUSPAO was itself committed to the psywar effort in that it was formally responsible for it in Vietnam, it was also independent of the units themselves. Its report drew evidence from a number of sources to assess effectiveness – one, for instance, being the priority which Hanoi placed at that time on the assassination of the head of the Chieu Hoi defection programme (8).

The JUSPAO report quoted several Viet Cong documents that seem to give some indication of the effectiveness of psywar. For example:

A report by the Propaganda and Indoctrination Section of the Ba Ria Province Party Section, March 13, 1966

. . . we failed to have any clear plans to deal with enemy Open Arms activities or soldiers who deserted.

For example, in Lon Phuoc the people received and kept enemy documents concerning Chieu Hoi, but our cadre did not do anything about it, or did not dare to indoctrinate them because the cadre were afraid their daughters would work for the enemy . . . .

From a VC analysis of desertion:

During the last year there were deserters from all agencies and units, especially before and after each battle. This is an important problem. The number of deserters is as follows:

Cadre A and B (party and group members)	5%
Cadre who surrendered and had weapons with them	4%
Cadre influenced by Chieu Hoi	5%

Reasons

Indoctrination and ideological guidance neglected
Some cadre were authoritarian and did not behave kindly to troops
Many were demoralized by bombardments, shellings and fierce attacks
Many influenced by Chieu Hoi

We must train our troops to hate the US.

Then in 1967, the VC developed a programme to combat the effects of Chieu Hoi.

Directive against desertion: from Troung Cien Current Affairs Committee to local units

. . . many deserted because they were shocked by casualties after battle . . . .

The Viet Cong document goes on to recommend a study of the deserter, including biographical data, his mood immediately prior to his desertion, his family situation and attitudes, what kinds of things he collects, how he gets on with his colleagues. The document also explored other effects on morale such as 'melancholy music', and techniques to counter the Americans' propaganda, like daily indoctrination and the 'publicizing of contempt' for those who do desert. Three types of deserter were recognized and the Viet Cong had different fates for each: the revolter – who was killed; the ralliers – who were re-educated with guarantees to their dependents; those who returned to their families – who were returned to their unit.

The JUSPAO report also contains evidence of what it calls a 'thought control' plan of the VC to prevent desertion. This included a study of all suicides and self-injury among the VC to see how this form of battle avoidance could be checked.

JUSPAO's aim was clearly to marshal evidence which shows how obsessed the VC were at that time with the desertion rate and how seriously they took the task of combating it. By implication the Chieu Hoi programme at least, and other psychological methods, were getting through as intended.

Many other anecdotes were also given:

For example, they [the psyop troops] published a diary of one of the ralliers – only to find that the boy's mother came looking for him – she had read the diary and wanted to find out if he was OK; in another case, the psyop troops gave away gifts, like psyop soap and later found that the gifts had all been confiscated; to the psyop people this meant that the soap must have been judged effective by the VC; at another time, a defector turned himself in with a weapon; he was then made to broadcast to his company by name and within 24 hours 88 more men had defected; in other cases, the Chieu Hoi teams had posed as tourists in a local beauty spot and had gone to a local family to ask for water; they had spoken to the parents in general terms about defection – having found out where their boy was they then directed broadcasts into his areas naming him personally, mentioning his relatives by name. Two days later he defected.

Certainly General Westmoreland was convinced of the value of psychological warfare. In October 1967, for example, he wrote to Ellsworth Bunker, the US Ambassador in Saigon, applauding the psyop personnel there and saying that current arrangements were working, and arguing for more men to be supplied (9).

One study of the Korean War concluded that 80 per cent of all Communists who gave themselves up voluntarily had been influenced by psychological warfare (10); and a British account of events in Malaya noted that the Communists had at one stage taken to flying their own planes low over the jungle, broadcasting the message to their own men not to believe enemy propaganda (11).

This study, by F. H. Lakin, assessed effectiveness in the British tradition by measuring behaviour. Table 32, for example, gives some idea of the proportion of defectors who were influenced to surrender through psychological warfare (12).

Table 32 Percentage of 432 Malayan deserters induced to surrender by type of inducement

Period of surrender	*Read leaflets*	*Heard broadcasts*	*Believed promises*	*Were otherwise influenced to defect*
1949–51	63	0	42	26
1952	70	0	42	27
Jan–June 53	77	10	59	52
Jul–Dec 53	84	53	55	37
1st quarter 54	88	48	58	47
2nd quarter 54	96	61	61	50
3rd quarter 54	97	73	63	47
4th quarter 54	95	77	36	33

A parallel study by the American Operations Research Office also found that psywar was a 'definite causal factor in both disaffection and surrender' (13).

On the other hand, it has been argued on different occasions (though sometimes by the same organizations) that psychological warfare has little effect on an enemy. A study by Lessing Kahn and Julius Segal, carried out in Korea for the Operations Research Office in 1953, concluded that 'leaflet and radio broadcasts had minimal effect . . . in altering enemy predispositions' (14). And Harris Peel, the Fort Bragg representative of the United States Information Agency, the government-sponsored world-wide propaganda outfit, referred to psywar in

Vietnam as 'amateur' (15). (It has to be remembered that the USIA and military psywar essentially do the same job and so there is a jealous rivalry between them.) This criticism comes from the belief that truthful information is what people want, especially in an emergency. Credibility takes ages to build up but pays off. Peel cites as evidence of this the natural reactions of many people in the world to believe only the British Broadcasting Corporation and Voice of America when even their own countries are in turmoil. Harris Peel's point is well made, but it is surely also true that the BBC or the VOA cannot hope to cover every little skirmish in a war like that in Vietnam which went on for years. Rather it would seem that active, tactical psychological warfare, employing all sorts of subterfuge and other ruses, has its place, as does a high credibility strategic radio source like the BBC. They need not conflict.

But just as there are different ways that propaganda can achieve credibility so its effects will be felt in different ways. Usually scientists and soldiers have concentrated on defection rates since these are the most objective measure for all to see. At one time in Vietnam, however, there were plans to use other, more subtle, psychological techniques to assess effectiveness. Some of these were thought up by James Dodson, of CRESS, and produced in an undated document which was never approved for publication (16). The techniques included giving tests to Vietnamese which measured the cohesion of a village, placed their hopes and aspirations for the future on a 'ladder', and measured how frustrated they were with their life (using the Rosenweig Picture Frustration Test). Dodson's reasoning was that the more successfully psychological warfare techniques sapped the morale of the civilians, the less cohesive villages would become (on a sociometric test there would be fewer mutual choices and more isolates), the less ambitious their hopes for the future would be (they would expect to climb only a few rungs up the ladder), and the more frustrated they would feel (as revealed in the stories they told about the pictures in the test).

I do not know whether such tests were ever given. Even without them, however, it seems fair to say that the effectiveness of psychological warfare is beyond dispute though the exact when and how is far less certain. No doubt its enthusiasts claim too much for it – and equally its critics devalue it too much. At any rate most armies want more of it, rather than less.

To end this section, however, we must consider a wider and conceivably even more important question, especially now that we know the outcome of the Vietnam War. This relates to the general approach of psywar – that basically one should not stress the ideological side of affairs but concentrate instead on more personal, psychological matters. Did that really work in the long run? Perhaps it helps to sustain one's own morale and to shatter that of the enemy when things are going badly for him; but when they are going badly for those using the psywar, as happened towards the end of the war, perhaps this approach merely makes total collapse that much more likely. Can psywar ever hope to transform a *legitimate* political grievance into something else that can then be manipulated away? Isn't it misleading oneself even to try to do so? Isn't psywar, in this sense, encouraging soldiers to think in apparently value-free social science terms when in fact the political context cannot be ignored?

There is one other issue, too. There is some evidence, as was mentioned in chapter 1, that a bad analysis of the state of Viet Cong morale and motivation led President Johnson to step up the bombing in 1966 in the belief that it would quickly end the war. We can see now that it may instead have prolonged it. Psywarriors may well have had some major effects on policy. But these effects were not always the ones that were intended.

25 Conclusions: Military Psychology, Political Psychology and Politics

We live at a time when the military uses of science are of immense concern to everyone. The apocalyptic potential of nuclear explosion and the dangers of accident inherent in many chemical and biological weapons mean that, more than ever, war is too important to be left to the generals. These issues, however, have been with us for a quarter of a century at least. To them, recently, have been added the proven excesses of government agencies – military and quasi-military: the CIA capers with LSD and other psychochemicals in the USA and its bizarre activities in many Third World countries; FBI buggings in the United States; the British Army and its brutal interrogation 'techniques' in Northern Ireland; the Nixon administration itself and the Watergate scandal. Most of these incidents involved the use of techniques, originally devised for military deployment against enemy countries, against either neutral powers or domestic political opponents. Surveys show that the armed forces are as popular as ever but that the military authorities are not. No doubt this has something to do with the decline in respect for authority in general due to wider changes in society, but the behaviour of a number of authorities, or what has gone in their name, must – at least in part – be responsible.

With such a background it is perhaps understandable that so many books about science and its applications to military affairs in the last ten years have been written with a marked anti-military slant. The shadow of the Vietnam War, of course, has also darkened almost everyone's motives. Researching and writing this book, I felt very keenly this kind of 'anti-military' reaction myself on several occasions. As I mentioned in chapter 1, so many of the studies seem silly or provocative – they were asking for trouble. The 'Value of Life Questionnaire', for example, was just asking for the congressional inquiry that it eventually got. To produce tapes imitating tribal gods is another provocative example – a kind of psycho-technological inquisition that could easily backfire.

Taking the evidence in the book as a whole, however, and considering the more humdrum experiments alongside the controversial ones, is it fair to judge all military psychology as provocative and dangerous? Should, in fact, our view of military psychology be anything at all like the views we hold about the military uses of other sciences? Should all military psychology be condemned out of hand?

For this concluding chapter I have chosen to focus on several fairly specific issues which highlight, to me at any rate, some of the moral, political and professional controversies raised by the studies considered in the preceding chapters. The theme they share is the link between military psychology, political psychology and politics. I make no pretence that my list of issues is exhaustive; but I hope that it will stimulate others, closer perhaps to the material or the political–military background of specific ventures, to pose more such questions.

THE BOGEY OF 'MIND CONTROL'

Often, when the words 'psywar', 'psychac' or 'mind control' by the military are used, they are couched in such a way as to suggest some hidden power which the initiated have over the unfortunates on the receiving end of these mysteries. Headline phrases like 'the mind benders' are familiar currency. The techniques seem to be regarded invariably as unquestionably effective, but the account I have given in this book, deliberately stressing the 'nuts and bolts' of every technique where possible and giving some idea of its success rate, will I hope dispel this bogey. There is nothing mystical or mysterious about psywar. There *have* been times when certain individuals have behaved as though psychological techniques could wreak terrible damage – Donald Tepas's approach to harassment, discussed on page 411, is one which comes to mind in this context, written up as it is, as though Tepas himself believes his techniques give him extraordinary powers. So to do Commander Narut's disclosures at the NATO conference in Oslo – on how to train assassins (pp. 248–9) – invoke a science fiction flavour. The CIA disclosures in 1977, about attempts by the agency to counter Russian and Chinese brainwashing techniques twenty years earlier, are the most flamboyant of all. But even their attempts to change the behaviour of spies or statesmen by chemicals or hypnosis almost invariably failed.

Interfering with 'the mind' in most cases translates more prosaically into attempts to understand and control the *behaviour* of men in war.

Though many projects – besides the CIA capers – have not worked, one or two have done so very successfully (for example, the effects of coordination in squads and some of the defection programmes). The majority, however, seem to have produced a modest but significant improvement in military productivity – and one that is, in comparison with the scientific effort in other fields, *cheap*.

So I hope that as well as persuading the reader of the enormous range of military psychology, this book helps dispel some of the vague but potent anxieties which have grown up and confuse people about the nature of 'psywar'.

WAR PSYCHOLOGY VERSUS PEACE PSYCHOLOGY

Another question which struck me time and again as I was preparing the material for this book was the irony of the situation which the profession of psychology now finds itself in. For years there have been at least two groups of psychologists who have interested themselves in psychological aspects of war and peace. But they have approached the subject from totally different standpoints and appear to have been unaware of (or at least not to have taken account of) the work of their colleagues.

On the one hand, there have been the social psychologists and ethologists who looked at the nature of aggression and violence, its biological and instinctive roots. They have mainly considered war in the evolutionary context – as 'intraspecific aggression' – and have studied such things as primate aggression, 'territorial imperatives', ritual fighting and so forth to see what lessons these patterns of behaviour might have for the conduct of warfare. Their aim appears to have been to try to understand the springs of war so that we might better bring about peace. Similarly, it is now twenty years since the inception of the *Journal of Conflict Resolution* in 1957 which embodies the aims of another group of scientists, psychologists prominent among them, who view wars and other conflicts (like industrial disputes) in terms of a certain number of games, 'zero-sum situations' or dilemmas which seek to characterize the essential psychological features of conflict in a controlled and simplified way – again in the hope that this may provide understanding and perhaps produce systems for preventing conflict of the escalation of small 'disturbances' into something far worse.

Neither of these approaches is necessarily invalidated by the material presented here. Both fields of study have produced information that is of great intellectual interest. But it is arguable that either has really thrown light on wars to an extent where control of the course of a war, or the people fighting in it, is even remotely possible.

It is therefore not entirely inappropriate to ask whether the somewhat more practical researches of war-making reported here can tell the peace-researchers anything. Again, I make no general claims, but what follows are just five examples of areas where I think a joint input might be fruitful – or at any rate where the lack of such collaboration might be expected to hinder the truth.

Leadership. Conflict research has tended to regard leadership as a role, suggesting that a leader's personality matters less than the constraints put upon him by the situation. Yet the military research, reported here in chapter 7, takes almost the opposite view, regarding personality as very important. We have also seen that in a unit like a battalion the top four or five leaders perform very different functions (and that those functions change as the battle approaches, begins and develops). Very little is known about the interplay of leaders and how this affects the course of a battle or even of an entire war; it could be a fruitful area of research.

The 'weapons effect'. Theories of deterrence tend to regard more lethal weapons as more of a deterrent. Yet more than once in this book there is the suggestion that though this may be broadly true there are wide variations. Most important, the *use* of any particular weapon may depend on the context of *other* weapons already in use or available for use later. This appears to affect how lethal a weapon is regarded and may tell us something about the nature of escalation.

Risk-taking. Conflict research has looked into risk-taking in the theoretical context in quite some detail. But it has rarely been concerned with the actual risks soldiers have to take. The risks reported here are perhaps relatively small ones – whether to say a photograph does or does not show a target, for example – but they are far more real than the theoretical risks taken in 'conflict games'. There seems to be plenty of scope for cross-checking between the two to see whether the more theoretical studies of the conflict researchers are anything like the truth.

Survival research. This is clearly a fertile field for military psychologists. Can one say that a nation's view of how well it will survive a nuclear war will affect its willingness to engage in such conflict? The civil defence efforts of the Russians in early 1977 created a flurry of excitement in some western observers which suggests that survival capacity might well be important. The survival research reported here may therefore need to be more widely known in order that the conflict researchers may take account of it in their calculations.

Fighters. The mass of research on what makes a good fighter offers scope for comparison with ethological studies to see whether man's behaviour corresponds at all to the picture we get of warlike animals and the part which the more aggressive individuals play in a society. (Marc Pilisuk quotes evidence to show that, in Vietnam, units with highly aggressive individuals encountered the enemy more often – but as a result other men tended to try to avoid these units (1). So what is the overall effect of aggressive people on wars? And there have been no studies, for example, of whether these aggressive individuals get promoted and what effect this has if they do.)

As I say, these may not be the most important points to be made about the studies reported in this book, and their relationship with other branches of psychology concerned with war and peace. But it does seem clear that the three fields – of military psychology, ethology and conflict studies – can learn a lot from one another and have been prevented from doing so by the reluctance of the first group to mix their findings with the others.

Nor is this the only general problem which psychologists as a professional group will have to face. I suspect that, like me, many psychologists have been unaware of the vast amount of research that has been going on. They may well have thought that psychology's only concern with war and peace was to reduce the probability of war and increase that of peace; or to help find ways of resolving conflicts. This, we can now see, is a mistaken view. Like all other sciences psychology has been used, and abused, for military purposes.

MILITARY PSYCHOLOGY, POLITICAL PSYCHOLOGY AND SECRECY

In the introduction I discuss some of the effects of secrecy – notably the way it can breed bad social science. Having studied the evidence of the intervening pages, several other points are worth adding here.

In the first place, one might ask how it is that all this research has been possible without psychologists themselves drawing wider attention to it. Clearly, psychology as a discipline has no structure which enables it to search out these issues and give them a full airing. The American Psychological Association, the largest organization of its kind in the world, does have one of its twenty-odd divisions devoted to military psychology – Division 19. But this is a perfect example of the way one part of the profession has talked to itself while others, who should have been interested in this research, contented themselves discussing related issues without apparently inviting the military specialists to have their say.

For example, in the fifteen years since 1962, Division 19 has held at least ninety-one symposia or lectures at the annual conferences of the APA. Invariably these have been on topics in the first sections of this book – selection, marksmanship, training, leadership, human engineering, that sort of thing. Only six of these ninety-one symposia referred to foreign area research, Vietnam studies, or psywar. A handful looked at survival studies, none at all referred to captivity.

But, and this is most important, in only *three* symposia were non-military psychologists invited to make known their views. Yet during precisely the same time, Division 9 of the APA – the Society for the Psychological Study of Social Issue – held a dozen seminars with such titles as 'Models of revolutionary behaviour' (1969) and 'The psychological aspects of war in Vietnam' (1966) yet did *not* invite participants from military institutions. In fifteen years' research documented in the *American Psychologist*, the house journal of American psychologists, I could find only three meetings where the speakers were listed as coming from both military and non-military institutions. When you add to that the fact that only one or two of the symposia and papers on the ethical problems of psychology have considered military research, and that this compares very badly indeed with the attention given to race, women's studies, testing in general and the disadvantaged, you can see how blinkered psychologists have been on this issue.

It is not as if the clues have not been there. For instance, *The Pentagon*

Papers were published in 1971 and these contained *specific* discussion of psychological aspects of the Vietnam War on no fewer than 109 of the book's 677 pages (2).

A book published in Britain in spring 1977 underlines a different aspect of this blinkered approach of psychologists. The book was called *The Technology of Political Control*, at least one of its authors was/is a psychologist, and it seems to demonstrate that even when psychologists and allied scientists have been interested in this general area they just do not have the wherewithal to find out anything new (3). *The Technology of Political Control* looked at a small number of counter-insurgency techniques including a short section on interrogation and one on sensory deprivation. Yet all the prime sources for this were either established (and so fairly ancient) texts or newspaper clippings. Even here the scientists had not really discovered any new material, or even any new links, for themselves.

Besides the straightforward secrecy in this area, then, psychologists appear to have been guilty of an extraordinary compartmentalization in their thinking that has already had an effect on the way the subject is – or rather is not – understood. For instance, Senator Fulbright has said that, in this field of the psychological aspects of war and peace, psychology's role is to provide a 'new dimension of self-understanding':

> We have got to understand, as we have never understood before, why it is, psychologically and biologically, that men and nations fight; why it is, regardless of time or place or circumstances, that they always find *something* to fight about; why it is that we are capable of love and loyalty to our own nation or ideology and of venomous hatred toward someone else's. We have got to understand whether and how such emotions satisfy certain needs of human nature and whether and how these needs could be satisfied in a world without war.
>
> Only on the basis of an understanding of our behaviour can we hope to control it in such a way as to ensure the survival of the human race (4).

Yes – but . . . the crucial point now, surely, is that military psychology has gone far enough, apparently without people like Senator Fulbright realizing it, to provide some understanding of the way people behave in wars – but behaviour of a quite different order from that which the senator would like to see. The research is *not* leading us towards being able to control the way we slide into wars or to understand why we are capable of loyalty and hatred. It is leading us towards being able (however unsatisfactorily) to dislodge the loyalty of enemy figures and to understand how, for example, the presence of an aggressive man in a platoon or squad changes that unit's performance. Senator Fulbright posed the problem as if the scientific questions and the political ones could be kept apart. But, as we have seen, the research just does not allow us that luxury. The statement of the senator's is now ten years old, but it would still hold, I would judge, for what many people hope for from psychology. This is the effect of the lack of publicity in this field – the expectations most people have about it are quite out of touch with the way it is going.

Some of the defection studies particularly suggest, or are designed to suggest, that political views are determined by psychological factors. One reaction to this is 'so what', but another – very prevalent – is as a result to devalue any political stance once it can be 'explained away' Social change then becomes a matter of social engineering rather than of politics.

This is the nub: that all psychological research, when carried out by or for the military, is likely to look hostile – as was the case with Project Camelot. Yet in some ways the more purely scientific aims of Camelot were worthwhile. But this would only be so, provided that the results of the information were made available to all. And this brings us back to secrecy. Much of the information reported in this book is of use only if it is possessed exclusively. Surely, if the tribes know that their tribal differences are being exploited by a common enemy, this must change their perception of the way they respond to the exploitation of those differences? In the first chapter I have drawn attention to some of the problems of secrecy. But we can now see that not only does it lead to bad science but perhaps most important of all it has allowed military psychology to turn itself into a crude, *ad hoc*, political psychology, one that is dominated by attempts to control specific situations rather than attempting to achieve a more general understanding of the psychology of warfare and peace with a view to everybody being clearer as to what it is all about.

Two other points about secrecy are worth making briefly. The first concerns the non-military use of findings from military research. Armed forces are in general zealous to have it known that the peaceful spin-offs of their research are always a major consideration. Yet in psychology there would seem to be a long way to go on this score. Due to secrecy many of its more useful findings have not been picked up as widely as they deserve. The difference between leadership and command, for example, would seem to be a useful distinction which could be transferred from a military context to an industrial one (and might be especially suitable for civil service departments where there is a permanent staff under more temporary political masters). Or again the concept of slack on naval ships might well be introduced on a more systematic scale into many enterprises to ensure productivity stays at a stable level.

A final drawback with secrecy is that, as recent events have shown only too well, it is extremely difficult to maintain if political divisions are involved. And it often follows that the process of revelation does incalculable harm to the country that has carried out the secret enterprise – witness the Camelot fiasco and the ramifications of Philip Agee's CIA *Diary* (5). Might it not be far better for military–political research to be carried out openly, or at least to be published openly, so that it can be freely reviewed and critized and its implications fully aired?

The harm that secrecy does brings us to the two most important questions we must attempt to answer before finishing the book: do the studies reported here make war more likely? and what is an abuse?

DO THE STUDIES REPORTED IN THIS BOOK MAKE WAR MORE OR LESS LIKELY?

There can be no categorical answer to this question, either way. In practice we have to address ourselves to a set of more specific issues.

A less lethal alternative. To begin with it is always worth remembering that a chief characteristic of the military uses of psychology is that in many cases they do not involve killing; civic action, heart-and-minds, for example, often involve the creation of drainage systems, new bridges, etc., rather than destruction. Such activities are designed, in the short run at least, to relieve rather than exacerbate suffering. And in so far as these studies make clear the risks involved in various options, the overall risk of inadvertent catastropic war is reduced. Other studies have shown how to stop people taking risks (groups, for instance, seem to be more conservative) and how to be more specific about their guesses in balancing pros and cons – all these, it must be concluded, would probably reduce the likelihood of war.

On the other hand, as I have mentioned several times already, the creation of less lethal alternatives to major weapons may not be the advance it seems. By increasing the steps to total war, less lethal weapons may actually lessen the gradient and thus make it more likely. (The experiments in which this has happened are admittedly artificial but the possibility has to be considered.)

Dehumanization. Major political changes in the world are likely always to be the main causes of large-scale conflict. Other things being equal, however, the studies reported here where the main aim has been to demean the enemy in the eyes of soldiers and politicians are likely to increase the chances of war by decreasing respect for that enemy and his political system. And, of course, in so far as many of the 'psychological' methods involve the encouragement of conflict between tribal groups (perhaps sub-divisions of a nation) then such conflict *within* countries may be at the expense of wars *between* countries. Less blood may not necessarily be shed, but a full-scale war, in which one's own side is involved, may be avoided. It is arguable whether this counts as a reduction of conflict.

The image of the enemy. The image of the enemy is very important – Dr Jerome Frank, in his book *Sanity and Survival*, devotes an entire chapter to this issue but makes several points which may now need to be modified (6). For instance, he argues from experimental studies that enemies who are unknown are more feared than those who are known. This may not say much about war. Civil wars, where even brothers may be enemies and thus know each other very well, are notorious for being bloody and more prolonged than other conflicts. Dr Frank suggested that getting to know other nations would help the cause of world peace. Clearly the context of that familiarity is important; as we have just remarked, the studies of potential enemies, discussed in Part Five of this book, in dwelling on undesirable characteristics and weaknesses, although making these potential enemies better known, may make war against them more likely for precisely the reason that they are regarded as adversaries who can be beaten and are worth beating.

On the other hand, studies of censorship show that people filter information coming in to suit their preconceptions. Particularly, we have a tendency to see the leaders of potential enemy nations as the real villains. Psychological studies showing such leaders as either typical of their nation, or less warlike than the rest, might be used to make one's own military leaders more cautious, and more realistic about potential threats.

Frank also makes the point that ideological conflicts (including holy wars) are among the most bitter (though it must be said that territorial claims have recently been as bitter as any). What could be more risky than scientists producing 'evidence' to show that a potential enemy's ideology, far from being the one that your own side supports, is not a real ideology at all but a psuedo-scientific trick? Yet this has been the message of many social science studies for the military.

Finally, Frank argues that 'Before wars break out military over-confidence is usually a characteristic of both sides and based on wishful thinking.' It is difficult to know how – or whether – psychological studies will change this. In providing more information about the men it should make military leaders more realistic; in its general tone of optimism it may make them even more (over-)confident.

The military mind. This too has been modified in recent years. It is still true that those attracted to a military career (or who have been in it for some time, having been conscripted) are more authoritarian than most, more conservative, more bureaucratic and likely to have a more negative view of human nature – to assume that people tend naturally to be selfish, aggressive, untruthful and so on. On the other hand, it has recently been shown that the military mind is *not* more ideological than the non-military mind; if anything, the studies show, it is more pragmatic (7). If true this has certain implications in the conext of our discussion. It would appear to follow, harping back to the points made earlier, that the psychological 'explaining away' of political positions would be particularly acceptable to the military mind. On the other hand, it would render suspect Frank's comment that the armed forces tend to be over-confident of their abilities before a war. This may be propaganda for their own men but if they are pragmatic they should not be taken in; nor – since they are not especially ideological – should they have a zealous dislike of the enemy any more than the rest of the population and which might lead them to foolhardy acts. In fact, the picture of the military mind that now obtains is one of generals as cautious but firm war-makers rather than as headstrong wild-men.

The real significance of this new formulation of the military mind may be that it explains so well why military–political psychology has grown up. Its non-ideological stance suited the generals so well. They simply see military–political psychology as a pragmatic way to deal with operational political difficulties which they have to face; others, as over the Camelot issue, see the whole business as posing fundamental problems about human freedom. Military–political psychology may be a crucial area where the generals differ from the rest of us.

Civil rights implications. *The Technology of Political Control* has been referred to earlier in the context of discussion of how well informed psychologists are (and are not) on the general topic of military psychology. The book also exemplifies another important issue: the use of counter-insurgency techniques not against legitimate enemies but against *domestic* political opponents. It should be clear from all sections of this book that there are many techniques which could be used in such a way: the psychological analysis of political leaders, the manipulation of captivity psychology, exploitation of the psychology of lying, manipulation of the need to conform, the psychology of cell structure, not to mention sensory deprivation, brainwashing and hypnosis. It is the thesis of *The Technology of Political Control* that techniques like these (it covers only the last three in the list above) were specifically developed for domestic use as part of a wider argument that western (capitalistic) democracies can only survive with such brutal help. Clearly the authors are wrong in this – most of the techniques were developed for use abroad. The threat of their use domestically remains, of course; but we should be wary of attributing sinister motive where there is none (this is usually the accusation levelled against military authorities). Rather than allege conspiracy where none exists, it would be far more useful for scientists like the authors of *The Technology of Political Control* to think up ways to ensure that the civil rights implications of counter-insurgency weapons are publicized and then to devise ways to prevent domestic use. A conference on this topic with both the military authorities and civilian psychologists present might be a useful starting point.

The military uses of psychology thus have no simple single effect making war either more or less likely. The discussion, even when confined to more specific questions like those just examined, is by no means conclusive. In order to try to be a little more helpful, however, I will now try to come at the same question from another direction – by asking: what is an abuse?

WHAT IS AN ABUSE?

To begin with, helping rifle marksmen or missile guiders to shoot straight is not, as I see it, an abuse. If you are going to have a rifle or a missile in the first place, there is no sense in having an inaccurate one. The greater the ambiguity in a situation, the more likely there are to be unintended consequences and the greater the unintended and unnecessary suffering. So I would regard as legitimate any psychological device which made *existing* hardware more precise, or safer to handle – in the long run I think this would keep suffering to a minimum. So most of chapter 3, on artillery, target detection, camouflage, navigation, code breaking and so forth, would seem to me legitimate military *uses* of psychology.

Similarly, the studies of the allocation of men to units and of leadership also seem on balance to do more good than they do harm. If it is true, as it appears to be, that the intelligent avoid combat then at least armies should be certain of their facts before moving men around. Some might say that the natural state of things is good because it keeps the intelligent alive longer as a national resource after hostilities are over. Others might argue that if the intelligent were nearer the action the authorities might be forced to conclude matters more quickly, thus saving lives. A third view might be that intelligence should not affect who gets to the front. But either way it is better to have the information than to remain in ignorance.

With leadership, too, in so far as it opens up a vague concept for study, increased knowledge must be for the general good. Better leaders are, by definition, those men who win battles but also keep their own men alive. Better leaders do not commit atrocities, do not go in for unnecessary suffering.

Much stress research also seems legitimate. The precise figures for risks associated with various combat specialities surely help rotation policies to be effective and so preserve the sanity of men in high risk positions. Witness the difference in breakdown rates between the Korean and Vietnam Wars, achieved partly as a result of such studies.

Then there is a second area of research about which, it seems to me, it is very difficult to generalize, where it all depends on circumstance. I would include in this the research on what makes a man a good fighter, what makes a man a good bomb disposal operator, how you combine different personalities in squads to make them better fighting units, the use of animals in a military context.

With this range of studies there opens up the possibility, small though it may be, that the mere possession of an ability to do something makes it more likely that that something will be more frequently carried out. If, for example, one could select with some accuracy the men who can defuse bombs and thus train them more quickly, does this, however slightly, make them more expendable in the eyes of their superior officers? Were this to happen it might affect the delicate interplay between terrorist and army. At one stage in the Ulster troubles, for example, the British Army bomb disposal men got so good at disposal and the security on explosives so tight, that the IRA were forced to use chemicals that were more accessible but less stable – thus posign *more* of a threat to innocent people. Is this good? The same type of argument is mentioned by Pilisuk in discussing good, aggressive fighters. He notes how, in Vietnam, units with very aggressive individuals in them met more trouble than units without such men (and that as a result other soldiers tried to boycott these especially aggressive units). With this kind of study perhaps the 'abuse' (it does not seem quite the right word for the criticism I am about to offer) is that the impact of these findings on the course of conflict has never been examined. On balance it seems that such studies are legitimate but that better tabs should be kept on their impact. I return to how this might be done a little later.

A third group of subjects includes research into atrocity makers, captivity behaviour, interrogation techniques, brainwashing. Here the problem is the traditional military one – that a weapon can usually be both offensive and defensive. It is surely legitimate, because humanitarian, that researchers should study who commits atrocities in the hope of being able to stop further brutality. Not merely because by definition atrocities involve unnecessary killing but also because publicity about atrocities can be bad for morale. It leads to cover-ups which sometimes do more damage than the atrocity itself.

But each of this set of subjects can be turned around and used in a far more bloodthirsty way. Atrocity research, as we have seen (page 246) can be turned on its head to train assassins; interrogation techniques (page 274 ff.) can produce lasting damage psychologically; an understanding of behaviour in captivity can help guards terrify prisoners so that they give away information; it is claimed that brainwashing can change a man's political beliefs almost against his will (though as we have seen this claim is often exaggerated). Some people object to these procedures flatly on moral grounds. Many others, I suspect, would say it all depends on circumstances – the relevant criterion being the need to know something. *If* torture of one of the enemy will save many lives of your side, at what point – if ever – does it become legitimate to torture him? Does that kind of situation ever arise? Probably these psychological techniques are controversial because they are new and their effects on the personality are long-lasting. Perhaps more than any others the circumstances under which such methods, when they are used, should be studied so that eventually some guidelines may be layed down. I return to this issue in just a moment, in discussing a way to monitor studies of this sort; but I confess I do not hold out much hope for general agreement. Wars being what they are, it is an area almost bound to remain controversial. Armies will always need information.

I have little doubt in my own mind that Part Five of this book – chapters 17 to 24 – contains a great deal of material that many will view not merely as an abuse of psychology but as bad psychology in any case. Taken individually, many of the studies may seem reasonable enough in themselves. It is reasonable to try to find out who would make the best stability operations officer (if you are going to have stability operations). It is reasonable to try to induce defection of the enemy and to devise methods to do it, to take advantage of the enemy's psychological weakness, be they civilians or military groups in the jungle. But far more than any others in this book these studies commit three cardinal errors.

Methodologically, for example, they often leave a great deal to be desired. I have drawn attention (on page 337 ff.) to Andrew Molnar's brief look at insurgencies, in which in a very few pages he drew enormous conclusions from scanty evidence. Later, between pages 346 and 351, I showed how the methodology of morale studies in Vietnam affected presidential decisions but were themselves based on very small samples, provided contradictory conclusions and how, in one case at least, a study that did not fit the prevailing 'wisdom' was suppressed, in a quite blatant abuse of scientific procedure. Partly as a result, as Pilisuk notes, Third World leaders were at one point getting tired of poor quality US social scientists asking the same questions time and again, each one to come to a different conclusion, this confusion leading to yet more study.

We are back, inevitably, at secrecy. No need to repeat the discussion we have already had, but I will simply add that secrecy is often due as much to the shabby motives of the poor scientist carrying out the research, as to any large-scale conspiracy of silence by 'the authorities'.

The most important abuse of the material in Part Five, however, is of course, as I said in the last section, that more than any other type of study, it has made war more likely. It is impossible to say how much. Some of the arguments are, perforce, artificial as with 'the weapons effect'. Others are intuitive, as with the reasoning that emphasizes the demeaning effect of much military–political psychology. But the worry is there, more than anywhere else, that a secret, bad science is lulling military thinking into a posture it may some day regret.

Let us be clear: it must be fair for any nation to use its resources as it best sees fit to ensure its own survival. Arguments about internationalism, though they may offer the best long-term solution, do not appear to be making much headway in the present decade. But unlike almost any other weapon, military–political psychology has developed, or been allowed to develop, without any safeguards at all. At the end of the day it may turn out to be a weak weapon, as Adam Roberts suggested with regard to South African psychac teams and the psyops outfits in Vietnam (8). But it is unlikely that the commitment recorded here will disappear overnight (in fact, as the final section shows, there are plenty of plans for military psychology in the future). The edifice of military–political psychology has been allowed to grow not merely unchecked but also un-thought-out. Michael Klare argued in his book on US military planning, *War Without End*, that the provision of advanced scientific gadgetry in Vietnam bred a new generation of 'scientific mercenaries' – using a term with overtones of an ever-ready, swashbuckling attitude to war (9). No one knows whether the development of military–political psychology has had this effect on either psychologists or soldiers. I have tried in this book to put some order into things; but I am well aware that there is a long way to go yet.

Before we consider, in a final section, what the future may hold, can we just make a start on a suggestion for some kind of institution that might help monitor these future developments and put past ones into context? Senator Fulbright, in his preface to Dr Jerome Frank's book, *Sanity and Survival: Psychological Aspects of War and Peace*, suggested that international organizations like the United Nations should be brought more into things. I rather feel a different type of institution is called for.

From the way the various sorts of psychologist fail to keep in touch with one another's work, which this chapter has shown to have been the state of things for many years, the first thing to do seems to be to introduce one type to the other. Psychologists from division 19 of the APA seem to be fairly familiar with the research reported in this book up to, but excluding, Part Five. Division 9 psychologists seem ready to ask wider questions, though perhaps not the most important ones since the work which falls in the most controversial area – military–political psychology – seems to have had less of an airing than the rest. Together they might begin a progress of sorts.

But there is more to do than this. What should the attitudes of psychologists be towards secrecy? Should they carry out classified research knowing that it may have the kinds of effects sketched here? Knowing that quality in the social sciences is particularly vulnerable to secrecy. For the psychiatrists particularly, what advice should they offer on conscription – and on psychiatric discharges? Should they, for instance, abide by military rulings or by those of the psychiatric profession if these differ?

A few years ago the Group for the Advancement of Psychiatry (GAP) produced informative booklets about 'forceful indoctrination' and the hazards of shelter living in a post-nuclear age and this sort of service might well continue.

An institute – not attached to the UN or some similar body but set up by psychologists themselves in the manner of the British press council – might be a start towards being able to carry out these functions. It might be attached to a university (but should be independently funded, perhaps like some of the peace research institutes). It could carry out its own research. Its aim should *not* be to stop all military psychology altogether but to try to put some order into the field, as I have begun to try to do here, so that the public is kept aware of the general trends in research and development. It could try to define abuses better than I have been able to do in the hope of their more efficient prevention. Organizations like the North American Congress on Latin America, the British Society for Social Responsibility in Science, or Amnesty International might provide other models. It would, I am sure, help the image of psychology and psychologists if they were to take the initiative.

Whatever happens on the institution front, however, one thing is certain – there *will* be a military psychology in the years to come and it is unlikely to be very much less controversial than it has been in the past. So the need for an institution of some sort will not go away.

FIGHTERS OF THE FUTURE

How can I be so sure that the kinds of studies reported here will continue? Because psychologists have learned one thing: that if you wish to be part of the future you must help plan that future so that it turns out the way you want it. There is now an active field of forward-looking studies exploring the ways in which psychology can be applied to the armed forces between now and the end of the century. So the wheel turns.

Perhaps the single most comprehensive document on the future is that produced by social scientists at the US Combat Developments Command, and entitled *Man and the 1990 Environment* (10). In typically precise army fashion this document projects, first in general terms, the nature of society in the 1990s in four areas – technology, economy, sociology, politics. This, however, is followed by twenty-three specific implications of these general changes, each of psychological importance, and each of which has a number of policy implications for the army. Here is a list of some of the most intriguing:

People will be considerably wealthier in the decade of the 1990s;

In 1990 men will seek vocational situations conducive to their personal need;

Leadership styles will change;

A majority of the military manpower will be from an urban environment;

People may live in an essentially 'non-private' environment;

Religion in 1990 will be more important but less visible;

Drugs will be more prevalent in the 1990 society;

To the extent that Government is successful in accommodating the public welfare, youth of military age will be more content and more adaptable to army training and living;

The importance of national goals may diminish;

Reduction of the role of the military–industrial complex will aid the image of the army.

These implications themselves give a good idea of future army pre-occupations with which psychologists will be concerned. Let us examine the military significance of a few of them in detail.

The overall flavour of the document may strike some as too optimistic but this is intentional, says CDC – 'It seems that goals are better motivators for the future than dire threats.' But it is not without its sinister overtones either – for instance, it predicts that 'The major portion of the army's effort [will be] devoted to an accomplishment of "quasi-civilian functions".' Whether this means the army will make more use of civilians on army installations or something else – like taking over some police functions – is not made clear.

In spelling out the implications of the changes listed above, intriguing possibilities for army development are revealed. For example, the document says that, as a result of men in the future being better educated and more mobile and with a greater tendency to look upon themselves as professionals in whatever they do, experience and ability will come to count more than age or time served. In consequence, says the document, 'people are becoming increasingly responsive to rank structure' and adds that the army will have to devise new forms of commissioning and promotion. It specifically predicts 'lateral entry' to the higher ranks – i.e. specialists will be able to enter the army for the first time *above* junior officer rank (maybe straight in as the equivalent of a colonel).

Leadership, says the document, will change – becoming more 'intellectual'. By this is meant the fact that detailed orders from leaders will not be needed so much because men will have to be fully fluent in the operation and functioning of complex machinery. So leadership will be less concerned with these kinds of detail – more concerned with the general direction of strategy. Command structures will also become less rigid – specialists may assume temporary leadership when their particular knowledge makes this appropriate.

Together with this shift in the various characteristics of work, there will, says the document, be a move towards the 'disposable society' in which many everyday objects will be used once and then thrown away rather than cleaned, or renovated to be used again. This will change attitudes towards such things as maintenance, thinks CDC: it may prove harder to persuade men that military equipment is complex and expensive and so has to be maintained rather than replaced. A humdrum prediction, but one which could, in a 'conventional' war particularly, prove vital if it turns out to be true.

The army specialists expect society (or at least American society) to have solved many of its economic problems by the 1990s and as a result predict that pay will be less of a motivating force. In addition, more people will live in cities and the spread of income will decrease so that more people will be earning roughly the same. One result of this, say the CDC psychologists, is that people will become more *ethical*. Everyone will tend more to share middle-class values and as a result crime will drop and there will be fewer discipline problems in the army of the future. On the other hand, because of this ethical interest the army predicts that orders will have to be explained more – especially the reasons for unpopular orders.

Religious changes mean that it will be harder to identify conscientious objectors on traditional grounds. The chief problem here could be public relations. Some conscientious objectors might seek the psychiatric way out – but as we have seen that avenue may not always be open and imposes strains on psychiatrists. Drugs will be more accepted, says the document, making a far from radical point. But it goes on rather more controversially to add that as a result the use of drugs by the army *as a weapon* may come to be more accepted by the public in general.

Finally, the document addresses itself to the breakdown in nationalistic feelings and what it predicts will be a growth in internationalism. One consequence of this, the CDC people foresee, is that it may get much harder for the army to warn its country of developing threats; its information techniques may need to be re-thought to cope with this.

A fairly wide-ranging set of concepts about the future, not all of which (like the predicted growth in internationalism) would gain universal acceptance. But they give a good flavour of some of the fresh controversies which military psychology may have to face in the years to come.

Three other ideas need to be added to these before we round off our discussion, only briefly. First, it is likely to become a matter of routine (if it is not already) for many armies to bring in psychologists at the planning stages of all hardware, whether they be guns, tanks, aircraft or merely smaller pieces of equipment like helmets, boots and so on, to ensure that problems of perception, comfort and other 'human factors' are taken into account. This is not trivial for it means that psychologists will always be aware of weapons developments and general army thinking and so will be able to suggest more imaginative uses of psychology as and when they arise.

Second, armies will seek to extend men's perceptual limits with 'psychotechnology'. We saw on page 104 how links are being sought between the pilot's brain and a missile, via a computer. The Army Research Institute has an entire work unit area devoted to this type of advancement. Scientists would like to change pilots' blink patterns to help them read maps better. The electrical problems of the brain are being investigated to see whether man's ability to distinguish enemy planes and weapons can be improved; ways to circumvent the effects of fatigue are being explored; and the responses of the autonomic system are being manipulated through conditioning so that subtle changes in body chemistry can in the future act as danger signals to the soldier in the field.

For the moment, however, a third psychological aspect of the future is likely to occupy psychologists' imaginations. This is whether likely political and social changes in the years ahead will affect the overall willingness of men and women to fight at all. This is not something the military are ever keen to discuss openly but it does crop up repeatedly at staff colleges and similar institutions. There appears to be a general feeling – a worry almost – that with the growth of weapon power and the increasing de-personalization of war, there are fewer and fewer situations worth dying for. One of the most open discussions of this which I have seen is that by James Scovel, a lieutenant colonel in the US Infantry, in a special paper he wrote called 'Motivating the US soldier to fight in future limited wars' (11).

After surveying animal aggression, the history of warfare and some experimental studies on conflict, Scovel concluded that the basic reasons why men fight do not change – war is a response to threat and/or territorial ambitions. These are, he says, biologically determined and will continue to operate. Nevertheless, he does foresee certain social developments which suggest that a change in the general level of these motivations could well occur.

He is one, for instance, who subscribes to the view that there will be (is) a reduction in the number of causes regarded as worth dying for. This is partly due to wider education and the increase in the number of independent small states which may make small wars more likely but the involvement of third party powers less probable. But as a result, he says, the egalitarian type of leadership style developed, for example, in Vietnam by NCOs who were of a similar educational background to their men will be inappropriate. He appears to favour a return to the German model of the Second World War in which there was a 'hard core' of officers and men imbued with ideals of 'toughness', manly comradeliness and group solidarity (*Gemeinschaft* as the Germans called it). This will be all the more necessary, Scovel maintains, because the conventional notions of masculinity are changing. (He does not explore the corollary to this, that women may be more willing and able to fight. In Britain there were plans in 1977 to arm the WRAC. And in the same year an American research project found that women actually do better on some army tests, so that, with suitable training, they might make better soldiers than men in some circumstances.)

This 'hard core' would also help define, he says, what measure of variation could be allowed in the armed forces in terms of behaviour. In Vietnam, for instance, there was tolerance of AWOLs who had 'seen their share', non-productive jobs for the twice wounded and 'milk-run' patrols. Scovel writes as though this sort of thing was too prevalent, due partly to social changes in the composition of the army, and that the only way to reduce its incidence is not by fiat from above but by an informal but very real 'hard core' setting an example everyone else has to live up to.

On the other hand, says Scovel, there will be forces acting against non-conformity. As a result of people in the West having smaller families and at the same time there being a stable or rising birth rate, the proportion of first-borns in the society is increasing. And, says Scovel, there is plenty of evidence to show that first-borns are more conforming (and incidentally better fighters) than others.

Unions also are likely to grow in importance, in western armies at least, and their presence will affect soldiers' psychology. By the summer of 1977, for example, unions or similar organizations existed in the armies of Sweden, Norway, Denmark, Holland, West Germany and their introduction was being discussed in the USA. So far their existence has not influenced the *operational* direction of armies, that is to say, the unions have not expressed a view on which wars are worth fighting and which not. Their interest has almost solely been in working conditions and pay. (And in which they have been very successful. In Holland the union negotiated a 1200 per cent pay rise over a ten-year period at a time when, according to NATO, the Dutch army was above average in efficiency.) But who can tell whether the picture will remain as equitable as it has so far been?

Cautious optimism, then, is the prevailing mood among military psychologists in regard to the continued availability of men to recruit as soldiers and their willingness to take part in war.

Nevertheless, the continued production of psychological knowledge about soldiers, their weapons (thus affecting their capabilities) and their enemies, must inevitably have a continued *political* significance. Military psychologists, other psychologists, even politicians, have appeared unwilling to recognize this in the past. I hope that this book will go some way to change that in the future.

Appendix I: 70 Classified Documents encountered in research

The following is a chronological list of seventy classified documents which I encountered while working on this book. Full details are not given in all cases because they were not always available – to prevent me knowing the precise source of many documents, individuals in certain institutions made it their practice to tear off the title pages. The original classification of the research is given – in some cases this research may have changed its classification since, or been de-classified altogether. But the purpose of the combined list is to show what *type* of research is classified, so the original label has been left for this reason. Some of the research was only seen in 'abstract' – usually with a sanitized (unclassified) title. Where this occurs, it is indicated. Where the sanitized title was so vague as to be meaningless to outsiders it has been omitted altogether.

S = Secret; C = Classified; R = Restricted; L = Limited; Co = Confidential; F = For official use only.

1949

'Conference on methods for studying the psychological effects of unconventional weapons', Rand Corporation. (C)

1951

'Implications and summary of a psywar study in South Korea', Human Resources Research Institute, Maxwell Air Force Base, Alabama (May). (S, later R)

'The detection of malingering', American Institutes of Research, Technical Report 947 (December). (Co)

'Study of ineffective soldier performance under fire in Korea'. (F)

1952

'Survey of opinions of officers and senior NCOs in Korea: integration', American Institutes of Research, Technical Report 953 (March). (Co)
'Survey of opinions of officers and Senior NCOs in Korea: Foreign nationals', American Institutes of Research, Technical Report 952 (March). (Co)
D. O. Hebb *et al.*, 'The effects of isolation upon attitudes, motivation and thought', Fourth Symposium, Military Medicine I, Defense Research Board, Canada. (S)
D. O. Hebb and W. Heron, 'Effects of radical isolation upon intellectual function and the manipulation of attitudes', Fourth Symposium, Military Medicine, Defense Research Board, Canada. (S)

1953

'Military aspects of psywar', Department of Army, Washington, DC, Authorization of the Psychological Warfare Board to make sure that experience from the Second World War, from Korea and from the developing Cold War was integrated. The board had five members, the President's adviser on psywar and four others.
'Regular army officer integration', Army Research Institute, Research Memo 53–30 (September). (Co)
'Tests for marginal personnel', Army Research Institute, Research Memo 53–28 (September). (Co)
'Area manual for China' (not classified but) 'Psyop implications'. (C)
Joseph W. Hasell and Michael Leyzorek, 'A preliminary study of combat-information-handling' (unclassified title), Operations Research Office, Johns Hopkins University, Bethesda, Md. (Co)

1954

'An examination of single-call-sign procedure' (unclassified title), AFFE, carried out by Operations Research Office, Johns Hopkins University, Bethesda, Md. (Co)
'Communist vulnerabilities to the use of music in psywar'. (S)

1955

Rodney A. Clark, 'Leadership in rifle squads on the Korean front line', HumRRO Technical Report 21 (September). (F)

'Delays in handling and transmission of combat information' (unclassified title), Operations Research Office, Johns Hopkins University, Report T-56 (AFFE). (Co)

1957

A. R. Ashkenasy, 'An analysis and synthesis of information on mass defection organized for planning and operational use', subcontracted by Psychological Research Associates (July). (Co)

1958

William C. Bradbury, 'Motivation of Chinese Communist soldiers: basis for research in support of military psywar', HumRRO Staff Memo (May). (F)
Sam Meyers and W. C. Bradbury, 'Political behavior of Korean and Chinese POWs in the Korean conflict: a historical analysis', HumRRO Technical Report 50 (August). (F)

1959

'Research on psychological and political effects of military postures' (unclassified title), Defense Science Board Working Group of the Advisory Panel on Psychology and Social Science, Department of Defense, Washington, DC, (July). (Co)

1962

'Psyops vulnerabilities in the Soviet Union', (Project Exploit USSR): (1) Orientation and summary; (2) Component social systems of the Soviet Union); (3) Synthesis and overview. In all, a 5-volume typescript (March). (C)
Hilton Bialek *et al.*, 'Psychological techniques for facilitating and countering interrogative processes; exploratory effort concerned with a study of interrogation processes in survey activities, conceptualization and pilot studies', HumRRO Research Memo (May). (F)
John Anspacher, 'Psychological strategy and tactics in countering insurgency', Foreign Service Institute, Department of State. (F)
'Cold war psyops in developing countries'. (C)
'Five steps in the operation of a revolution' (translation of an official Viet Cong document captured in Vietnam). (C)
'Essay on aspects of air power in morale-oriented tactics'. (S)

1963

J. E. Barmack, 'Behavioral science research relevant to military psyops', (unclassified title), Institute of Defense Analysis Research and Engineering Support Division, S–102 (April). (S)
F. F. Medland and L. V. Gordon, 'Leadership assessment of Cuban enlisted men and officers in the US Army,' Army Research Institute, Research Study 63–3 (June). (Co)
'A study of provocation and harassment against US citizens in Soviet/Satellite countries', Security Committee, US Intelligence Board (July). (C)
'Psyops against Communist China: a background reading list', CIA. (F)

1964

'Research in the behavioral and social sciences relevant to counter-insurgency and special warfare' (unclassified title), Memo from the Director of Defense Research and Engineering Support Division to the Assistant Secretary (Research and Development) of the Army, Navy and Air Force and Director of ARPA (dated 2 September 1964). (S)

1965

'Psywar in a tactical nuclear war: 1968–72' (unclassified title), A sub-study of Project Oregon Trail, US Army Combat Development Command, Special Warfare Agency (January). (C)
'Training birds for field surveillance: phase II', US Army Land Warfare Laboratory, Technical Report 65–226 (April). (S)
Calvin Green *et al.*, 'Problems affecting special forces performance in Vietnam', Army Research Institute, Research Study 65–4 (June). (Co)
L. P. Willemin *et al.*, 'Prisoner behavior in simulated interrogation', USAPRO, Paper to the 11th US Annual Army Human Factors Research and Development Conference (October). (C)
R. R. Kramer, 'Some effects of stress on rifle firing', Paper to the 11th Annual US Army Human Factors Research and Development Conference (October). (C)
Louis Willemin *et al.*, 'Prisoner-of-war behavior in simulated interrogations', Army Research Institute, Research Study 65–7 (December). (F)
Clifford P. Hahn and staff, 'Psychological phenomena applicable to the development of psychological weapons', American Institutes for Research, Contract AF 08(635)–4238 (December). (L)
'Psywar in a tactical nuclear war'. (C)

1966

'Psychological reactions to weapons: a factor analysis', prepared by Ohio State University Research Foundation for the Directorate of Armament Development (January). (C)
'Human factors considerations for the atomic/nuclear demolition system', ESL Information Report No. 249 (February). (Co)
Harold Martinek and James Thomas, 'The effects of image resolution and lighting orientation on intelligence', Army Research Institute, Technical Research Note 166 C (February). (Co)
John Tiedermann, 'An evaluation of three work methods in special devices search and analysis', Army Research Institute, Technical Research Note 168 (S) (June). (S)
M. Dean Havron and Randolph Berkeley Jr, 'The role of civil affairs in marine corps operations', Human Sciences Research Incorporated (June). (R)
R. G. Smith, 'Problems in measuring the performance of incapacitated troops', paper presented to the 11th Annual US Army Human Factors Research and Development Conference (October). (C)
Emilie Rapoport, 'Psychological personnel barriers', Remote Area Conflict Center. (C)
'Strong men series', Defense Intelligence Agency, Washington, DC. (S)

1967

'Development of a pigeon ambush detection system', Animal Behavior Enterprises Inc., Contract DA 18-001-AMC-860 (January). (Co)
O. K. Hansen *et al.*, 'Method for developing a laboratory model of an image interpretation system', Army Research Institute, Technical Research Note 185 (C) (June). (Co)
John Tiedermann and Richard Decey, 'Measurement error in special devices search and analysis', Army Research Institute, Technical Research Note 190 (S) (June). (S)
'Psychological operations role in establishing a sense of nationhood', International and Civil Affairs Directorate (August). (Co)
Cedric Smith and Joseph Coates, 'Olfaction and its potential application in personnel detection', Institute for Defense Analysis, Research Paper P-198 (December). (L)
'Role of women in Vietnam'. (F)
'Characteristics of selected societies relevant to US military interests: North Vietnam'. (C)
'Psychological study of regional/popular forces in Vietnam'. (F)

1968

W. H. Helme and I. Orleans, 'Personnel systems of the Imperial Iranian Army – research requirements', Army Research Institute, Research Study 68–1 (C) (March). (Co)

John Tiedermann and R. F. Dean, 'Field evaluation of the BERL RIB method in special devices search and analysis', Army Research Institute, Technical Research Note 196 (S) (April). (S)

'Project night life – human factors summary' (unclassified title), US Army Material Command, TR RH–TR–68–2 (May). (C)

B. A. Wilson, 'Psychological trends, capabilities and vulnerabilities: Africa south of the Sahara, Asian Communist countries, East Asia and Indian Ocean, Latin America' (October). (S)

Jack Sternberg *et al.*, 'Human performance with selected night vision devices: methodology and analysis of preliminary data', Army Research Institute, Research Study 68–6 (C) (December). (Co)

John Parsons *et al.*, 'Americans and Vietnamese: a comparison of values in two cultures', Human Sciences Research Incorporated, Report–68/10–Ct (November). (L)

1969

Edmund Fuchs, 'Characteristics of stockade prisoners – summary of major findings', Army Research Institute, Research Study 69–3 (March). (F)

Francis Medland, 'Research program for selection and performance evaluation in overseas security operations assignments', US Army BESRL, Research Study 69–1 (March). (L)

1970

Joyce House and Stanley Cohen, 'Current procedures in selection, training and utilization of voice processors', Army Research Institute, Research Study 70–4 (April). (Co)

1971

Isaak Orleans, 'Feasibility of using FTX Gobbler Woods for criterion evaluation of psyop officers', Army Research Institute, Research Study 71–1 (May). (F)

Joyce House and Stanley Cohen, 'Quantitative evaluation of current procedures in voice processing', Army Research Institute, Technical Report 1174 (July). (Co)

No date given

'Programs of education and recreation for oriental POWs'. (C)
'The needs of the Vietnamese peasant'. (F)
'Dancing dolls', Conduction Corporation, for US Army Materiel Command, Missiles Divisions, Aberdeen Proving Ground. (C)
'Bibliography of psywar background materiel available in the Pentagon'. (R) Includes: Document no. 6: Lt Col O. R. Schoof, 'The use of psychology to prevent armed conflict with Russia'; Document no. 8: 'Psychological aspects of air force weapons'.

Appendix II: Psychological Think-Tanks

A short background account of the main research organizations referred to throughout this book is given below. Remember, though, that I counted 146 research organizations in this field in the Preface (page 16).

Human Resources Research Office (*HumRRO*)

HumRRO was founded in 1951 when the army concluded that 'human factors' were going to be a more important aspect of military affairs and as a way to bring university scientists into military research. Headquartered in Alexandria, just outside Washington, DC, HumRRO has seven divisions scattered across the continental United States – Kentucky, California, Georgia, Alabama and Texas. These combined divisions are staffed by 275 employees, 90–100 of whom are psychologists with an MSc or PhD.

The office's six main research areas are: man as an individual performer; as a member of an organization; as a leader; as a member of one culture operating in another setting; the technology of instruction; the management of human systems. HumRRO also specializes in devices for training – simulators, models as part of miniaturization systems and other gadgets such as mock ups, job performance tests, self-instructional taped courses, scrambled books and so on. According to HumRRO's own propaganda, these devices have produced, on average, a 10–15 per cent reduction in training time and a 20 per cent drop in attribution rate after such training.

In October 1969, following a series of student demonstrations, HumRRO was severed from George Washington University, of which it had till then always formed a part, and re-constituted as an independent non-profit research organization. Today, 60–65 per cent of its research is sponsored by the military, and about 3 per cent is classified. The annual budget is approximately $6 million and some thirty-five technical reports are published every year.

Center for Research in Social Systems (CRESS) – formerly the Special Operations Research Office (SORO)

Founded, as SORO, in 1956 as part of the American University in Washington, DC, CRESS later became the US Federal Contract Research Center primarily responsible for social science research in counter-insurgency, psychological warfare and civic action. To begin with SORO directed most of its attention at the Soviet bloc and was mainly concerned with propaganda, its production, its dissemination and research into its effectiveness. During the early 1960s, however, counter-insurgency became the dominant aspect of its work and several leading researchers from HumRRO (e.g. Charles Windle and Theodore Vallance) moved over, the focus of SORO's concern then becoming the Third World.

SORO was the home of Project Camelot and following the extremely adverse publicity associated with the discovery of the project and its somewhat doubtful aims (see pp. 30–31) it was re-organized as CRESS. The Camelot episode, and student demonstrations which ensued, resulted in CRESS separating from the American University (much as HumRRO was later to separate from George Washington). CRESS also then became linked with American Institutes for Research.

CRESS now consists of two divisions, CINFAC, the Cultural Information Analysis Center, and SSRI, the Social Science Research Institute. CINFAC originally called the Counter-insurgency Information Center, exists to 'provide a rapid-response capability system which can effectively store and retrieve raw data as well as completed studies in counter-insurgency, emphasizing the social, psychological and economic sciences'.

SSRI was established to carry out in depth studies of 'unconventional warfare, psychological operations, military assistance programs and other studies and evaluations of foreign cultures'. It seems to specialize in handbooks and bibliographies.

US Army Research Institute for the Behavioral and Social Sciences (USARIBSS)

First known as USAPRO, the US Army Personnel Research Office, when it was founded after the Second World War, it changed its name to BESRL, the Behavioral Sciences Research Laboratory in the early 1960s and then to ARI, as it is now called, in 1974. Its headquarters are in Washington, a short taxi ride from the Pentagon. It shares the building with many other research organizations including ARPA and the US Air Force research labs (see below).

In some ways similar – and a rival – to HumRRO, it has 250 full-time professionals, two labs (and two more just coming into service), many field units and a special laboratory in Korea. Like HumRRO its interests are in training (including computer-assisted instruction), the performance of crews together, tactical training, soldier socialization. If anything it is more hardware oriented than HumRRO and slightly less geared to counter-insurgency work. It is my impression that it does more classified work than HumRRO though I could easily be wrong. It seems to fit more closely into the mainstream army.

For years it has been run by Jay Uhlaner, who edited *Psychological Research in National Defense Today*, a good, though very technical overview of military psychology for the specialist (I have often referred to it in this book). Uhlaner is much involved in working on future developments of the army, particularly the growth of computers, the changing sociology of army life, changing attitudes and so forth.

Rand

Rand (the name is an acronym for *R*esearch *an*d *D*evelopment) is probably the world's most famous think-tank. It is also America's oldest. It arose out of the partnership which grew up in the Second World War between university scientists and soldiers, the air force acting as a kind of midwife to establish it in 1946. Rand HQ in Santa Monica, California, looks like a mini-university except that security restrictions are much in evidence. It also maintains an office in Washington, along the street from the White House, and though much smaller than the California base, security is tight there too.

It is not a predominantly psychological think-tank like HumRRO or ARI and even among social scientists, economists and political scientists overshadow the rest. But in the Cold War it was among the first organizations to look at the more psychological side of things – like censorship policy, the effects of translation and mis-translation, propaganda and so on. It was also the home for such 'way out' ideas as 'psychological inspection' for atom bomb preparation (assessing the truthfulness of enemy figures). Then in the early 1960s, ARPA commissioned Rand to study the motivation and morale of underground organizations in Vietnam – again a predominantly psychological approach, albeit one not carried out at Rand by psychologists for much of the time. Apart from this study, which gave rise to some highly controversial conclusions, I have referred to Rand's work on thirty-one occasions, mainly on the question of tactics rather than the nuts and bolts of psywar.

Operations Research Office/Research Analysis Corporation (ORO/RAC)

ORO, at Johns Hopkins University, did the early work identifying fighters, and good fighting units, in Korea. RAC developed as an independent non-profit research organization when Johns Hopkins grew unhappy with its military links. Often RAC is said to be to the army what Rand is to the air force; it concerns itself with all fields of operational research not just psychology. But in psychology its interests, so far, have been motivating soldiers; guerrillas; pre-testing methods for psychological warfare and evaluation of their effectiveness; theories of propaganda and study of Communist uses of propaganda. Located in McLean, Virginia, the home of several other smaller think-tanks, like the Human Resources Research Inc. (see below).

Office of Naval Research (ONR)

The psychological sciences division of ONR is one of eight such divisions within the office which has its headquarters in Arlington, Virginia, a couple of miles down the street from the building housing ARI, ARPA and the others. There are nine professionals in the Washington office, but there are three field stations – in Chicago, Boston and Pasadena in California. And there is also a psychologist as one of ONR's dozen or so technical staff at the US Embassy in London, England, part of a branch that has been there since the Second World War.

The reason the HQ staff is so small is that ONR contracts out most of its work to university departments. The office also has good links with the US Navy's *Medical Neuropsychiatric Research Unit* at San Diego, California. This is a 'center of excellence' for military research in the medical and psychiatric sciences.

This unit has undertaken research in Arctic studies; small groups in remote areas; problems of rank; problems of combat stress and mental illness; the determinants of military effectiveness.

ONR's technical man in London in the psychological sciences seems to busy himself with surveys of military psychology in Europe (the office produces a monthly: *European Scientific Notes*).

Other US Laboratories

The *Air Force Office of Scientific Research*, at Arlington and in the same building as ARI and ARPA, maintains a life-sciences section with four divisions on psychological areas: habitat–crew environments; problems in flight; human engineering; personnel systems technology. Run for ages by Charles Hutchinson, a well-known figure in the field.

The air force seems not to have concentrated its efforts on any one institution. Rand, as we have seen, does much air force work; but so does the *AF Human Resources Laboratory*, at Brooks Air Force Base in Texas. This looks at the wider aspects of military education, recruitment, allocation to combat and so forth. The *Advanced Projects Research Agency* (ARPA) is specifically designed to sponsor projects which will not come to fruition for ten to fifteen years. We have seen that in the psychological sciences they are encouraging biofeedback, brain links to weapons and so forth. Run by a scientist with the rank of Lieutenant Colonel.

Establishments in the United Kingdom

British science does not, in general, have anything to compare with the USA's independent think-tanks and this applies certainly in psychology. The British Aircraft Corporation has a human factors group looking at problems associated with flight; and Loughborough University and the University of Wales Institute of Science and Technology both offer degrees in human factors (called, in Britain, ergonomics). But of course these do not deal by any means exclusively with military work.

The largest laboratory in this field of a military nature is the *Army Personnel Research Establishment* (APRE) at Farnborough. APRE became an independent unit in 1965 to carry out research into 'selection and training, ergonomics, environmental physiology, clothing and equipment research and trials'. It was created by combining the human factors divisions of the *Army Operational Research Establishment* (AORE) at West Byfleet with the *Clothing and Equipment Physiological Research Establishment* (CEPRE) The two posts were not fully integrated until August 1972.

APRE is responsible to the Chief Scientists (Army) through the Deputy Chief of Staff (Army) at the Ministry of Defence. The research programme is determined by the Army Human Factors Research Committee and reviewed by a scientific advisory panel set up by the Army Personnel Research Committee of the Medical Research Council. APRE is the sole army organization concerned with human factors research and has seventy-five scientists and supporting staff, fifteen industrial staff, six military officers and thirty other ranks. Its five sections are manpower studies; experimental psychology; applied physiology and field trials; weapons and vehicle ergonomics; a human factors unit.

The British do not give a detailed breakdown of their spending on human factors research but I understand that it is in the the region of £500000 – and certainly much less than £1 million.

The British research thrust appears to have been in four areas: (a) a concern with unusual environments, especially the stress of working in a hot, tropical climate. The services have been fortunate in having, in London, the London School of Tropical and Hygienic Medicine which has helped in a number of studies; (b) the problems arising from an ever smaller but increasingly professional army – the question of waste, drop-out rates, attrition. One by-product of this has been the centralization of selection at Sutton Coldfield. The British approach is in general more physiological than social, as is perhaps the case in the USA, and this is shown in area (c) in which the British have been concentrating their efforts on remote-controlled devices for use in future wars and the human factors associated with that. The final area (d) has been research into vehicles so that they are easier to operate, quicker and safer in the event of chemical or biological warfare.

While these are the main thrusts, there has been other research, most notably into vigilance but also into isolation, discipline and sleep deprivation. The psychologists hope to get into the problems of command but so far that has not yet happened. In general psychology has only been recognized as a central aspect of weapons planning since the integration of the human factors divisions of AORE and CEPRE mentioned above. Psychiatry, however, had been accepted somewhat earlier.

The psychiatrists, in fact, carry out other more contentious work. It is they, for example, who train soldiers to resist interrogation. According to General McGhie, chief army psychiatrist, all British psychiatrists are cleared to top security levels and help screen for sensitive posts. They also advise the Special Air Services and other special forces of the effects of drugs, hypnosis and isolation as used in interrogation.

Whereas the army has all its psychologists in one place, the RAF has *Science 3*: selection and training; *Science 4*: social science studies; *Brampton*: training command; and the *Institute of Aviation Medicine* (also at Farnborough), which contains the RAF's human factors group. It mainly looks at the effects of fatigue on stress and thus carries a higher classification rating than APRE. The RAF is more oriented to social psychology than the British Army (for instance, it does sociometry tests and research).

In the Royal Navy all psychological research comes under a senior psychologist and the work is divided as follows: selection; conditions of service; physiology and medicine; ergonomics; training. According to Ted Elliott, senior naval psychologist, the navy accepts psychology more than, say, the RAF, and is especially well organized for studying things like hazardous missions. Studies of training and equipment for missions like this are carried out at the *Naval Applied Psychology Unit* at Teddington, Middlesex. The kinds of research done there include stress resistance – the British approach is usually to overload people on certain tasks, on the basis that this is a realistic stress even if it is not the same as combat stress. Elliott has himself done quite a lot of research into isolation in, for example, Polaris submarines.

The Naval APU (near the National Physical Laboratory) also carries out research on the uses of computers, particularly computer-assisted instruction.

Other countries

In other countries the organization of research into military pscyhology (psychological warfare excepted) seems to parallel the British situation rather more closely than the American – there are few if any private research enterprises, most work being done in government and/or military installations.

In West Germany, for example, the *Bundesministerium der Verteidigung* in Bonn is responsible for psychological studies of military organization, ergonomics, personality assessment and social psychology in the Federal

Armed Forces and hospitals. But two thirds of the 120 psychologists are, according to Dr Friedrich Steege, head of the unit, involved in selection and allocation, both of men and of officers. Officer candidates are selected at special centres in Cologne, Hanover, Dusseldorf, Munich, etc. The *Forschunginstitut für Soziologies der Universität Köln* has carried out many studies into group psychology in the German forces. These include the effects of cohesion on military attitudes and behaviour; relations between enlisted men and officers; the soldier and his free time; how conscripts differ from volunteers; the effects of the German brand of radical politics on soldiers; the prestige of the army; recruitment, propaganda, etc.

In Australia, the *Navy Psychological Service* was founded in 1949 and consists of about a score of psychologists. Its main research areas include sonar detection displays; soldier attitude surveys; the ratings system; individual differences in the learning of complex tasks.

Both the Air Force and Navy in Australia are committed to the validation of selection tests, presumably as part of their problems with a small professional armed service, susceptible to waste and attrition. Plans are afoot for a 'research cell' to expand the research into computer studies, psychometrics studies and the amalgamation of psychologists in all the services so that a better coordination of research can be attained. Visits to US and British establishments are being sought, too. Approaches have been made to ARI for help with research on 'culture-fair testing and counter-insurgency'. Staffing is a problem, however, and this makes the development of some areas of research uncertain.

References

U = Unclassified title of a classified paper.

Chapter 1 Introduction

1. Clifford P. Hahn and staff, 'Psychological phenomena applicable to the development of psychological weapons', American Institutes for Research, US Department of Defense Project 2516, Contract AF 08(635)–4238, December 1965. (Distribution limited)
For a similar study see: Joseph Weitz, 'An exploratory study of psychological weapons', Contract AF 08(635)–5220, 18 June 1965–17 August 1966.
2. Clay George, 'Training for coordination within rifle squads', Paper given to the 12th Annual US Army Human Factors Research and Development Conference, Fort Benning, Georgia, October 1966.
3. Robert E. Lubow, 'High-order concept formation in the pigeon', *Journal of the Experimental Analysis of Behavior*, vol. 21, 1974, pp. 475–83.
See also: Peter Watson, 'How the Israelis train pigeons to spy on the Arabs', *Psychology Today* (UK edition), vol. 1, no. 1, April 1975, where the links between Lubow and the US Army's Land Warfare Laboratory were first pointed out. Dr Lubow later denied some aspects of this account but in his book, *The War Animals* (Doubleday, 1977), he does concede that several studies have been carried out to investigate just this sort of possibility.
4. R. Huey Wright and Kenneth Michels, 'Human processing of olfactory information', US Air Force Aeronautical Systems Division, Wright–Patterson Air Force Base, Ohio, March 1963.
See also: 'Soviet Union trying to unlock parapsychology as a weapon', *International Herald Tribune*, 20 June 1977; and: 'Mind field', *Guardian*, 11 August 1977.
5. There are several papers in the book on this topic. Three examples are: S. M. Fallah, 'A bibliography on witchcraft, sorcery, magic and other psychological phenomena and their implications on people in SE Asia', prepared for the Remote Area Conflict Information Center and then

released to the library at Fort Bragg, 7 October 1965. It includes such things as Chinese ancestor worship in Malaya, Burmese folk tales, SE Asian birth customs.
Keyser Bauer, 'Anthropological–cultural influences on weapons use, choice and maintenance', Paper prepared for the US Army Materiel Command Human Engineering Laboratory, Aberdeen Proving Ground, no date.
Stuart Howard and William D. Hitt, 'Intercultural differences in olfaction', Battelle Memorial Institute for the Project Agile Report for ARPA, 2 May 1966. This explores the idea of smell bombs.
6. Louis Guttman, 'Social problem indicators for the period 12–14 November 1974 (from the continuing survey)', Israel Institute of Applied Social Research, PO Box 7150, Jerusalem.
7. 'The strange tale of Commander Narut', *Sunday Times* (London), 6 July 1975.
8. F. A. Geldard (ed.), *Defence Psychology: Proceedings of a symposium held in Paris*, Pergamon Press, 1962.
F. H. Lakin, 'Psychological warfare research in Malaya 1952–55', Army Operational Research Establishment, Ministry of Defence, UK, Paper given to the 11th Annual US Army Human Factors Research and Development Conference, October 1965.
'Research in the behavioral and social sciences relevant to counterinsurgency and special warfare' (U), memo from the Director of Defense Research and Engineering to the Assistant Secretary (Research and Development) of the Army, Navy and Air Force and Director of ARPA, 2 September 1965. (Secret)
9. Charles Windle and Theodore Vallance, 'The future of military psychology: paramilitary psychology', *American Psychologist*, vol. 19, 1964.
10. 'Mind-guided missiles', *Sunday Times* (London), 'Spectrum', 7 July 1964.
11. 'Propitious and non-propitious dates in the Vietnam calendar (1964–67)', Center for Research into Social Systems (CRESS). This document also refers to Cambodia.
Another untitled document I have also explores the abuses of personal feelings possible at New Year's day, weddings, anniversaries of death. See chapter 22 on special psywar techniques.
12. 'Military psychology, propaganda and psychological warfare', vol. 1, US Department of Defense, July 1968, 609 pp. A report bibliography.
13. Ritchie P. Lowry, 'Toward a sociology of secrecy and security systems', *Social Problems*, vol. 19, no. 4, 1972, p. 437.
14. 'The Viet Cong Motivation and Morale Project', by the Rand Corporation, gathered some 2000 interviews with NLF prisoners, defectors and refugees from free-fire zones. There were countless documents produced for this project but some of the more pertinent and controversial are:
J. J. Zasloff, 'Origins of the insurgency in South Vietnam, 1954–60: The role

of the Southern Vietminh cadres', RM–5163–2–ISA/ARPA, May 1968.
Leon Goure and C. A. H. Thomson, 'Some impressions of Viet Cong vulnerabilities: An interim report', RM–4699–ISA/ARPA, August 1965.
J. J. Zasloff, 'Political motivation of the Vietcong: The Vietminh regroupees', RM–4703/2–ISA/ARPA, May 1968.
W. P. Davison, 'Some observations on Viet Cong operations in the villages, Rand, RM–5267/2–ISA/ARPA, May 1968.
Stephen T. Hosmer, 'Viet Cong repression and its implications for the future', R–475/1–ARPA, May 1970.
15. Personal interview with Dr Shlomo Breznitz at the International Nato Conference on Dimensions of Stress and Anxiety, Oslo, July 1975.
16. 'The Rand Papers: The secret study which lies at the heart of the war', *Ramparts*, vol. 11, no. 5, November 1972.
17. Quoted in Lowry, reference 13 above, p. 442.
18. Lowry, reference 13 above, p. 441.
19. 'Behavioral sciences and the national security, Report no. 4', testimony before the House Subcommittee on International Organizations and Movements of the Committee on Foreign Affairs, Washington, DC, 8 July 1965.
Irving Louis Horowitz (ed.), *The Rise and Fall of Project Camelot*, MIT Press, 1967.
20. See particularly: Mitchell Berkun *et al.*, 'Psychological and physiological responses in observers of an atomic test shot', *Psychological Reports*, vol. 4, no. 5, December 1958.
Mitchell Berkun *et al.*, 'Human psychophysiological response to stress: successful experimental simulation of real-life stress', Human Resources Research Office (HumRRO) Research Memorandum, December 1959.
Mitchell Berkun, 'Urinary responses to psychological stress', Paper given to the Society for Psychological Research, Denver, October 1962.
21. Lewis A. Coser, 'The dysfunctions of military secrecy', *Social Problems*, vol. 11, Summer 1963, pp. 13–22.
22. 'Human factors considerations for the atomic/nuclear demolition system', ESL Information Report No. 249, February 1966. (Confidential)
23. B. A. Wilson, 'Psychological trends, capabilities, and vulnerabilities: Africa south of the Sahara, Asian Communist countries, East Asia and Indian Ocean, Latin America', October 1968. (Secret)
24. Emilie Rapoport, 'Psychological personnel barriers', Remote Area Conflict Information Center, 1966. (Classified)
25. 'Strong men series', Defense Intelligence Agency, Washington, DC, 1966. (Secret)
26. 'Essay on aspects of air power in morale-oriented tactics', 1962. (Secret)
27. 'Communist vulnerabilities to the use of music in psywar', 1954. (Secret)
28. Hilton Bialek *et al.*, 'Psychological techniques for facilitating and countering interrogative processes: exploratory effort concerned with a study of the

interrogation process in survey activities, conceptualization and pilot studies', HumRRO Research Memorandum, May 1962. (For official use only)
29. John Anspacher, 'Psychological strategy and tactics in countering insurgency', Foreign Service Institute, Department of State, 1962. (For official use only)
30. 'Psyops against Communist China: a background reading list', CIA, 1963. (For official use only)
See also: 'Area manual for China', 1953. (*Not* classified)
'Pysops implications', 1953. (Classified)
'A background reading list on psyops with special emphasis on Communist China', Department of State, External Research Staff, Research Paper 153, May 1964.
31. Major Alexander (US officer), 'Precis 6: Psyops unit – general', Training Report, Senior Officers' Psyops Course, Royal Air Force Old Sarum, Salisbury, Wiltshire, UK, 14–18 February 1972.
Lt Col B. Johnston, Principal General Staff Officer Information Policy, 'Military information policy in low intensity operations', quoted in detail in preceding reference.
32. For example: E. K. E. Gunderson, 'Personal history correlates of military performance at a large Antarctic station' (NM. N, RU 64–22), National Technical Information Service Microfiche: AD 609542.
33. Joel T. Campbell *et al.*, 'Public opinion and the outbreak of war', *Journal of Conflict Resolution*, vol. 9, 1963, p. 318.
34. See reference 6 above.
35. Personal communication from the Home Office Research Unit, Autumn 1975.
36. Peter G. Bourne, *Men, Stress and Vietnam*, Little, Brown & Co., 1970.
37. See, for example: Robert Boldt and Dennis Seidman, 'Aptitude distribution in the combat arms', PRB Technical Research Report no. 1119, April 1960.
Robert Boldt, 'Combat allocation and future combat tasks: Status report fiscal year 1962', USAPRO Research Study 62–2, August 1962.
38. Robert Vineberg, 'Human factors in tactical nuclear combat', HumRRO Technical Report 65–2, April 1965, p. 60, table 2 and references.
Franklin Del Jones and Arnold W. Johnson, Jr, 'Medical and psychiatric treatment policy and practice in Vietnam', Walter Reed Army Medical Center, Washington, DC, 1966.
39. 'Gallagher proposes study of ending Navy Department's project: Group Technology', by correspondence inserted in the *Congressional Record* by the Hon. Cornelius Gallagher, Representative of New Jersey, 2 March 1971, pp. E1295–E1302.
40. See reference 5 above.
41. J. McGuffin, *The Guinea Pigs*, Penguin Books, 1974.

42. A report of some of his conclusions appeared in *The Times* (London), 9 June 1973.
See also: McGuffin, reference 41 above, pp. 121–8.
43. Andrew Molnar, Jerry Tinkler and John Leitoir, 'Human factors considerations of undergrounds in insurgencies', CRESS, December 1966.
44. 'Development of a psychiatric/psychological mechanism to enhance national security', HQ Department of Army, Office of the Deputy Chief of Staff for Personnel, Washington, DC, 20310.
In April 1964, a Department of Defense research project was mooted to improve the screening of 'non-prior service personnel', though it was delayed until fiscal year 1966. Then a programme entitled 'Dependable performance' was begun. This information is contained in a document held in the Pentagon and which contains several annexes, as follows:
Annex A: 'Present use of psychiatric and psychological specialists and mechanisms by Department of Army to enhance national security', document no. AR 611/15.
Annex B: 'The possibility of expanding provision of AR 611/15 – The selection and retention criteria for personnel in nuclear weapons positions to include persons referred to and retained in highly sensitive positions'. This annex also contains details of an Enlisted Personnel Management Systems Document, AR 600/200: 'Procedures and selection criteria for meeting requirements of designated organizations and agencies for personnel with specific qualifications'.
Annex C: 'Expanded employment of psychiatric and psychological mechanisms through the Department of Army to enhance national security'.
Annex D: 'Development of a psychological mechanism'.
45. Jeffrey Maxey *et al.*, 'Identification of the potential characteristics, aptitudes and acquired skills involved in human detection of mines', Hum RRO Technical Report 73–18, August 1973.
46. Personal communication from British army psychiatrists and psychologists. A shortened account appeared as 'How to choose a hero', *Sunday Times* (London), 'Spectrum', May 1974.
47. See reference 45.
48. 'The selection of cryptography students', Report no. 434. (No agency or authors given)
49. Helen Ross, *Behaviour and Perception in Strange Environments*, Allen & Unwin, 1974, chapter 4.
50. Division 2 of HumRRO is known as 'ARMONITE' – Human Factors in Armor Operations under conditions of limited visibility. The division has produced at least two dozen reports but for an introductory review see: 'Collected papers prepared under work unit Armonite: Human Factors in Armor Operation under conditions of limited visibility', HumRRO Professional Paper 12–68, May 1968.

Chapter 2 The Individual Soldier: Selection and Training

1. S. A. Stouffer *et al.*, *The American Soldier: Combat and its aftermath*, vol. I of *Studies of Social Psychology in World War 2*, 4 vols., Princeton University Press, 1949.
2. S. L. A. Marshall, *Men Against Fire*, Infantry Journal Press, Washington, DC; William Morrow & Co., 1947 and 1950.
3. R. A. C. Ahrenfeldt, *Psychiatry in the British Army in the Second World War*, Routledge & Kegan Paul, 1958.
4. S. L. A. Marshall, 'Commentary on infantry operations and weapons usage in Korea, winter 1950–51', Operations Research Office, Johns Hopkins University, 1951.
5. The Personnel Research Branch (PRB) documents number at least thirteen, several of them very long. The most useful are:
Frank E. Manning *et al.*, 'Interviews with Korean casualties, a pilot study', Personnel Research Section (PRS) Report no. 886, April 1951.
M. Dean Havron *et al.*, 'The effectiveness of small military units: Part I The measurement of infantry rifle squad effectiveness; Part 2 Handbook: Construction and use of small unit field tests; Part 3 The prediction of field performance', PRS Report no. 980, September 1952.
M. Dean Havron *et al.*, 'An interview study of human relations in effective infantry rifle squads', PRS Report no. 983, January 1953.
6. An entire division of HumRRO was given over to its 'FIGHTER' project – 'Factors related to effectiveness and ineffectiveness of individuals in combat', and at least forty-one papers have been produced on this subject. Of interest in the Korean context are:
(a) Robert Egbert *et al.*, 'Incidental observations gathered during research in combat units', Information Report AD–478–562, October 1953.
(b) Robert Egbert, 'Profile of a fighter', *Infantry School Quarterly*, October 1954.
Jerald Walker, 'Subsequent army careers of effective and ineffective combat soldiers', Paper given to the Western Psychological Association, Spring 1957.
(c) Robert Egbert, Tor Meeland *et al.*, 'Fighter I: An analysis of combat fighters and non-fighters', Technical Report 14, December 1957.
(d) Tor Meeland and Mitchell Berkun, 'Sociometric effect of race and of combat performance', *Sociometry*, vol. 21, no. 2, 1958.
(e) Robert Egbert, Tor Meeland *et al.*, 'Fighter I: A study of effective and ineffective combat performers', Special Report 13, March 1958.
(f) Mitchell Berkun, 'Inferred correlation between combat performance and some field laboratory stresses', Research Memo (Fighter II), November 1958.
7. Eric Gunderson, 'Body size, self-evaluation and military effectiveness', *Journal of Personality and Social Psychology*, vol. 2, 1965, p. 902.
8. See reference 6(f) above.

9. Yehuda Amir and Shlomo Sharon, 'Birth order, family structure and avoidance behavior', *Journal of Personality and Social Psychology*, vol. 10, 1968, p. 271.
Robert Helmreich, 'Birth order effects', *Naval Research Reviews*, vol. 21, no. 2, February 1968.
10. E. P. Torrance, 'The development of a preliminary life-experience inventory for the study of fighter-interceptor combat effectiveness', Research Bulletin AFPTRC–TR–54–89, Lackland Air Force Base, San Antonio, Texas, December 1954.
11. Roland Radloff and Robert Helmreich, *Groups Under Stress: Psychological Research in Sealab II*, Appleton–Century–Crofts, 1968.
12. See, for example: Abraham Birnbaum, 'Collection of combat evaluations based on overseas manoeuvres', HumRRO Research Memo 56–17, 1956.
Frederick O. Caretta, 'Improvement of the fighter-factor inventory', HumRRO Research Memo 56–22, 1956.
13. A. U. Dubuisson and W. A. Klieger, 'Combat performance of enlisted men with disciplinary records', USAPRO Technical Research Note 148, June 1964.
14. Louis Willemin, 'Development of a combat aptitude rating for combat arms', HumRRO Research Memo 55–18, July 1955.
Louis Willemin *et al.*, 'Identifying fighters for combat', PRB Technical Research Report 1112, August 1958.
See also chapter 1, reference 37 above, and:
Victor Fields, 'Differential classification and optimal allocation of personnel in the military services', in J. J. Uhlaner (ed.), *Psychological Research in National Defense Today*, US Army BESRL Technical Report S–1, June 1967.
Samuel H. King *et al.*, 'Improving assignment and allocation of personnel to combat units', PRB Research Studies nos. 58–1 and 59–2, April 1958 and May 1959.
15. Hilton Bialek and Michael McNeil, 'Preliminary study of motivation and incentives in basic combat training', HumRRO Technical Report 68–6, May 1968.
16. Joseph Olmstead, 'The effects of "quick kill" upon trainee confidence and attitudes', HumRRO Technical Report 68–15, December 1968.
17. Boyd Mathers and John F. Hayes, 'Criterion-referenced performance testing in combat arms skills', US Army Research Institute, Contract DACH 19–73–C–0016, UCS/Matrix Co., Systems Research Division.
M. Stephen Sheldon, 'Evaluation of training in simulated environments', in J. J. Uhlaner (ed.), *Psychological Research in National Defense Today*, US Army BESRL Technical Report S–1, June 1967.
18. Interview with Lt Col Austin W. Kibler, ARPA, Washington, DC, 22 January 1974.

'See also: Peter Watson, 'The ultimate war game', *Sunday Times* (London), 'Spectrum', 26 May 1974.
19. Personal briefing with Mr E. Elliott, Senior Royal Naval Psychologist, London, 24 July 1974; and personal briefing, Royal Air Force, Kinross, Scotland, 19 June 1974.
See also: I. D. Fauset, 'The quantitative evaluation of aircraft flight simulators', Ministry of Defence, UK, Paper presented to the West European Association for Aviation Psychology, Cambridge, 1971.
20. Marksmanship – and its training – comprised at one time two divisions of HumRRO. The main document is:
(a) James Dees *et al.*, 'An experimental review of basic combat rifle marksmanship: Marksman phase 1', HumRRO Technical Report 71–4, March 1971.
See also:
(b) Joseph Hammock, 'Rifle marksmanship as a function of manifest anxiety and situational stress', HumRRO Technical Memo, 1954.
(c) R. R. Kramer, 'Some effects of stress on rifle firing', unclassified abstract of classified paper presented to the 11th Annual US Army Human Factors Research and Development Conference, October 1965.
See also: David Vielharber *et al.*, 'Pretraining correlates of trainfire marksmanship', Army Hospital, West Point, Technical Report 16, January 1965.
Carl Lauterbach *et al.*, 'Personality–leadership correlates of trainfire marksmanship of new West Point cadets', Army Hospital, West Point, Research Report no. 22, August 1965.
Publications of HumRRO Division 4, listed on pp. 183–4 of the HumRRO *Bibliography of Publications* (1971), also includes papers studying: personalized stocks; the effects on aiming of the loop sling; the effects of different types of target; the effects of aiming point on accuracy.
21. See references 16 and 20(a) above.
22. See references 16 and 20(a) above.
23. See reference 20(a) above.
24. Joseph Hammock and Albert Prince, 'A study of the effects of manifest anxiety and situational stress on M–1 rifle firing', HumRRO Staff Memo, October 1954.
25. See reference 20(c) above.
26. W. O. Evans, 'Problems and approaches to the measurement of troop performance', US Army Medical Research and Nutrition Laboratory, Fitzimmons General Hospital, Denver, Colorado, Paper given to the 12th Annual US Army Human Factors Research and Development Conference, October 1966.
27. R. G. Smith, 'Problems in measuring the performance of incapacitated troops', Paper given to the 12th Annual US Army Human Factors Research and Development Conference, October 1966.
J. Whitehouse, 'Pharmacological enhancement of performance', US Army Materiel Command, *circa* 1963.

28. Leon Goldstein and Seymour Ringel, 'Survey of human factors problems in missile and communication systems', USAPRO Research Memo 60–17, October 1960.
Norman Walker and Janet Burkhardt, 'Combat effectiveness of human operator control systems', Norman Walker Associates, 4809 Auburn Avenue, Suite 201, Bethesda, Md (no date but early 1970s).
Norman Walker, 'The use of tracking tasks as indicators of stress: present position' (same publishing details as previous reference).
29. This subject received publicity from the middle of 1976 through to summer 1977, first due to the official report: *Foreign and Military Intelligence: Book I Final Report of the Select Committee to Study Governmental Operations with Respect to Intelligence Activities*, US Senate Report no. 94–755, 26 April (legislative day 14 April) 1976.
A series of documents was also made available to John Marks under the Freedom of Information Act. See, for example: 'Private institutions used in CIA effort to control behavior', *New York Times*, 2 August 1977.
'Lengthy mind-control research by CIA is detailed', *Washington Post*, 3 August 1977.
30. James Dees and George Magnon, 'A survey of opinions about the bayonet in the US Army', HumRRO Technical Report 69–13, June 1969.
31. See reference 30 above, p. 9, table 12.

Chapter 3 Artillery

1. Charles Bray, *Psychology and Military Proficiency: A history of the applied psychology panel of the National Defense Research Committee*, Greenwood Press, 1948; reprinted by Princeton University Press, 1950, pp. 155–6.
2. See reference 1, pp. 166 ff.
3. See reference 1, pp. 184 ff.
4. See reference 1, pp. 198 ff.
5. See reference 1, pp. 210 ff.
6. Robert Foskett *et al.*, 'A review of the literature on the use of tracer observation as an anti-aircraft firing technique', HumRRO Technical Report 68–11, September 1968.
7. Personal briefing with Lt Col Austin W. Kibler, ARPA, Washington, DC, 22 January 1974.
8. Edward W. Frederickson *et al.*, 'Methods of training for the engagement of aircraft with small arms', HumRRO Technical Report 70–2, February 1970.
9. See reference 8, p. 7.
10. See reference 8, p. 18.
11. See reference 8, p. 35.

12. Bernad Lyman, 'Visual detection, identification and localization: an annotated bibliography: HumRRO Technical Report 68–2, February 1968.
Robert Baldwin, 'Relationship between recognition range and the size, aspect angle and color of aircraft,' HumRRO Technical Report 73–2, February 1973.
John Whittenburg *et al.*, 'A field test of visual detection and for real and dummy targets', HumRRO AF 637244, April 1959.
'Magnification and observation time in target identification in orbital reconnaissance', *Human Factors*, vol. 7, no 6, December 1965.
Robert Foskett, 'Determination of ground-to-aircraft distance by visual techniques', HumRRO Technical Report 69–22, 1969.
Michael McCluskey *et al.*, 'Studies on training ground observers to estimate range to aerial targets', HumRRO Technical Report 68–5, May 1968.
13. Edward W. Frederickson, 'Shape perception judgements as a function of stimulus orientation, stimulus background and perceptual style', HumRRO Technical Report 79–24, December 1970.
14. A. Dean Wright, 'The performance of ground observers in detecting, recognizing and estimating range of low-altitude aircraft', HumRRO Technical Report 66–19, December 1966.
Edward W. Frederickson, 'Aircraft detection, range estimation and auditory tracking tests in a desert environment', HumRRO Technical Report 67–3, March 1967.
Robert Baldwin *et al.*, 'Aircraft recognition performance and crew chiefs with and without forward observers', HumRRO Technical Report 70–12, August 1970.
15. Fred Attneave, 'The relative importance of parts of a contour', US Air Force Resource Center Research Bulletin no. 51–8, 1951.
16. William Beva and Edward Turner, 'Assimilation and contrast in the estimation of number', *Journal of Experimental Psychology*, vol. 67, 1964, pp. 458–62.
17. W. E. K. Middleton and G. G. Mayo, 'The appearance of colors in twilight', *Journal of the Optical Society of America*, vol. 42, 1952, pp. 116–21.
Perhaps more accessible is: 'The effects of colour and rain on lookouts', *Ergonomics*, vol. 8, no. 2, 1965, *E. C. Poulton, et al.*, p. 163.
The Office of Naval Research is also looking into the effects on colour vision of breathing different mixtures of air and oxygen.
18. Richard W. Coan, 'Factors in movement perception', *Journal of Consulting Clinical Psychology*, vol. 28, 1964, pp. 394–402.
19. Robert Gottsdanker, 'How the identification of target acceleration is affected by modes of starting and of ending', *British Journal of Psychology*, vol. 52, 1961, pp. 155–60.
20. D. A. Robbins *et al.*, 'Effects of seasonal variation on personnel detection in an evergreen rainforest', Part III of 'Jungle vision', Research Report no. 3,

US Army Tropic Test Center, Fort Clayton, Canal Zone, May 1965. (This project also included jungle acoustics.)
21. 'Camouflage and deception: Report bibliography', US Army Combat Development Command, Special Warfare School, May 1969.
'Detectability ranges of camouflaged targets', *Human Factors*, vol. 14, no. 3, June 1972.
22. H. C. Muffley, 'A biochemical approach to camouflage for army application', US Army Weapons Command, Research and Engineering Division, Rock Island, Ill., January 1967.
23. R. V. Atkinson, 'Reversible uniform', ARPA Research and Development Field Unit, January 1965.
(Also discussed at personal briefing with Lt Col Austin W. Kibler, ARPA, Washington, DC, 22 January 1974.)
24. 'Color regional atlas' (a few details given in reference 21 above).
25. Michael McCluskey *et al.*, 'Studies on training ground observers to estimate range to aerial targets: Experiment I', HumRRO Technical Report 68–5, May 1968.
26. See reference 25, Experiment IV.
27. See reference 25, Experiment V.
28. Edward W. Frederickson and Robert Donohue, 'Auditory and visual tracking of a moving target', HumRRO Technical Report 70–4, March 1970.
29. Edward W. Frederickson *et al.*, 'Aircraft detection, range estimation and auditory tracking tests in a desert environment', HumRRO Technical Report 67–3, March 1967.
30. See references 28 and 29 above.
31. Elmo Miller, 'Prompting and guessing in tank identification', HumRRO Technical Report 70–21, December 1970.
32. Robert Baldwin *et al.*, 'Aircraft recognition performance of crew chiefs with and without forward observers', HumRRO Technical Report 70–12, August 1970.
33. See Reference 32 above.
34. For this section I have used three basic documents:
(a) Bruce Bergum and I. Charles Klein, 'A survey and analysis of vigilance research', HumRRO Research Report 8, November 1961, re-issued November 1964.
(b) A. H. Birnbaum *et al.*, 'Summary of BESRL surveillance, research' BESRL Technical Research Report 1160.
(c) D. R. Davies and G. S. Tune, *Human Vigilance Performance*, Staples Press, 1970.
Other individual references are given in the text.
35. James Miller *et al.*, 'ONR's role in human factors engineering', *Naval Research Reviews*, vol. 20, no. 7, July 1967, pp. 1–11.

36. See reference 34(c) above, pp. 58ff.
37. See reference 34(c) above, pp. 83–6.
38. See reference 34(c) above, p. 84.
39. See references 34(c) above, pp. 138ff, and 34(a), pp. 13ff.
40. See reference 34(c) above, pp. 95ff.
41. See reference 34(a) above, pp. 17ff.
42. See reference 34(c) above, chapter 5.
43. Personal briefing with Lt Col Austin W. Kibler, ARPA, Washington, DC, 22 January 1974.
44. Bruce Bergum and Donald Lehr, 'Vigilance performance as a function of environmental variables', HumRRO Research Report 11, May 1963.
See also reference 34(a) above, chapter 4.
45. T. O. Jacobs, 'Training for modern combat operations', Paper given to the 12th Annual US Army Human Factors Research and Development Conference, October 1966 (where there was a special section on night operations).
46. Howard Olson *et al.*, 'Recognition of vehicles by observers looking into a searchlight beam', HumRRO Technical Report 49, July 1958.
47. Nicholas Louis, 'The effects of observer location and viewing method on target detection with the 18-inch tank-mounted searchlight', HumRRO Technical Report 91, June 1964.
48. See reference 47 above, pp. 12ff.
49. Alfred Kraemer, 'An evaluation of flash localization performance with the fire control system of the M48 tank', HumRRO Technical Report 78, June 1962.
50. David Easley *et al.*, 'The effects of interruption of dark adaptation on performance of two military tasks at night', HumRRO Technical Report 69–20, December 1969.
51. T. A. Garny, 'Human factors considerations for the use of guided missiles during periods of limited visibility, US Army Human Engineering Laboratory, Aberdeen Proving Ground, Paper given to the 12th Annual US Army Human Factors Research and Development Conference, October 1966.
52. The US Army Research Institute for the Behavioral and Social Sciences has a special work unit, the Human Performance Capability (Technological Base) Unit. Principal investigators are Aaron Hyman, technical area chief, and M. Kaplan, supervisory project director. The unit is studying 'Overt and covert responses in perception' and 'Enhancing perception through selective reinforcement'. Dr Hyman also worked on a project with J. J. Sternberg called 'BESRL's field-laboratory studies in human performance experimentation', May 1969.
It was Dr Sternberg who, in 1965, produced a paper on 'Effects of conditioned aversive stimuli on behavior – DANGER SIGNALS', US Army

Briefing Supplement. Related studies may be found itemized in a bibliography of the US Army Materiel Command, Electronics Research and Development Laboratory, Fort Monmouth, New Jersey. This includes: 'Longitudinal studies of hearing sensitivity patterns in monkeys under normal and stress conditions'.
'Unaided auditory localization of stationary and moving sound sources.'
A. H. Humphrey, 'Principles of concealment of the individual soldier by camouflage', Paper presented at symposium on ambush detection at the Land Warfare Laboratory, Aberdeen Proving Ground. (The LWL is also exploring, with the help of perceptual psychologists, the use of mirrors as camouflage.)
J. A. Swets *et al.*, 'Further experiments on computer-aided learning of sound identification', TR–789–2, April 1964.
W. H. Teichner *et al.*, 'Visual after images as a source of information', TR–1303–3, 11 June 1964.

Chapter 4 Special Skills

1. George Magnon, 'Detection and avoidance of mines and boobytraps in South Vietnam', HumRRO Consulting Report, June 1968.
2. J. Carlock and B. Bucklin, 'Human factors in mine warfare: an overview of visual detection and stress', Paper prepared for the TTCP Panel 0–1, Working Group 8, Mine Warfare Study Group Seminar, October 1971.
3. F. Jeffrey Maxey *et al.*, 'Identification of the potential characteristics, aptitudes and acquired skills involved in human detection of mines', HumRRO Technical Report 73–18, August 1973.
4. Personal briefing with psychologists at Royal Victoria Hospital, Netley (the Ministry of Defence mental hospital near Portsmouth), 3 May 1974. Part of this account first appeared as 'How to choose a hero', *Sunday Times* (London), 'Spectrum', 12 May 1974.
5. A. H. Birnbaum *et al.*, 'Summary of BESRL surveillance research', US Army BESRL Technical Research Report 1160, September 1969 (revised April 1974).
6. See reference 5 above.
7. H. N. Hepps *et al.*, 'Conducting helicopter reconnaissance experiments with air cavalry elements', RAC McClean, Virginia, Paper presented to 10th Annual US Army Human Factors Research and Development Conference, October 1964.
8. John Powers III *et al.*, 'Training techniques for rapid target detection', US Army BESRL Technical Paper 242, September 1973.
9. R. Kause and J. A. Thomas, 'Effect of training on coordinate determination of SLAR imaged features', US Army RIBSS Technical Research Note 235, April 1973.

10. L. R. C. Haward, 'Respiratory effects of stress during ejector seat operations', *Flight Safety*, vol. 1, 1968, pp. 8–15.
11. R. H. Stacy and O. S. Fedoroff, 'The Soviet pilot and psychology', *Air University Review*, vol. 19, no. 3, 1968.
For details of twenty-four highly technical psychological studies of RAAF pilots and aircrew, see:
'Abstracts of completed and current research projects', Psychology Service Report RAAF, Directorate of Psychology, Department of Air, Canberra, ACT 2600.
12. Personal briefing with Lt Col Austin Kibler, ARPA, Washington, DC 22 January 1974. An account of this first appeared as 'Mind-guided missiles', *Sunday Times* (London), 'Spectrum', 7 July 1974.
13. James Miller *et al.*, 'ONR's role in human factors engineering', *Naval Research Reviews*, vol. 20, no. 7, July 1967.
14. Division 6 of HumRRO – codename LOWENTRY – has produced at least eleven research papers culminating in:
Robert Wright and Warren Pauley, 'Survey of factors influencing army low-level navigation', HumRRO Technical Report, 71–10, June 1971.
15. See reference 14 above, pp. 72ff.
16. R. E. F. Lewis, 'A further study of pilot performance during extended low-speed, low-level navigation', Defence Research Board, Department of National Defence, Canada, November 1962.
17. See reference 14 above.
See also: E. W. Martin, 'Tree-top navigation on a reconnaisance mission', Paper given to the 10th Annual US Army Human Factors Research and Development Conference, October 1964.
18. Robert Wright, 'Some comments on the display of cartographic information for very low-level flight', HumRRO Professional Paper 13–67, March 1967.
19. Ed Moon Edmons and Robert Wright, 'The effects of map scale on position location', HumRRO Technical Report 65–9, September 1965.
20. See reference 14 above, pp. 64ff.
21. See reference 14 above, pp. 63ff.
22. George Wheaton *et al.*, 'The effects of map design variables on map user performance', AIR Final Technical Report R67–3, April 1967.
23. See reference 22 above.
24. See reference 14 above, pp. 12ff.
25. See reference 14 above, pp. 13ff.
26. See reference 14 above, pp. 31ff.
27. Dorothy Adkins *et al.*, 'An exploratory study of terrain orientation', Personnel Research Section Report no. 992, December 1952.
28. Clyde Brictson and Joseph Wulfeck, 'Human factors research on visual

aids to night carrier landing', *Naval Research Reviews*, vol. 20, no. 2, February 1967, pp. 8–15.
29. 'The selection of cryptography students' (no author), BESRL Report no. 434, December 1943.
For a modern, less psychological account of code breaking, see:
David Kahn, *The Code Breakers*, Macmillan, 1967.
Related: A. E. Castelnova *et al.*, 'Individual differences in transcribing voice radio messages embedded in atmospheric noise', USAPRO Technical Research Note 137, October 1963.
30. T. O. Jacobs, 'Training for modern combat operations', Paper given to the 12th Annual US Army Human Factors Research and Development Conference, October 1966.

Chapter 5 Soldiers in Groups

1. Quoted in Shelford Bidwell, *Modern Warfare: a study of men, weapons and theories*, Allen Lane, 1974.
2. Rodney Clark, 'Analysing the group structures of rifle squads in combat', Paper presented to the American Psychological Association Convention, Cleveland, Ohio, 1953.
3. Rodney Clark, 'Leadership in rifle squads on the Korean front line', HumRRO Technical Report 21, September 1955.
4. 'Potential determiners of infantry rifle squad effectiveness', PRB Research Memo 54–49.
'Effectiveness of small military units: parts I–III', PRB Report 980, September 1952.
5. See, for instance:
(a) Albert Pepitone, 'Enhancing team performance: reduction of hostility in groups to enhance team performance', Office of Naval Research Procurement Request and Approval, 23 July 1969, Contract NONR 551 (27).
(b) Arnold Buss and Norman Portnoy, 'Pain tolerance and group identification', *Journal of Personality and Social Pscyhology*, vol. 6, no. 1, 1967.
6. H. Baker, 'Effects of aggressing alone or with others', *Journal of Personality and Social Psychology*, vol. 12, 1969, p. 80.
7. See reference 5(b) above.
8. Clay George, 'Training for coordination within rifle squads', Paper presented to the 12th Annual US Army Human Factors Research and Development Conference, October 1966.
More generally – and more accessible: William C. Biel, 'Planning for team training in the system', in J. J. Uhlaner (ed.), *Psychological Research in National Defense Today*, BESRL TR S–1, June 1967, chapter 17.
9. Harold Gerard *et al.*, 'Conformity and group size', *Journal of Personality and Social Psychology*, vol. 8, 1968, p. 79.

10. Peter Bourne, *Men, Stress and Vietnam*, Little, Brown & Co, 1970, chapter 10.
11. See reference 4 above.
12. See reference 8 above.
13. See reference 8 above, Experiment IV.
14. See reference 8 above, Experiment VI.
15. Adie McRae, 'Interaction content and team effectiveness', HumRRO Technical Report 66–10, June 1966.
16. L. T. Alexander *et al.*, 'The effectiveness of knowledge of results in a military system training program', *Journal of Applied Psychology*, vol. 46, no. 3, 1962.
17. See reference 5(a) above.
18. James Dees, 'Squad performance as a function of the distribution of a squad radio', HumRRO Technical Report 69–24, December 1969.
19. See for example: R. L. Brown *et al.*, 'Recognition thresholds and accuracy for differing body regions as a function of electrode number and spacing', HumRRO Professional Paper 3–67, January 1967.
R. L. Brown *et al.*, 'A differential comparison of two types of electropulse alphabets based on locus of stimulation', HumRRO Professional Paper 32–67, June 1967.
Douglas Holmes *et al.*, 'Determinants of tactual perception of finger drawn symbols: reappraisal', HumRRO Professional Paper 37–68, November 1968.
20. For a number of studies on this topic see 'Collected papers presented under work unit AAA: Factors affecting efficiency and morale in anti-aircraft artillery batteries' (various authors), HumRRO Professional Paper 33–69, November 1969.
21. Thomas Myers and Francis Palmer, 'Crew description dimensions and radar crew effectiveness', Paper presented to the American Psychological Association Convention, September 1955.
22. See reference 20 above, paper entitled: 'Sociometric choices and group productivity among rader crews'.
For the general application of social science finding to naval ships: T. B. Sutherland, 'Application of recent sociological surveys to personnel management aboard ship', *Naval War College Review*, vol. 19, no. 8, 1967, pp. 8–50.
23. Philip Federman *et al.*, 'Communication as a measurable index of team behavior', Applied Psychological Services, Science Centre, Wayne, Pa, Contract no. N61339–1537, October 1965.
N. Egerman, 'Effects of team arrangement on team performance', *Journal of Personality and Social Psychology*, vol. 3, 1966, p. 541.
24. 'Digitial simulation of the performance of intermediate size crews', Project for the Group Psychology and Engineering Psychology Program of

the Office of Naval Research by Applied Psychological Services Inc., July 1972. (No author given)
25. Kent Crawford and Edmund Thomas, 'Organization climate and crew disciplinary rates on navy ships', *Armed Forces and Society*, vol. 3, no. 2, Whiter 1977, pp. 165–83.
M. T. Synder *et al.*, 'Techniques for developing systems to fit manpower resources', Paper given to the 14th Annual US Army Human Factors Research and Development Conference, October 1968.
Richard Sorenson, 'Amount of assignment information and expected performance of military personnel', US Army Research Institute for the Behavioral and Social Sciences, Report no. 1152, February 1967.

Chpater 6 Allocating and Inducting Men

1. See chapter 2, reference 14 above.
2. *Marginal Man and Military Service: A review*, Department of Army, USA War Office, December 1965.
3. See reference 2 above, chapter 6.
4. See reference 2 above, chapter 7.
5. See reference 2 above, chapter 9.
6. See reference 2 above, chapter 9, pp. 134ff.
7. See reference 2 above, chapter 10, pp. 170ff.
8. See reference 2 above, chapter 13.
9. See reference 2 above, chapter 13, p. 214.
10. See reference 2 above, chapter 13, p. 215.
11. See reference 2 above, chapter 13, pp. 217–18.
12. See reference 2 above, chapter 13, p. 218.
13. See reference 2 above, chapter 13, p. 218.
14. A. U. Dubuisson and W. A. Klieger, 'Combat performance of enlisted men with disciplinary records', USAPRO Technical Research Note 148, June 1964.
15. J. G. Bachman and J. Johnston, 'The all-volunteer force: not whether but what kind?', *Psychology Today* (USA), October 1972, pp. 113–16.
16. Clark C. Abt, 'National opinion and military security: research problems', *Journal of Conflict Resolution*, vol. 9, 1965, p. 334.
17. Ira M. Frank and Frederick S. Hoedemaker, 'The Civilian Psychiatrist and the Draft', *American Journal of Psychiatry*, vol. 127, no. 4, October 1970.
18. Robert H. Ollendorf *et al.*, 'Psychiatry and the draft', *American Journal of Orthopsychiatry*, vol. 41, 1971, p. 85.
19. Robert Lieberman, 'Psychiatric evaluations for young men facing the draft: a report of 147 cases', *American Journal of Psychiatry*, vol. 128, 1971, p. 147.

See also: Edward Shils, 'A profile of the military deserter', *Armed Forces and Society*, vol. 3, no. 3, 1977, pp. 427–32.
D. Bruce Bell and Beverley W. Bell, 'Desertion and antiwar protest: Findings from the Ford Clemency Program', *Armed Forces and Society*, vol. 3, no. 3, 1977, pp. 433–44.
20. Carl Kline *et al.*, 'The young American expatriates in Canada: alienated or self-defined?', *American Journal of Orthopsychiatry*, vol. 41, 1971, p. 74.

Chapter 7 Military leadership and the Skills of Command

1. Shelford Bidwell, *Modern Warfare: A study of men, weapons and theories*, Allen Lane, 1973, pp. 109–15.
2. See reference 1 above, pp. 112ff.
3. One of a series of papers stored in the Fort Bragg library prepared as course papers by serving officers attending staff courses at the war college.
4. See reference 1 above, pp. 112ff. The Bion techniques are now in use with other British services, particularly the RAF.
5. See reference 1 above, p. 116.
6. Jay L. Otis *et al.*, 'Pyschological requirements analysis of company grade officers', Department of Army, Personnel Research Section, Program 5176, 29 December 1950.
7. See reference 6 above, pp. 50ff.
8. William S. Turley, 'Origins and development of Communist military leadership in Vietnam', *Armed Forces and Society*, vol. 3, no. 2, 1977, p. 219.
9. 'OFFTRAIN' produced at least fifteen separate research papers, but see especially:
(a) Carl J. Lange *et al.*, 'A study of leadership in army infantry platoons', HumRRO Research Report 1 (OFFTRAIN II), November 1958.
(b) Carl J. Lange and T. O. Jacobs, 'Leadership in army infantry platoons: study II', HumRRO Research Report 5 (OFFTRAIN III), July 1960.
(c) Frank L. Brown and T. O. Jacobs, 'Developing the critical combat performance required of the infantry rifle platoon leader', HumRRO Technical Report 70–5, April 1970.
10. T. O. Jacobs, 'Leadership in small military units', Paper for the 4th International Congress on Applied Psychology, The Hague, Holland, September 1967.
11. Rodney A. Clark, 'Leadership in rifle squads on the Korean front line', HumRRO Technical Report 21, September 1955. (For official use only)
12. Personal communication, July 1975.
13. Personal communication, July 1975.
14. Aaron J. Spector *et al.*, 'Supervisory characteristics and attitudes of subordinates', *Personnel Psychology*, vol. 13, no. 3, 1960, pp. 301–16.
For an imaginative approach to training, see Paul D. Hood, 'The use of

follower stooges for field evaluation of leadership ability', Paper to the American Psychological Association, Cincinnati, Ohio, 1959.
15. D. Kipnis *et al.*, 'Leadership problems and practices of petty officers', Bureau of Naval Personnel, Technical Bulletin 61–13, August 1961.
16. Robert Ziller, 'Leader assumed dissimilarity as a measure of prejudicial cognitive style', *Journal of Applied Psychology*, vol. 47, no. 5, October 1963, pp. 339–42.
17. Richard Snyder, 'Communication and leadership roles', Paper given to the West Coast Society for Small Group Research, April 1955.
Richard Snyder and Carl H. Rittenhouse, 'The influence of cognitive dissonance of sequential decisions', Paper for Western Psychological Association, 1957.
Eugene H. Drucker, 'The differences in the power, authority and responsibility of leaders on team performance', US Army Armor Human Research Unit (undated).
18. Paul R. Bleda *et al.*, 'Enlisted men's perception of leader attributes and satisfaction with military life', *Journal of Applied Psychology*, vol. 62, no. 1, 1977, pp. 43–9.
See also: Morris Showel and Christian W. Peterson, 'A critical incident study of infantry, airborne and armored junior non-commissioned officers', Staff Memo, US Army Leadership Human Research Unit, Presidio of Monterey, California, July 1958.
19. Yehuda Amir *et al.*, 'Peer nominations as a predictor of multistage promotions in a ramified organization', *Journal of Personality and Social Psychology*, vol. 10, 1968, p. 271.
20. Paul D. Hood *et al.*, 'Research on the training of non-commissioned officers: a summary report of pilot studies', HumRRO Technical Report 65–17, December 1965.
21. Meredith P. Crawford, 'A review of recent research and development on military leadership, command and team function', HumRRO Research Memo 7, September 1964.
Also: T. O. Jacobs, 'A program of leadership instruction for junior officers', HumRRO Technical Report 84, June 1963.
HumRRO have also produced a textbook on leadership – one not confined solely to military purposes:
T. O. Jacobs, *Leadership and Exchange in Formal Organizations*, December 1970. It includes chapters on: motivation; small group processes; the organizational context of leadership; leader behaviour and organizational effectiveness.
22. William H. Helme *et al.*, 'Dimensions of leadership in a simulated combat situation', US Army BESRL, Technical Research Report 1172, July 1971.
23. See also: William H. Helme *et al.*, 'Prediction of officer behavior in a simulated combat situation', US Army RIBSS, Research Report 1182, March 1974.

24. See reference 23 above, pp. 6ff.
25. See reference 23 above, pp. 10ff.
26. Shepard Swartz and Arthur Floyd Jr, 'Improving tactical training for tank commanders: test development and performance assessment', HumRRO Technical Report 82, March 1963.
See also: Shepard Swartz, 'Tank crew effectiveness in relation to the supervisory behavior of the tank commander', HumRRO Technical Report 68–12 September 1968.
27. Frank L. Vicino *et al.*, 'Conspicuity coding of updated symbolic information', US Army RIBSS Technical Research Note 152, May 1965.
28. Frank Vicino and Seymour Ringel, 'Decision-making with up-dated graphic v. alpha-numeric information', US Army RIBSS Technical Research Note 178, November 1966.
29. Seymour Ringel *et al.*, 'Human factors research in command information processing systems – summary of recent studies', US Army BESRL Technical Research Report 1158, July 1969.
More accessible is: 'Threat and decision-making in a simulated combat game', *Behavioral Science*, vol. 11, 1966, p. 167.
30. Michael Strub, 'Perception of military event patterns in a two-alternative prediction task', US Army BESRL Technical Research Note 221, February 1970.
31. James D. Baker, 'Human factors experimentation within a tactical operations system (TOS) environment', US Army BESRL Research Study 68–4, October 1968.
32. James D. Baker *et al.*, 'Certitude judgements in an operational environment', US Army BESRL Technical Research Note 200, November 1968.
Edgar Johnson, 'Numerical encoding of quantitative expressions of uncertainty', US Army RIBSS Technical Paper 250, December 1973.
33. Ralph E. Strauch, 'The operational assessment of risk: a case study of the *Pueblo* mission', Rand Paper R–691–PR, March 1971.
34. See reference 29 above, pp. 12ff; also reference 31 above.
35. Howard C. Olson, 'Improvement in performance on a leadership game as a result of training in information handling', HumRRO Professional Paper 24–69, June 1969.
36. Theodore R. Powers and Arthur J. DeLuca, 'Knowledge, skills and thought processes of the battalion commander and principal staff officers', HumRRO Technical Report 72–20, July 1972.
Joseph A. Olmstead, 'Leadership at senior levels of command', HumRRO Professional Paper 5–68, February 1968.
37. See reference 36 (Powers and DeLuca) above, p. 26.
38. See reference 36 (Powers and DeLuca) above, pp. 12ff.
39. See reference 36 (Powers and DeLuca) above, p. 24, figure 6.
40. L. L. Lackey *et al.*, 'The effects of command position upon evaluation of

leader behavior' HumRRO Technical Report 72–32, November 1972.
41. Joel T. Campbell and Leila S. Cain, 'Public opinion and the outbreak of war', *Journal of Conflict Resolution*, vol. 9, 1965, p. 318.
42. Personal communication at Nato Conference on the Dimension of Stress and Anxiety, Oslo, July 1975.
43. Hazel Erskine, 'Polls: atomic weapons and nuclear energy', *Public Opinion Quarterly*, vol. 27, 1963, p. 155. (One of several articles on the social and psychological aspects of nuclear fall-out.)
44. Jerome Laulicht, 'Leaders and voters' attitudes on defence and disarmament', *Public Opinion Quarterly*, vol. 28, 1964, p. 357.
Compare with: Frank Morris, 'The general or admiral of military service: an analysis of his role as a public opinion leader in civil–military relations', Contract AF33608, August 1965.
More specifically: Lt Col Edward Kelley, 'The president and the people – strategic psyop in the early phase of the Vietnam War', US Army War College Research Paper, USAWC, Carlisle Barracks, Pennsylvania, 14 January 1971. Highly critical of President Johnson's role.
45. Richard Teevan, 'A motivational correlate of Vietnam protest group members', Office of Naval Research Report no. 20, 1966.
H. Gergen, 'Aging, time perspective and preferred solutions to international conflicts', *Journal of Conflict Resolution*, vol. 9, 1965, p. 177.
For a more general discussion about changing public attitudes to the Vietnam War, see: Milton J. Rosenberg *et al.*, *Vietnam and the Silent Majority*, Harper & Row, 1970.
46. Hazel Erskine, 'Is war a mistake?', *Public Opinion Quarterly*, vol. 34, 1970, p. 134.
47. See reference 45 (Teevan) above.
48. Information for this section given in a personal interview with Professor Louis Guttman in Jerusalem, November 1964.
But see also: 'Social problem indicators of the period 12–14 November 1974 (from the continuing survey)', Israel Institute of Applied Social Research, PO Box 7150, Jerusalem, Israel. The charts are taken from this paper.
Note: The British Ministry of Defence also has six questions per month in an NOP poll in which it can survey public opinion on any politico-military issue (personal interview with D. Bennison, psychologist, Ministry of Defence, 16 April 1974).

Chapter 8 Whales as Weapons? Animals in a Military Context

1. See, for example: 'Tuning in on porpoises that "talk" and "work" with people' (unsigned article), *Science Digest*, vol. 22, February 1973.
2. See reference 1 above, p. 10–F.

3. Variously: C. A. Bower and R. S. Henderson, 'Project Deep Ops: deep object recovery with pilot and killer whales', Naval Undersea Center (NUC) Technical Publication 306, 1972.
M. E. Canboy, 'Project Quick Find: a marine mammal system for object recovery', NUC Technical Publication 268, 1972 (using sea-lions).
S. H. Ridgway, 'Studies on diving depth and duration in *Tursiops truncatus*', Proceedings of the Conference on Biological Sonar and Diving Mammals, Stanford Research Institute, Menlo Park, California, 1966.
F. G. Wood and S. H. Ridgway, 'Utilization of porpoises in the man-in-the-sea program', Office of Naval Research Report ACR–124, 1967.
4. See reference 1 above, p. 10–F, column 3.
5. Kenneth H. Norris, 'Studying porpoises in their natural habitat', *Naval Research Reviews*, vol. 21, no. 8, August 1968, p. 1–5.
6. For example: W. E. Evans, 'Uses of advanced space technology and upgrading the future of oceanography', AIAA Paper no. 70–1273, 1970.
7. General summary on the navy's marine mammal programme, unsigned, undated factsheet published by the Office of Naval Research, Washington, DC.
8. Cedric M. Smith and Joseph M. Coates, 'Olfaction and its potential application in personnel detection', IDA Research Paper P–187, December 1967. (Limited availability)
9. L. V. Kruchinskii and D. A. Fless, 'Strengthening of olfaction in police dogs', *Zhur. Vyssh.* Deiatel'nosti (trans.) vol. 9, no. 2, 1959, pp. 266–72 (in reference 8 above, page 37).
10. See reference 8 above, pp. 14–15.
11. See reference 8 above, p. 21.
12. Personal interview with Professor Robert Lubow, Tel Aviv, 9 July 1977.
13. E. Carr-Harris and R. Thal, 'Mine, booby trap, trip wire and tunnel detection in dogs: final report', US Army Land Warfare Laboratory Technical Report, July 1969.
14. Robert Lubow, 'High-order concept formation in the pigeon', *Journal of the Experimental Analysis of Behavior*, vol. 21, no. 3, 1974, pp. 475–83.
15. See reference 14 above, especially the references.
16. B. F. Skinner, 'Pigeons in a Pelican', *American Psychologist*, vol. 15, 1960, pp. 28–37.
17. In chronological order the papers listed by Professor Lubow (some of which, he says, are still not de-classified) are:
'Training birds for field surveillance: Phase II' (Unsigned), Final Report TR 65–226 US Army Land Warfare Laboratory, Aberdeen Proving Ground, 15 April 1965. (Secret)
'Development of a pigeon ambush detection system', Animal Behavior Enterprises Inc., Final Report, Contract no. DA 18–001–AMC–860, 4 January 1967. (Confidential)

T. A. McHaffied, 'The use of dogs in searching for pre-scented objects', Summary of RAF work on detection of squalene-scented flight recorders (undated but probably about 1969).
Robert Lubow and E. E. Bernard, 'The use of birds in selected military applications', for Israeli Ministry of Defence, January 1972.
Robert Lubow, 'Survey of local bird population', for Israeli Ministry of Defence, *circa* 1972/3.

Chapter 9 In the Danger Zone: Testing Men Under Stress

1. Albert J. Glass, 'Psychiatry in the Korean Campaign (Instalment I)', *US Armed Forces Medical Journal*, vol. 4, 1953, pp. 1387–1401.
2. Shelford Bidwell, *Modern Warfare: A study of men, weapons and theories*, Allen Lane, 1974.
3. Franklin Del Jones and Arnold W. Johnson Jr, 'Medical and psychiatric treatment policy and practice in Vietnam', final draft of paper prepared at the Walter Reed Army Medical Center, Washington, DC, 20 October 1973.
4. Quoted in reference 3 above.
5. Richard Kern and Howard McFann (from HumRRO), 'Can our troops be battleproofed?', *Army Information Digest*, vol. 20, no. 2, December 1965, pp. 24–7.
6. See reference 5 above, p. 25.
7. James G. Miller, 'The development of experimental stress-sensitive tests for predicting performance in military tasks', PRB Technical Research Report 1079, October 1953.
8. Wiley R. Boyles, 'Aviator performance under stress', HumRRO Professional Paper 27–67, June 1967.
Wiley R. Boyles, 'Background and situational confidence: their relation to performance effectiveness', HumRRO Professional Paper 22–68, June 1968.
'Stress testing of special mission personnel', *Human Factors*, vol. 7, no. 6, December 1965.
Wiley R. Boyles, 'Measures of reaction to threat of physical harm as predictors of performance in military aviation training', HumRRO Professional Paper 15–69, May 1969.
(In Britain Dr L. R. C. Haward has been using hypnosis to get parachutists and pilots who have to use ejector seats to relive their stress and to teach them how to cope. See: 'Respiratory effects of stress during ejector seat operations', *Flight Safety*, vol. 1, 1968, p. 85. The paper also contains some useful references.)
9. Jerald N. Walker and Tor Meeland, 'Relationship of life history, family background and intelligence data to performance in situations employing

height, fire, distraction, shock, dark and noise as sources of stress', HumRRO paper for Western Psychological Association Meeting, spring 1956.
10. Kan Yagi *et al.*, 'Development of a verbal measure for use in stress study', HumRRO paper for Western Psychological Association Meeting, 1959.
11. Mitchell M. Berkun, 'Psychological and physiological criteria for stress simulation research', HumRRO paper for Third Annual Symposium on Performance Reserve, Human Factors Society of Los Angeles, 18 June 1963.
12. Peter G. Bourne, *Men, Stress and Vietnam*, Little, Brown & Co, 1970; see especially chapters 6 and 7.
13. V. H. Marchbanks, 'Effects of flying stress on 17-hydroxycorticosteroid levels', *Journal of Aviation Medicine* vol. 29, 1958, pp. 670–82.
14. See reference 12 above, chapter 7.

Chapter 10 The Psychological Effects of Weapons

1. The four papers are:
(a) J. C. Naylor, 'Proposed method for determining the psychological effects of weapons', Ohio State University.
Related: R. A. Terry, 'Toward a psychological index of weapons effectiveness. Part I. field studies', Air Force Contract no. 086353693.
(b) Monte Page *et al.*, 'Psychological effects of non-nuclear weapons: bibliography with selected abstracts, vol. 1', University of Oklahoma Research Institute. (Unclassified)
(c) A. E. Dahlke and Ron Burnett, 'Psychological reactions to non-nuclear weapons: problems and potential of the experimental approach', University of Oklahoma Research Institute.
(d) Monte Page *et al.*, 'Prior art in the psychological effects of weapons systems', University of Oklahoma Research Institute.
(This first symposium of the psychological effects of non-nuclear weapons was held at Eglin Air Force Base on 29 April 1964. The bibliography is dated August 1964.)
2. See reference 1(a) above and references therein.
3. See reference 1(a) above and references therein.
4. See reference 1(a) above and references therein.
5. See reference 1(a) above, table 2.
6. See reference 1(a) above, table 3.
7. See reference 1(a) above, table 4.
8. See reference 1(a) above.
9. 'Study of AGF battle casualties', HQ Army Ground Forces, Washington, DC, Plans Section, 1946.
10. Quoted in references 1(a–d) above.
11. D. L. Mills *et al.*, 'Exploratory study of human reactions to fragmentation weapons', Stanford Research Institute, WSLRM66, 1961.

12. W. S. Vaughan and P. G. Walker, 'Psychological effects of patterns of small arms fire', Psychological Research Associates Report 57–16, 1957.
For more up-to-date but less detailed work see: J. M. Keyser, 'Weapons preferences in South Vietnam', HELTM 9–64, Aberdeen Proving Ground, Md, June 1964.
13. See reference 12 above.
14. See references 1(a–d) above.
15. I. Atkin, 'Air-raid trauma in mental hospital admissions', *Lancet*, vol. 2, 1941.
16. See: Monte Page, Clinton Goff and J. D. Palmer (eds.), 'The psychological effects of non-nuclear weapons: a bibliography with selected extracts', vol. 1 (unclassified), University of Oklahoma Research Institute Technical Report, August 1964, p. 6.
17. F. C. Ikle, *The Social Impact of Bomb Destruction*, University of Oklahoma Press, 1958.
18. I. L. Janis, 'Psychological impact of air attacks: survey and analysis of observations on civilian reaction during World War Two', Rand RM–93, 1949.
19. Ignacio Matte, 'Observations of the English in wartime', *Journal of Nervous and Mental Disease*, vol. 97, 1943.
20. See reference 1(d) above which includes a re-analysis of the Morale Division's material.
21. P. E. Vernon, 'Psychological effects of air raids', *Journal of Abnormal and Social Psychology*, vol. 36, 1941, pp. 457–76.
More up-to-date is D. Kuobovi's work with children in Ma'alot and Qiryat Shmona in Israel, both of which towns were attacked by Palestinian terrorists in the early 1970s. See her: 'Individual treatment by the teacher of emotionally disturbed children', *Megamot*, vol. 11, no. 2, 1971 (a mimeographed paper); and her book, *Therapeutic Teaching at a Time of National Crisis*, 1973. This shows how she developed and adapted Old Testament stories to the current tensions to help the children 'abreact' – that is live through their fears in their imaginations in such a way as to prevent these fears being 'bottled up' and showing themselves later as neurotic symptoms.
22. See reference 1(d) above and:
Alexander George, 'Human factors in air-to-ground interdiction operations in the Korean war', Rand RM–659, 1951.
23. Leonard Berkowitz, 'Weapons as aggression-eliciting stimuli', *Journal of Personality and Social Psychology*, vol. 7, no. 2, 1967, pp. 202–7.
24. See, for example, Stanley Milgram, *Obedience to Authority*, Harper & Row; Tavistock, 1974. Though many studies have grown out of this particular work, all supporting this general point.
25. *Health Aspects of Chemical and Biological Weapons*, Report of a World Health Organization Group of Consultants, WHO, Geneva, 1970.

26. See reference 25 above, pp. 46–51.
27. R. G. Smith, 'Problem of measuring performance of incapacitated troops', US Army Edgewood Arsenal, Paper given to 12th Annual US Army Human Factors Research and Development Conference, October 1965. (Classified)
Also see: S. Jackson, 'Effects of chlorpromazine, amphetamine and meprobromate on performance in human subjects under delayed auditory feedback', Draft CDRL Report, Edgewood Arsenal, Munitions Command (no date but late 1960s).
28. See reference 25 above, Annex 6: 'Psychosocial consequences of chemical and biological weapons', pp. 121ff. and references.

Chapter 11 Combat Psychiatry

1. Franklin Del Jones, 'Psychological adjustments to Vietnam', Speech to Medical Education for National Defense (MEND), 18 October 1967.
2. Personal briefing with Tom Brown, psychologist, Royal Victoria Hospital, Netley, Hampshire, 3 May 1974.
3. Col Robert Bernstein (Medical Corps), 'Getting to the fight – a review of physical and emotional problems encountered in moving troops into combat', Student Thesis, US Army War College, Carlisle Barracks, Pa, 6 March 1964.
4. References 1 and 3 above.
5. W. O. Evans, 'Problems and approaches to measurement of troop performance', US Army Medical Research and Nutrition Laboratory, Paper to the 12th Annual US Army Human Factors Research and Development Conference, October 1966.
6. See reference 1 above, pp. 3ff.
7. Shelford Bidwell, *Modern Warfare: A study of men, weapons and theories*, Allen Lane, 1974.
8. Franklin Del Jones and Arnold W. Johnson Jr, 'Medical and psychiatric treatment policy and practice in Vietnam', Paper produced at the Walter Reed Army Medical Center, Washington, DC.
9. See references 1 and 3 above.
10. See reference 7 above.
11. Meyer Teichman and D. Kligger, 'Interpersonal relationships among soldiers during wartime', Paper presented to the Nato Conference on Dimensions of Stress and Anxiety, Oslo, 29 June–3 July 1975.
12. See reference 7 above.
13. See reference 1 above.
14. Quoted in Alber Glass, 'Psychiatry in the Korea campaign (Instalment I)', *US Armed Forces Medical Journal*, vol. 4, 1953, pp. 1387–1401.
15. See reference 7 above.

16. See reference 7 above.
17. See reference 8 above, p. 17ff.
18. Peter G. Bourne, *Men, Stress and Vietnam*, Little, Brown & Co. 1970, chapter 1.
19. See reference 18 above.
20. R. S. Anderson *et al.* (eds.), *Neuropscyhiatry in World War II*, US Government Printing Office, Washington, DC, 1966.
21. Fatigue and Stress Symposium, Operations Research Office, Johns Hopkins University, 1952.
22. See reference 3 above, pp. 13ff.
23. See reference 20 above.
24. Quoted in reference 18 above, chapter 1.
25. See: Edward Shils, 'A profile of the military deserter', *Armed Forces and Society*, vol. 3, no. 3, spring 1977, pp. 427–33.
Also: D. Bruce Bell and Beverley Bell, 'Desertion and anti-war protest: findings from the Ford Clemency Program', *Armed Forces and Society*, vol. 3, no. 3, spring 1977, pp. 433–45. (This has a comprehensive reference list on military desertion.)
26. See reference 18 above, chapter 1.
27. See reference 25 above.
28. See reference 25 above.
29. See reference 8 above, p. 9 and table 1.
30. See reference 18 above, chapter 1.
31. See reference 8 above, pp. 20ff.
32. Robert E. Strange, 'Effects of combat stress on hospital ship psychiatric evacuees', in Peter G. Bourne (ed.), *The Pyschology and Physiology of Stress (with reference to special studies of the Vietnam War)*, Academic Press, 1969, chapter 4, pp. 75–92.
Frank Hayes, 'Psychiatric aeromedical evacuation of patients during Tet and Tet II offensive, 1968', *American Journal of Psychiatry*, 1968, p. 503.
33. See reference 8 above, pp. 3–6.
34. See reference 8 above, pp. 22–3.
35. See reference 8 above, pp. 18–19.
36. Gary L. Tischler, 'Patterns of psychiatric attrition and of behavior in a combat zone', in Peter G. Bourne (ed.), *The Psychology and Physiology of Stress*, Academic Press, 1969.
37. Meyer Teichman, 'Affiliative behavior among soldiers during wartime', *British Journal of Social and Clinical Psychology*, vol. 16, 1977, pp. 3–7.
38. Personal communication to colleagues at the *Sunday Times*, October–November 1973.
39. See reference 8 above, pp. 23–7 and: Norman E. Zinburg, 'Rehabilitation of heroin users in Vietnam', *Contemporary Drug Problems*, spring 1972, pp. 263–94.

Chapter 12 Atrocity Research

1. Neil Sheehan, 'Should we have war crime trials?', a major review of thirty-three war books in the *New York Times Book Review*, 28 March 1971.
2. H. V. Dicks, *Licensed Mass Murder: A socio-psychological study of some SS killers*, Tavistock, 1972.
3. (a) S. H. Haley, 'When the patient reports atrocities', *Archives of General Psychiatry*, vol. 30, no. 2, 1974, pp. 191–6.
(b) Peter Watson, 'My Lai: what makes a killer?', *Psychology Today* (UK edition), vol. 1, no. 3, June 1975, p. 70 (similar to reference (a) but with thirty-one ex-Vietnam veterans).
(c) Peter Watson, 'The soldiers who become killers', *Sunday Times* (London), 'Spectrum', 8 September 1974.
4. See reference 3(b) above, table p. 70.
5. Peter G. Bourne, *Men, Stress and Vietnam*, Little, Brown & Co, 1970, chapters 6 and 7, pp. 63–102.
6. Reference 3(b) above, table p. 70.
7. Reference 3(a) above, pp. 195–6.
Also see: Robert J. Lifton, 'Beyond atrocity', *Saturday Review*, 27 March 1971, pp. 23–6 for an entirely different type of analysis.
8. 'Gallagher proposes study of ending Navy Department's project: Group Technology', *Congressional Record*, 2 March 1971, pp. E1295–E1302.

Chapter 13 Captivity

1. F. H. Lakin, 'Psychological warfare in Malaya 1952–55', Army Operational Research Establishment, Ministry of Defence, UK, Paper presented to the 11th Annual US Army Human Factors Research and Development Conference, October 1965.
2. W. C. Bradbury and A. D. Biderman, *Mass Behavior in Battle and Captivity*, Aldine, 1962.
3. V. A. Kral, 'Psychiatric observations under severe chronic stress', *American Journal of Psychiatry*, vol. 108, September 1951, p. 185.
4. For instance: Bruno Bettelheim, 'Individual and mass behavior in extreme situations', *Journal of Abnormal and Social Psychology*, vol. 38, 1943, pp. 417–52.
5. Stewart Wolf and Herbert Ripley, 'Reactions among allied prisoners subject to three years of imprisonment and torture by Japanese', *American Journal of Psychiatry*, vol. 104, 1947, pp. 180–93.
6. J. E. Nardini, 'Survival factors in American POWs of the Japanese', *American Journal of Psychiatry*, vol. 109, 1952, pp. 241–8.
7. Edgar H. Schein, 'Patterns of reaction to severe chronic stress in American army POWs of the Chinese', in *Methods of Forceful Indoctrination: Obser-*

vations and interviews, GAP Symposium no. 4, New York, Group for the Advancement of Psychiatry, Publications Office, July 1957, pp. 253–69.
8. H. D. Strassman *et al.*, 'A prisoner-of-war syndrome: apathy as a reaction to severe stress', *American Journal of Psychiatry*, vol. 112, 1956, pp. 998–1003.
9. Charles V. Ford, 'The *Pueblo* incident: psychological responses to severe stress', Paper given to Nato Conference on Dimensions of Stress and Anxiety, Oslo, July 1975.
For another recent account of being a POW, see: Charles A. Stenger, 'Life-style shock: the psychological experience of being an American POW in the Vietnam conflict', Veterans Administration, Central Office, Washington, DC, December 1972.
10. See reference 9 (Ford), above, pp. 9ff.
11. The best account is: Craig Haney, Curtis Banks and Philip Zimbardo, 'A study of prisoners and guards in a simulated prison', *Naval Research Reviews*, vol. 26, no. 9, September 1973, pp. 1–17.

Chapter 14 Interrogation

1. Peter Deeley, *Beyond Breaking Point: A study of the techniques of interrogation*, Arthur Barker, 1971.
2. Skaidrite Maliks Fallah, 'Certain aspects of interrogation techniques: a bibliography', CRESS, The American University, December 1967.
Robert Beezer, 'Research methods of interviewing foreign informants', HumRRO Technical Report 30, August 1956. (Four methods of interrogating East German refugees are compared.)
3. See, for instance:
(a) Cyril Cunningham (British Air Ministry), 'Communist indoctrination and interrogation techniques in Korea', *Science*, March 1952.
(b) Alexander Kennedy, 'The scientific lesson of interrogation', *Proceedings of the Royal Institute of Great Britain*, vol. 170, 1960, pp. 93–113.
(c) 'Physical and mental consequences of imprisonment and torture', Lectures presented at the Conference at Lysebu, near Oslo, Amnesty International, 5–7 October 1973.
4. See reference 3(b) above.
5. Personal interview with Portuguese psychiatrist in Lisbon, March 1975. See also: *Insight on Portugal: the year of the captains*, Deutsch, 1975. (The author was part of the Insight team which wrote this account of the Portuguese revolution.)
6. Most notably: D. O. Hebb *et al.*, 'The effects of isolation upon attitudes, motivation and thought,' Fourth Symposium, Military Medicine I, Defence Research Board, Canada, December 1952. (Secret)
7. D. O. Hebb and W. Heron, 'Effects of radical isolation upon intellectual

function and the manipulation of attitudes', Fourth Symposium, Military Medicine I, Defence Research Board, Canada, December 1952. (Secret)
8. See reference 7 above.
Also: Report no. HR63, Defence Research Board, Canada, October 1955. (Secret)
9. For example: W. H. Bexton *et al.*, 'The effects of decreased variation in the sensory environment', *Canadian Journal of Psychology*, vol. 8, 1954, pp. 70–76.
10. The earliest work was: Thomas Myers *et al.*, 'Experimental studies of sensory deprivation and social isolation', HumRRO Technical Report 66–8, June 1966.
11. Thomas Myers, 'Human reaction to monotony', Paper presented at a Symposium on the Effects of Reduced Sensory Stimulation at the American Association for the Advancement of Science, Philadelphia, 1971.
12. See reference 11 above, p. 8 passim.
13. See reference 10 above, chapter 8: 'Conformity to a group norm'.
14. See reference 10 above, chapter 9: 'Propaganda and attitude change'.
15. See reference 10 above, chapter 10: 'Conditioning of connotative meaning'.
16. See reference 11 above, pp. 2–5, tables 14 and 15.
17. See reference 11 above, pp. 6ff and appendix.
18. See reference 11 above, p. 11 and table 12.
19. See reference 11 above, p. 14.
20. See reference 11 above, p. 17, table 16.
21. See reference 11 above, pp. 17–18.
22. John Rasmussen (ed.), *Man in Isolation and Confinement*, Aldine, 1973, based on a symposium organized by Nato Science Committee Advisory Group on Human Factors held in Rome, 1969.
23. Tim Shallice, 'The use of sensory deprivation in depth interrogation', paper 2 in reference 3(c) above.
See also Dr Shallice's 'The Ulster depth interrogation techniques and their relation to SD research , *Cognitive Psychology*, vol. 1, 1973.
24. (a) *The Times*, 9 June 1973.
(b) John McGuffin, *The Guinea Pigs*, Penguin Books, 1974.
25. 'Building a better thumbscrew', unsigned article in *New Scientist*, 19 July 1973, p. 141.
Besides the international exchanges of psywarriors mentioned later (in chapter 20), British troops have attended interrogation courses at Fort Huachuca, Arizona; and the SAS and US Special Forces have attended similar courses at Bad Tolz, West Germany. (Information contained in library of the International Police Academy before it was disbanded.)
26. Tim Shallice, 'Solitary confinement – a torture revived?', *New Scientist*, 28 November 1974.

27. L. E. Hinkle and H. G. Wolff, 'Communist interrogation and indoctrination of "enemies of the state" ', *Archives of Neurology and Psychiatry*, vol. 76, 1956, pp. 115–74.
28. See reference 26 above, column 2.
29. 'Wakefield's grim secret', an Insight report in the *Sunday Times* (London), 26 October 1974.
30. J. Kubis, 'Instrumental, chemical and psychological aids in the interrogation of witnesses', *Journal of Social Issues*, vol. 13, 1957, p. 40.
Louis Gottschalk, 'The use of drugs in interrogation', in A. D. Biderman and H. Zimmer (eds.), *The Manipulation of Human Behavior*, Wiley, 1960.
See also: Jean S. Cameron *et al.*, 'Effects of meprobormate on moods, emotions and motivations', *Journal of Psychology*, vol. 65, 1967, pp. 209–21.
See reference 26 above for an account of similar practices in Uruguay.
31. L. R. C. Haward, 'Investigations of torture allegations by the forensic psychologist', *Journal of the Forensic Science Society*, vol. 14, no. 4, 1974, pp. 299–310.
32. See reference 24(b) above.
33. 19 May 1974 – though denied.
34. See chapter 2 of this book, reference 29.
35. A. D. Biderman, 'Sociopsychological needs and "involuntary" behavior as illustrated by compliance in interrogation', *Sociometry*, vol. 23, 1960, pp. 120–217.
36. L. P. Willemin *et al.*, 'Prisoner behavior in simulated interrogation', USAPRO, Paper given to the 11th Annual US Army Human Factors Research and Development Conference, October 1965. (Classified)
See also: Peter B. Field *et al.*, 'Strategies of hypnotic interrogation', Veterans Administration Hospital, Brooklyn, New York, Contract NONR 4294–(00), 29 May 1967.
Seymour Fisher, 'The use of hypnosis in intelligence and related military situations', Bureau of Social Science Research TR–4 SR177–D.
37. *Sunday Times*, 25 February 1973.
38. Augustus F. Kinzel, 'Body buffer zones in violent prisoners', American Journal of Psychiatry, vol. 127, July 1970, pp. 54–64.
39. Cyril Cunningham, 'Korean war studies in forensic psychology', *Bulletin of the British Psychological Society*, vol. 23, no. 81, October 1970, pp. 309–12.
40. See reference 31 above.
41. Personal interview with ex-PIDE psychiatrist, Lisbon, March 1975.
42. See reference 31 above.
43. J. A. M. Meerloo, 'Medication into submission: the dangers of therapeutic coercion', *Journal of Nervous and Mental Disease*, vol. 122, 1955, pp. 353–60.
44. Frank S. Horvath, 'Verbal and non-verbal clues to truth and deception

during polygraph examinations', *Journal of Police Science and Administration*, vol. 1, no. 2, June 1973, pp. 138–52.
45. In Arthur S. Aubry Jr and Rudolph R. Caputo, *Criminal Interrogation* (2nd edn.), C. C. Thomas, 1972, chapter 8 and 15 onwards.
46. See reference 1 above, pp. 216ff.
47. See reference 1 above, p. 219.
48. Jan Berkhout *et al.*, 'Autonomic responses during a replicable interrogation', *Journal of Applied Psychology*, vol. 54, no. 4, 1970, pp. 316–25.
49. Personal interview with Meyer Kaplan, Jerusalem Police HQ, November 1974.
50. This was used, at the request of the *Sunday Times*, by Dr Alan Smith, of Powick Mental Hospital, Worcester, to test the veracity of five witnesses in the George Davis case in Britain. Davis had been convicted of a crime which, he and others maintained, he did not commit. See Peter Watson, 'Lie detector okays George Davis alibi', *Sunday Times* (London), 14 December 1975.
(The Land Warfare Laboratory in Maryland is also looking at voice analysis in interrogation – see 'Current Tasks: US Army LWL, July 1973'.)
51. As for reference 49 above, but see also: 'Hypnosis used on witness', *Daily Telegraph* (London), 16 December 1974, for a similar technique used by police in New Jersey.
52. Richard I. Moren and Eugene H. Bocklyn, 'A limited language for obtaining combat information from POWs: a pilot study', HumRRO paper for presentation at the American Psychological Association meeting, September 1960.

Chapter 15 Hsi nao: 'Brainwashing'

1. 'Brainwashing: a guide to the literature', Society for the Investigation of Human Ecology Inc, Forrest Hills, New York, 1960. (Note that the Society for the Investigation of Human Ecology has since been shown to have been set up by the CIA – see references at 29, chapter 2 of this book. However, it still seems that this bibliography is valid.)
2. 'Communist psychological warfare (brainwashing)', Edward Hunter's evidence to the House of Representatives Committee on Un-American Activities, Washington, DC, 13 March 1958. This contains full reference to all Edward Hunter's books but his main work is:
Brainwashing in Red China, Vanguard Press, 1951, 1953, new enlarged edition, 1971.
3. Skaidrite Maliks Fallah, 'Certain aspects of interrogation techniques: a bibliography', CRESS, The American University, December 1967.
4. William H. Sargant, *Battle for the Mind: A physiology of conversion and brainwashing*, Doubleday, 1957.

For another odd insight into American feelings about the politico-military use of psychology at that time, see: Robert C. Tucker, 'Stalin and the use of psychology', Rand: RM–1441, 10 March 1955. This is an attempt to link Stalin, the Pavlovian revival in Russia, the Korean war and brainwashing. Unconvincing.
5. R. A. Bauer, 'Brainwashing: psychology or demonology', *Journal of Social Issues*, vol. 13, 1957, pp. 41–7.
6. For instance, Virginia Paley: *21 Stayed: The Story of the American GIs who chose communist China – who they were and why they stayed*, Farra, Strauss & Cudahy, 1955.
7. William E. Mayer, 'Why did many GI captives cave in?', *US News and World Report*, 24 February 1956.
8. *Methods of Forceful Indoctrination: Observations and interviews*, GAP Symposium no. 4, Group for the Advancement of Psychiatry, Publications Office, New York, July 1957.
Factors Used to Increase the Susceptibility of Individuals to Forceful Indoctrination: Observations and experiments, GAP Symposium no. 3, Group for the Advancement of Psychiatry, Publications Office, New York, December 1956.
9. See reference 7 above, pp. 58ff.
For an interesting but fairly academic examination of *Chinese* attitudes to this general topic, see: Paul Hiniker, 'Chinese attitudinal reactions to forced compliance: a cross-cultural experiment in the theory of cognitive dissonance', Research Program in Problems of International Communications and Security, Center for International Studies, MIT, May 1965. Hiniker concludes that Chinese attitudes are changed less by forced compliance than are those of Americans.
10. Cyril Cunningham, 'Korean war studies in forensic psychology', *Bulletin of the British Psychological Society*, vol. 23, no. 81, October 1970, pp. 309–12.
11. See reference 10 above, p. 309, col. 2.
12. Julius Segal, 'Factors related to the collaboration and resistance behavior among US Army POWs in Korea', HumRRO Technical Report 33, 1956.
13. J. McClintock, 'Reward level and game-playing behavior', *Journal of Conflict Resolution*, vol. 10, 1966, p. 98.
14. P. Tedeschi, 'Behavior of a threatener', *Journal of Conflict Resolution*, vol. 14, p. 1970, p. 69.
15. P. Tedeschi, 'Threatener's reactions to compliance and defiance', *Behavioral Science*, vol. 15, 1970, p. 171.
P. Tedeschi, 'Compliance to threats', *Behavioral Science*, vol. 18, 1973, p. 34.
Harvey A. Hornstein *et al.*, 'Compliance to threats directed against self and against an innocent third party,' Contract NONR 4294(00), 4 November 1966.

16. Morris Friedell, 'Laboratory experimentation in retaliation', *Journal of Conflict Resolution*, vol. 12, 1968, p. 357.
17. L. Pisano. 'Strategies to reduce physical aggression', *Journal of Personality and Social Psychology*, vol. 19, 1971, p. 237.
18. Kurt W. Back and Morton D. Bogdanoff, 'Buffer conditions in experimental stress', *Behavioral Science*, vol. 12, September 1967, pp. 84–90.
19. B. Rapoport, 'Comparison of Danish and US in a "threat" game, *Behavioral Science*, vol. 16, 1971, p. 456.
20. See reference 14 above, and: H. Kelly, 'Experimental studies of threats in interpersonal negotiations', *Journal of Conflict Resolution*, vol. 9, 1965, p. 79.
21. L. Piliavin, 'Effect of blood on reactions to a victim', *Journal of Personality and Social Psychology*, vol. 23, 1972, p. 353.
22. See chapter 14 of this book, reference 26 above.
23. Edgar Schein *et al.*, 'Distinguishing characteristics of collaborators and resisters among American POWs', *Journal of Abnormal and Social Psychology*, vol. 55, 1957, pp. 197–201.
24. Peter Watson, 'The New Secret Service', *Sunday Times* (London), 'Spectrum', 27 January 1974.
25. L. J. West, 'Psychiatric aspects of training for honorable survival as a prisoner of war', *American Journal of Psychiatry*, vol. 115, 1958, pp. 320–36.
26. These are outlined more fully in reference 24 above, based on interviews given the author mostly at Fort Bragg in January 1974.
27. See reference 24 above.
28. 'Hypnosis research: special method for resisting psychological warfare techniques', on-going project by Dr Martin Orne, Office of Naval Research. Contract no. 4331(00), at the Institute of Pennsylvania Hospital, Philadelphia, Pa, 19139.
29. See references 24 and 28 above.
Also: Martin T. Orne, 'The potential use of hypnosis in interrogation', in A. D. Biderman and H. Zimmer (eds.) *The Manipulation of Human Behavior*, Wiley, 1961, pp. 169–215.
T. H. McGlashan, F. J. Evans and M. T. Orne, 'The nature of hypnotic analgesia and placebo response to experimental pain', *Psychosomatic Medicine*, vol. 33, May/June 1969, pp. 227–46.
The Russians are – or were – regarded as behind the USA in this field; see: 'Soviet research and development in the field of direction and control of human behavior', Memo from Richard Helms, (then) Deputy Director for Plans, Central Intelligence Agency, to J. Lee Rankin, General Counsel, President's Commission on the Assassination of President Kennedy.
Allegations of hypnotically induced amnesia for spies is raised by Walter Bowart, *Operation Mind Control* (forthcoming).

Chapter 16 Living Where Others Would Die

1. See chapter 14 of this book, reference 11 above.
Also: 'Effects of controlled isolation on performance', seven papers by various authors, HumRRO Professional Paper 6–68, March 1968.
2. W. W. Haythorn and I. Altman, 'Alone Together', Naval Medical Research Institute, Naval Medical Center, Bethesda, Md, 1966.
3. W. W. Haythorn, 'Interpersonal stress in isolated groups', in J. E. McGrath (ed.), *Social and Psychological Factors in Stress*, Holt, Rinehart & Winston, 1970.
W. W. Haythorn, 'Project Argus – a program of isolation and confinement research', *Naval Research Reviews*, vol. 20, no. 2, December 1967, pp. 1–8.
4. Roland Radloff and Robert Helmreich, *Groups under Stress: Psychological Research in Sealab II*, Appleton-Century-Crofts, 1968.
5. E. K. E. Gunderson, 'Personal history correlates of military performance at a large Antarctic station' (NM.N.RU 64–22), National Technical Information Service Microfiche: AD 609542.
Also: E. K. E. Gunderson, 'Interpersonal compatibility in a restricted environment', in *Factors Affecting Team Performance in Isolated Environments*, American Psychological Association Symposium, Institute of Behavioral Research, Texas Christian University, Fort Worth, Texas, July 1968, pp. 11–25.
6. Robert Helmreich, 'The Tektite II human behavior program', Texas University, Report TR–14, 1971.
7. See reference 3 above.
Also: E. Wright, 'Personality factors in selecting civilians for isolated northern stations', (Defence Research Board, Canada), *Journal of Applied Psychology*, vol. 147, 1963, p. 24.
8. See reference 7 above.
9. See reference 7 above.
See also: R. Hammes (Department of Defense), 'Surival research in group isolation studies', *Journal of Applied Psychology*, vol. 49, 1965, p. 418.
10. Robert Helmreich, 'Navy diver training', *Journal of Applied Psychology*, vol. 57, 1973, p. 148.
11. Helen Ross, *Behaviour and Perception in Strange Environments*, Allen & Unwin, 1974, chapter 2, section 1, pp. 26–30.
12. A. D. Baddeley, 'Sea depth and manual dexterity of divers (Royal Engineers)', *Journal of Applied Psychology*, vol. 50, 1966, p. 81.
Reference 11 above, pp. 30–3.
13. 'Diver performance and cold', *Human Factors*, vol. 10, no. 5, October 1968.
14. See reference 11 above, pp. 92–5.
15. See reference 11 above, pp. 95–6 and 150–2.

Also: 'Skin communication', *Human Factors*, vol. 8, no. 2, April 1966, and no. 8, October 1966.
16. 'Human factors in long-duration space flight', Space Science Board, National Academy of Sciences, National Research Council, Washington, DC, 1972.
17. R. H. Gaylord, 'Nomination for inadequacy in Arctic duty', PRB Research Notes, no. 51–19, March 1951.
18. L. Harold Sharp and Bertha Harper, 'Selection of quartermaster personnel for Arctic assignment', PRB Research Report 999, February 1953.
19. E. K. E. Gunderson and P. D. Nelson, 'Attitude changes in small groups under prolonged isolation', US Navy Neuropsychological Research Laboratory, San Diego, Cal., 1962.
Similarly: 'Submarine escape and medical team size', *Human Factors*, vol. 19, no. 3, June 1972.
20. See reference 19 (Gunderson and Nelson) above.
21. E. K. E. Gunderson and P. D. Nelson, 'Adaptation of small groups to extreme environments', *Aerospace Medicine*, vol. 34, 1963, pp. 1111–15.
22. William F. Fox, 'Human performance in the cold', HumRRO Professional Paper 2–68, January 1968.
23. L. J. Peacock, 'A field study of rifle-aiming steadiness and serial reaction performance as affected by thermal stress and activity', US Army Medical Research Laboratory Report no. 231, 1956.
24. See reference 22 above, preface.
25. A. T. Kisson *et al.*, 'Modifications of thermo-regulatory responses to cold by hypnosis', *Journal of Applied Physiology*, vol. 19, 1964, pp. 1043–50.
26. Personal briefing with Lt Col Austin W. Kibler, ARPA, Washington, DC, January 1974.
See also: Donald Woodward, 'Self-regulation through biofeedback', *Naval Research Reviews*, vol. 25, no. 2, December 1972, pp. 14–22.
27. Gary L. Hart *et al.*, 'Attitudes of troops in the tropics: Final Report, Phase Two', prepared for US Army Natick Laboratory, Natick, Mass., Contract no. DA19–129–QM–2076(N).
More accessible: J. R. Allen, 'Training for a hot climate', *Ergonomics*, vol. 8, no. 4, p. 445.
K. D. Duncan, 'Artificial acclimatization for infantry in the heat', *Ergonomics*, vol. 9, no. 3, p. 229.
R. Ernest Clark and E. Ralph Dusek, 'Effects of climate, food, clothing and protective devices on soldier performance', in *Psychological Research in National Defense Today*, US Army BESRL Technical Report S–1, June 1967, chapter 10.
Related: William Montague and Richard Moren, 'Human factors in CBR operations; The effects of CBR Protection on the performance of selected combat skills in hot weather', HumRRO Technical Report 71, May 1961.

28. Robert Meyers, 'Is there a chemical thermostat in the brain?', *Naval Research Reviews*, vol. 20, no. 4, April 1967, pp. 1–7.
29. Eugene H. Drucker *et al.*, 'The effects of sleep deprivation on performance over a 48-hour period', HumRRO Technical Report 69–8, May 1969.
30. Laverne C. Johnson, 'Sleep and sleep-loss – their effect on performance', *Naval Research Reviews*, vol. 20, no. 8, August 1967, pp. 16–22.
31. D. Cannon *et al.*, 'Summary of literature review on extended operations', HumRRO Consult. Report, December 1964.
32. R. W. Russell, 'Psychological and pharmacological factors controlling gastrointestinal motility', US Army Medical Service, Research and Development Command, Office of the Surgeon General, Washington, DC, no date but *circa* 1964.
33. R. T. Wilkinson, 'Sleep deprivation: performance tests for partial and selective sleep deprivation', *Progress in Clinical Psychology*, vol. 8, 1968, pp. 28–43.
34. Donald F. Haggard, 'HumRRO studies in continous operations', HumRRO Professional Paper 7–70, March 1970.
Also: 'Body rhythms in unusual environments', *Human Factors*, vol. 8, no. 5, October 1966 (includes confinement, shelter living, time estimation, etc).
35. See reference 31 above.
36. Compare references 29 and 34 above.
37. James H. Bank *et al.*, 'Effects of continuous military operations on selected military tasks', US Army BESRL Technical Report 1166, December 1970.
38. See reference 37 above, page 3.
39. W. Phillips Davison, 'The human side of the Berlin airlift', Rand: P–1224, 3 December 1957.
40. See reference 39 above, pp. 11ff.
41. See reference 39 above, pp. 13ff.
42. Information for this section is based on a long document in the Pentagon library. The main section is headed 'Research into defection: data review on army personnel in certain categories'. The four annexes to the document are as follows:
Annex A: 'Present use of psychiatric and psychological specialists and mechanisms by Department of Army to enhance national security, document no. AR 611/15.
Annex B: 'The possibility of expanding provision of AR 611/15 – The selection and retention criteria for personnel in nuclear weapons positions to include persons assigned to and retained in highly sensitive positions'.
Annex C: 'Expanded employment of psychiatric and psychological mechanisms by the Department of Army to enhance national security'.
Annex D: 'Development of a psychological mechanism'.
See also 'Education for security', *Human Factors*, vol. 7, no. 4, August 1965.

43. Jerome D. Frank *Sanity and Survival: Psychological aspects of war and peace*, Barrie & Rockcliff, 1967.
44. See reference 43 above, p. 166 ff.
45. Robert Vineberg, 'Human factors in tactical nuclear combat', HumRRO Technical Report 62–5, April 1965.
46. Lewis C. Bohn, 'Psychological inspection', Rand: P–1917, 19 February 1960.
See also: Seymour Melman (ed.), *Inspection for Disarmament*, Columbia University Press, 1958, pp. 231–60.
'Psychology and policy in a nuclear age', *Journal of Social Issues* (entire issue), vol. 17, no. 3, 1961.
47. Robert J. Lifton, 'Psychological effects of the atomic bomb: the theme of death', *Daedelus*, summer 1963, pp. 462–97.
See also: Robert J. Lifton, *Death in Life: The survivors of Hiroshima*, Random House, 1967.
48. (a) Irving Janis, *Air War and Emotional Stress*, McGraw-Hill, 1951.
(b) Irving Janis, 'Psychodynamic aspects of stress tolerance', Paper given to Conference on Self-Control under Stressful Conditions', Bureau of Social Science Research, Washington, DC, September 1962.
John Hersey, *Hiroshima*, Doubleday, 1960.
49. See reference 48(a) above.
50. See reference 48(a) above.
51. See, for instance: A. H. Leighton, *The Governing of Man*, Princeton University Press, 1945, and *The Character of Danger*, Basic Books, 1963.
52. Edward A. Shils and Morris Janowitz, 'Cohesion and disintegration in the Wehrmacht in World War 2', *Public Opinion Quarterly*, vol. 12, no. 2, summer 1948, pp. 280–315.
53. See reference 46 above, pp. 38–54.
But also see: 'An approach to the study of social and psychological effects of nuclear attack', Human Sciences Research Incorporated, March 1963.
'Psychiatric aspects of the prevention of nuclear war', Group for the Advancement of Psychiatry, Report no. 57, 1964.
Scott Grier *et al.*, 'Kinship and voluntary organization in post-thermonuclear attack society', Human Sciences Research Incorporated, September 1965.
Neil J. Smelser, 'Theories of social change and the analysis of nuclear attack and recovery', Human Sciences Research Incorporated, January 1967.
V. R. L. Goen *et al.*, 'Analysis of national entity survival', Stanford Research Institute, November 1967.
William Brown, 'Recovery from a nuclear attack', US Office of Civil Defense, October 1971 (but based on a 1973 'scenario').
And especially: Horace D. Beach, *Management of Human Behavior in Disaster*, Emergency Health Services Division, Canada, 1967, particularly

chapter 4: 'Social problems in disaster'; chapter 5: 'Responses to warning and evacuation'; chapter 6: 'Entrapment and shelter living'.
54. See reference 45 above, pp. 50–51.
55. Irving Janis, 'Psychological effects of warnings', in *Man and Society in Disaster*, Basic Books, 1962.
For more recent, though more theoretical work, see: Shlomo Breznitz, 'Threat and false alarms as determinants of fear and avoidance of pain', US Army RIBSS, 31 August 1974.
56. Albert J. Glass, 'Psychiatry in the Korean campaign (Installment II)', *US Armed Forces Medical Journal*, vol. 4, 1953, pp. 1563–83.
57. See reference 45 above, part II: 'A model for estimating casualty rates for psychological attrition'.
58. See reference 45 above, pp. 63–5.
59. E. Paul Torrance, 'Psychological aspects of survival', Human Factors Operational Research Laboratories, HFORL Report no. 35, March 1953.
'Survival in unfair conflict', *Behavioral Science*, vol. 18, 1973, p. 313.
Richard Seaton, 'Hunger in groups: an Arctic experiment', QM Food and Container Institute for the Armed Forces, Chicago, no date.
60. I. Altman, 'Ecology of isolated groups', *Behavioral Science*, vol. 12, 1967, p. 169.

Part 5 – Introduction

1. Frank Kitson, *Low Intensity Operations: Subversion, insurgency and peace-keeping*, Faber & Faber, 1971.
2. James Davies, 'Smith's gentle mind benders', *Daily Express*, 8 March 1977.

Chapter 17 Human Factors in Guerrilla War

1. Andrew R. Molnar *et al.*, 'Human factors considerations of undergrounds in insurgencies', CRESS, The American University, December 1966.
As an example of a psychological profile of a country, see: J. L. Sorenson, 'Unconventional warfare and Venezuelan society', Weapons Planning Group, US Ordnance Tests Station, China Lake, California, November 1964.
Other examples of interest: S. R. Silverbury, 'Selected and annotated bibliography on civac and psyop in the Congo, Ethiopia and Nigeria', First Report produced for the Remote Area Conflict Center (then at Fort Bragg); this includes details on: 'Attitudes, hopes and fears of Nigerians'; 'Secret societies and prophetic movements in the Belgian Congo'; 'Suicide in Western Nigeria'; 'Omens in Ethiopia;' 'African nationalist and dissident groups in African capitals'; 'Congo psywar handbook'.
'Strong men series', Defense Intelligence Agency, 1966. (Secret)

'Congolese dissident leaders: who are they and what they represent', Research Memo RAF–48–64, Department of State, Bureau of Intelligence and Research, 1964. (Secret)
2. Gary T. Marx, 'Thoughts on a neglected category of social movement participant: the agent provocateur and the informant', mimeographed paper prepared at MIT and Harvard Center for Criminal Justice (undated but post-1973).
3. 'Communist manual of instructions of psycho-political warfare', Testimony of Kenneth Goff before the House of Representatives Un-American Activities Committee. Note the reservations about this document discussed in the footnote on p. 340.
4. For example, J. Rieselbach, 'Isolationist behavior', *Public Opinion Quarterly*, vol. 24, 1960, p. 645.
5. P. Overstreet, 'Extremism', *Public Opinion Quarterly*, vol. 31, 1967–8, p. 137.
6. For example: W. McEvoy, 'Content analysis of a super-patriot protest', *Social Problems*, vol. 14. 1966–7, p. 455.
See also reference 5 above and general studies like Sydney Verba (Harvard) *et al.*, *The Cross-National Program in Political Participation: An Overview*, Viking Press, 1972, includes sections on tribalism, women, regionalism, ethnicity, leadership, etc.
7. For example, R. Wright, 'Working class and the war in Vietnam', *Social Problems*, vol. 20, 1972–3, p. 133.
8. F. H. Lakin, 'Psychological warfare research in Malaya 1952–55', Army Operation Research Establishment, UK Ministry of Defence, Paper to the 11th Annual US Army Human Factors Research and Development Conference, October 1965.
See also: Frank Denton, 'Volunteers for the Viet Cong', Rand RM–5647–ISA/ARPA, September 1968 concludes that there are five main reasons for volunteering – economic-social frustration (48 per cent mentioned this); personal oppression by the government of South Vietnam (45 per cent), social justice (11 per cent), adroit VC recruitment (20 per cent), family–peer pressure (6 per cent). Compare this with:
'The Viet Cong: 5 steps in running a revolution', JUSPAO document (unsigned, undated) which highlights how 'captured documents consistently emphasize 5 revolutionary tasks'; (1) finding the enemy's weak points, exploited by (2) propaganda; (3) establishment of front groups; (4) training the most zealous to become new leaders; (5) the principles of escalation.
Viet Cong indoctrination is also covered in: David Elliott and Mai Elliott, 'Documents of an elite VC Delta unit: The demolition platoon of the 514th battalion. Part 4: Indoctrination and military training', Rand RM–5851–ISA/ARPA May 1969.
J. J. Kirkpatrick, 'Adjustment of Chinese soldiers to the communist demand

for ideological participation: an exploratory study based on the CCF in the Korean war', HumRRO staff memo, February 1969.
9. See especially 'The Rand Papers: the secret study which lies at the heart of the war', *Ramparts*, vol. 11, no. 5, November 1972, pp. 25–42.
10. Konrad Kellen, 'Conversations with enemy soldiers in late 1968/early 1969: a study of motivation and morale', Rand RM–6131–1–ISA/ARPA, September 1970, plus other Kellen references in reference 9 above.
11. J. J. Zasloff, 'Political motivation of the Viet Cong: The Vietminh Regroupees', Rand RM–4703/2–ISA/ARPA, May 1968 (original edition August 1966).
12. W. P. Davison, 'Some observations on Viet Cong operations in the villages', Rand RM–5267/2 ISA/ARPA, May 1968 (original edition July 1967).
See other Davison references in 9 above.
13. E. J. Mitchell, 'Inequality and insurgency: a statistical study of South Vietnam', Rand paper published in 1968, *Social Problems*.
14. Anthony Russo, 'Economic and social correlates of government control in South Vietnam, *Ramparts*, 1972.
See also: Anthony Russo, 'Looking backward: Rand and Vietnam in retrospect', in reference 9 above, pp. 40ff.
15. See among others: 'A selected bibliography of People's Republic of China foreign propaganda, information and culture programs with explanation on the agencies and individuals responsible for such programs', Psyop Special Intelligence Report, Issue SR 10–71, 7th Psychological Operations Group Army Post Office, San Francisco 96248, 7 December 1971.
Related: 'A background reading list on psyops with special emphasis on Communist China', US Department of State, External Research Staff, Research Paper 153, May 1964.
'The Chinese documents project, series III', No. 7 of *Studies in Chinese Communism*, Researched by the Human Resources Research Institute, Maxwell Air Force Base, Alabama, published by Air Force Personnel and Training Research Center, Lackland Air Force Base, Texas, January 1955.
16 (a). See among others: 'A handbook for mobile motivation teams (armed) in psychological operations', ST 33–154 US, Army Special Warfare School, Fort Bragg, North Carolina, March 1968.
(b) Lt Col Neil I. Leva, 'A comparison of the relative combat efficiency of insurgent forces and counter-insurgent forces', USAWC Research Element (Thesis), US Army War College, Carlisle Barracks, Pennsylvania, 9 March 1970.
17. See reference 1 above.
18. Lt Col Francis B. Kane Jr, 'Psychological warfare and revolution: a Cuban case study', USAWC Research Element (Thesis), US Army War College, Carlisle Barracks, 8 April 1966.

Chapter 18 The Growth of Special Operations

1. Information based on a personal visit to the International Police Academy in Washington in January 1974 and interviews with senior staff there and in the US State Department, to all of whom I owe a debt of thanks.
2. 'Psychology, social stability and internal state security of the Soviet bloc', mimeographed paper circulated privately by F. Rubin, a Polish psychologist now living in London. Includes translations of Russian, Czech and Hungarian articles on: 'Development of "class-antagonisms" formations'; 'The psychological factors in interrogation';' "Off-duty" behaviour of sensory deprived subjects'; 'The scientific problem of "alien ideological trends" '; 'Indoctrination of the Soviet armed forces'.
3. 'A new role in defense psychology: nation building research', SORO, The American University, Research Note 63–2, September 1963.
Also: Lorand Szalay *et al.*, 'Persuasion overseas (Republic of Vietnam emphasis),' SORO, 1965.
4. 'Psychological reactions to weapons: a factor analysis', Paper prepared by Ohio State University Research Foundation for the Directorate of Armament Development, January 1966. (Classified)
5. M. Dean Havron *et al.*, 'Constabulary capabilities for low-level conflict', Human Sciences Research Incorporated, prepared for Office of Naval Research Civil Affairs Branch, Research Report 69/1–Se, April 1969.
6. See chapter 17 of this book, reference 16(b).
7. 'Behavioral and social science research plan: 1967–71 (supporting military operations in the developing countries)'. Dated 7 February 1966, this unsigned document sets six major objectives including training of personnel on assignments, research on policies which influence foreign environments, psyops and civac research and the establishment of CINFAC, the cultural information analysis centre, plus twenty-nine specific research projects.
Related: Keyser Bauer, 'Anthropological-cultural influences on weapons use, choice and maintenance', US Army Materiel Command, Human Engineering Laboratories, Aberdeen Proving Ground, October 1965.
Military authorities also sponsored more general psychologically based studies the purpose of which, nevertheless, was designed to help understand what made the enemy tick. For example:
E. A. Weinberg, 'Soviet sociology', ARPA Contract no. 920F–9717, October, 1974.
Theodora Abel and Francis Hsu, 'Some aspects of personality of Chinese as revealed by the Rorschach test', Office of Naval Research, 1949.
Richard C. Teevan *et al.*, 'Fear arousing situations and the fear of failure', Technical Report no. 34, 1970.
R. Ulrich, 'How to prevent unwanted aggressive behavior', *Personnel Technology* S36K.

R. Taylor *et al.*, 'Variables influencing cooperative, escape and aggressive behaviour', TR no. 4, Office of Naval Research.
Marilla Svinicki, 'An exploratory study of methods for quantifying the analysis of human aggression', TR no. 3, June 1969, Office of Naval Research.
Alan Gonick, 'An analysis of climatic influences on aggression'.
S. Streufert and S. T. Ishibushi, 'American and Chinese perceptions of nations in conflict', TR no. 19, February 1969, Office of Naval Research.
More directly: Howard H. Sargeant, 'Soviet political warfare techniques – Propaganda in the 70s', National Strategy Information Center Inc., New York, 1972.

8. (a) Francis Medland, 'Research programme for selection and performance evaluation in overseas security operations assignments', US Army BESRL Research Study 69–1, March 1969. (Limited distribution)
(b) Also relevant: 'PSY-OP REASON: Psychological operations role in establishing a sense of nationhood', (U), International and Civil Affairs Directorate DCSOPS, August 1967. (Confidential)
'Personality factors in psyops', Report No. 8, 15th Psyop detachment, 7th Strategic Psyop Group, 10 January 1967.

9. See reference 8(a) above, p. 1 and p. 5.

10. Francis Medland and Calvin Green, 'Research on selection for combat and stability operations assignments', USAPRO Research Study 66–2, February 1966.

11. Anita Terauds *et al.*, 'Influence in intercultural interaction', CRESS Technical Report, August 1966.

12. Many anthropologists took exception to this move, see: Bryce Nelson, 'Anthropologists debate: Concern over future of foreign research', *Science*, vol. 154, 23 December 1966, pp. 1525–6.

13. Lt Col Howard J. Johnston (USMC), 'The tribal soldier: a study of the manipulation of ethnic minorities', *Naval War College Review*, vol. 19, no. 5, January 1967, pp. 98–144.

14. For the best account of this fiasco, see the collection of essays compiled by Irving Louis Horowitz, *The Rise and Fall of Project Camelot*, MIT Press, 1967 (revised 1974).

15. Robert J. Foster, 'Examples of cross-cultural problems encountered by Americans working overseas: an instructor's handbook', HumRRO, May 1965.
See also: Harley M. Upchurch, 'Towards the study of communities of Americans overseas', HumRRO Professional Paper 14–70, May 1970.

16. Arthur Niehoff and J. Charnel Anderson, 'Peasant fatalism and socio-economic innovation', HumRRO Professional Paper 33–67, June 1967.

17. Arthur Niehoff, 'Intra-group communication and induced change', HumRRO Professional 27–67, June 1967.

18. Arthur Niehoff, 'A quantitative approach to the study of directed cross-cultural change', HumRRO Professional Paper 40–68, December 1968.
Also of interest: Arthur Niehoff, 'Food habits and the introduction of new foods', HumRRO Professional Paper 9–67, March 1967.
Augusto Torres *et al.*, 'Social and behavioral impacts of a technological change in Colombian villages', AIR–E–40–4/68–FR, Prepared under contract for Agency for International Development, April 1968.
19. Dean Froehlich and Malcolm Klares, 'Advisor and counter-part activities in the military assistance program in the republic of China', HumRRO Technical Report 65–5, June 1965.
Dean Froehlich, 'An experimental criteria of cross-cultural interaction effectiveness: a study of military advisor and counterparts', HumRRO Professional Paper 38–68, December 1968.
Dean Froehlich, 'The military advisor as defined by counterparts', HumRRO Professional Paper 9–70, March 1970.
20. Eugene Rocklyn, 'The application of programmed instruction to foreign language and literacy training', HumRRO Professional Paper 8–67, February 1967.
21. Edward Stewart and John Pryle, 'An approach to cultural self-awareness', HumRRO Professional Paper 14–66, December 1966.
Jack Danielan and Edward Stewart, 'New perspectives in training and assessment of overseas personnel', HumRRO Professional Paper 19–67, April 1967.
Edward Stewart, 'Simulation exercises in area training', HumRRO Professional Paper 39–67, September 1967.
Edward Stewart, 'The simulation of cross-cultural communication', Hum RRO Professional Paper, 50–67 December 1967.
Arthur Hoehn, 'The design of cross-cultural training for military advisors', HumRRO Professional Paper 12–66, December 1966.
Fred Fiedler, 'The effect of culture training on leadership, organizational performance and adjustment', *Naval Research Reviews*, vol. 21, no. 7, 1968, pp. 7–13.
Alfred Kraemer, 'The development of cultural self-awareness: Design of a program of instruction', HumRRO Professional Paper 27–69, August 1969.
22. John Parsons *et al.*, 'Americans and Vietnamese: a comparison of values in two cultures', Human Sciences Research Incorporated, Research Report 68/10–Ct, November 1968 ARPA/Project AGILE T10–72–6. (Distribution limited to US Government agencies only).
See also: *Psyop-Polwar Newsletter* no. 11, 30 November 1969 for Vietnamese and US answers to open-ended sentence-completion tests.

Chapter 19 Selection and Training of Personnel for Counter-Insurgency

1. Personal interview with General Yarborough at his home near the Fort Bragg base, 10 January 1974.
2. 'Validation of counter-intelligence corps instruments' (unsigned), PRB Research Memo 54–19, May 1974.
3. Roy J. Jones and Berton Wonograd, 'Procurement of counter-intelligence corps trainees', HumRRO Special Report, 10 October 1957.
4. George Cantrell *et al.*, 'Application of a psychometric clinical approach to personnel selection for counterinsurgency duty', Personnel Research Laboratory (Aerospace Medical Division, Lackland Air Force Base), PRL-TR-64–24, October 1964.
5. Rudolph Berhouse, 'Research on combat selection and special forces manpower problems – status report', USAPRO Research Study 63–2, January 1953.
Francis Medland *et al.*, (USAPRO), 'Psychological factors in selection of special forces officers', Paper given to the 11th Annual US Army Human Factors Research and Development Conference, October 1965.
Francis Medland and Calvin Green, 'Research on selection for combat and stability operations assignments', USAPRO Research Study 66–2, February 1966.
For a comparison of special operations officers with the British SAS, see: Ken Hedges, 'Medical aspects of the SAS', *Journal of the Royal Army Medical Corps*, vol. 119, no. 2, 1973.
For a slightly different view see: 'Trouble-shooting behavior', *Human Factors*, vol. 8, no. 5, October 1966.
6. Personal interview with Tom Brown, navy psychologist, Royal Victoria Hospital, Netley, Hampshire, 3 May 1974, and Edward Elliott, Ministry of Defence, 24 July 1974.
7. Joseph Olsmtead *et al.*, 'Selection and training for SIAFs: final report', HumRRO Technical Report 72–2, February 1972.
8. 'Employment of US Army Psychological Operations Units in Vietnam', Army Concept Team in Vietnam, Final Report, ACTIV Project no. AGG-47F, Army Post Office, San Francisco 96384, 7 June 1969.
9. 'Precis 6: Psyops unit – General', Training Report, Senior Officers' Psyops Course, Royal Air Force, Old Sarum, Salisbury, Wiltshire, UK, 14–18 February 1972. This is a British restricted document devoted to the organization and equipment of a psyops unit – both at HQ and broken down into sub-sections. It includes details of 'consolidation psyops', counter-insurgency uses and uses in peacetime. It also includes details of deployment of psyops in the UK.
Similarly: 'Technical report of the senior officers' psyops course held at RAF

Old Sarum, 14–18 February 1972'. This course made it clear that the British psyops basis parallels the US Army doctrine.

Chapter 20 Psychological Warfare: Operational Organization

1. Peter Paret, 'A total weapon of limited war' based on a symposium in the German journal *Werwissenschaftliche*, 1959, includes details of the *Service Psychologique du FLN* and its French counter-part – the 5th section of the general staff.
'Guerrilla warfare and airpower in Algeria', Air University, Maxwell Air Force Base, March 1965. (NOFORN, a low classification which stands for, 'No foreigner allowed access').
R. J. Hill, 'French strategy and its political bases', Operations Research Divisions Department of National Defense, Ottawa, August 1966.
'FLN Psyops in Algeria' (unsigned), Defense Documentation Center Bibliography, May 1969.
2. F. H. Lakin, 'Psychological warfare research in Malaya 1952–55', Army Operational Research Establishment, Ministry of Defence, UK, Paper given to the 11th Annual US Army Human Factors Research and Development Conference, October 1965.
3. See chapter 19, reference 9 above.
4. See chapter 19, reference 9 above.
5. ' "Black propaganda" blue-pencilled', *Guardian*, 26 February 1977.
Also: 'The army's secret war in Ulster', *Sunday Times* (London), 13 March 1977.
6. See chapter 19, reference 9 above.
7, Frank Kitson, *Low Intensity Operations*, Faber & Faber, 1971, pp. 188–9.
Also: 'Four working papers on propaganda theory', an oldish document, the cover of which (with publication details) is missing from my copy, but which gives some details of Japanese, German, British and Russian concepts regarding the organization of propaganda.
8. *Guerra Psicologia*, Ministry of War, Brazil, 1956.
9. 'Vorster's men get psycho-war kit', *Sunday Times* (London), 3 April 1977.
10. Personal visit to the International Police Academy, Washington, DC, January 1974.
11. Information for this section is based on two visits to the psyop school at Fort Bragg. Interviews with: Col John D. Howard, director of the school, Majors Esworthy and Les Griffin, Capt. Hampton and others, to all of whom I am grateful for the time and help they so generously gave.
Also see: 'Psychological operations organization', HQ Department of Army, TOE–33–500H, 30 November 1970.
'4th psyop group – unit history', HQ Department of Army, 1969.
12. 'Employment of US Army Psychological Operations Units in Vietnam',

Army Concept Team in Vietnam, Army Post Office, San Francisco 96384, Final Report ACTIV Project no. AGG–47F, 7 June 1969.

Chapter 21 General Psywar Techniques

1. Charles Wang, 'Reactions in Communist China: An analysis of letters to newspaper editors', Human Resources Research Institute, Maxwell Air Force Base, Texas, Technical Report 33, published by Air Force Personnel and Training Research Center, Lackland AFB, Texas, June 1955 (part of 'Studies in Communism', Contract AF 33(038)–25075).
2. Wen-Hui Chen, 'The family revolution in Communist China', Human Resources Research Institute, Maxwell Air Force Base, Research Memo 35 (same publishing details as reference 1 above).
3. Alexander Askenasy and Richard Orth, 'A guide for field research in support of psychological operations,' CRESS, April 1970.
4. Most of these documents are put out from HQ of 7th Psychological Operations Group APO San Francisco 96248.
5. 'Exploitation of Viet Cong vulnerabilities: 1st supplement to JUSPAO guidance', Issue no. 9, JUSPAO Planning Office, Saigon, Vietnam, 5 August 1965.
Of related interest: 'Recommended psychological objectives for Vietnam with supporting tasks and themes', US Psychological Operations Committee, Saigon, 27 June 1963 (early but particularly clear).
For how the VC did it in return: 'The Viet Cong: 5 steps in running a revolution', full details given in chapter 17, reference 8 above.
'Implications and summary of a psywar study in South Korea', Human Resources Research Institute, Maxwell Air Force Base, Alabama, 15 January 1951. (Secret – downgraded to Restricted.) Old but shows how it was done.
6. Issue 23, pp. 41ff.
7. The following analysis is based on thirty-five of these *Psyop Intelligence Notes.*
8. At least half-a-dozen *Psyop Intelligence Notes*, dated 6 or 7 July 1970, are based on the de-briefing of just one North Korean pickpocket.
9. 'Psywar intelligence: technical questions for front line interrogators; field manual 30' (undated); includes list of thirty-two specific questions to be asked.
10. Col Emmett J. O'Brien, 'Defection: a military strategy for wars of liberation' (an individual research report), Military Intelligence, USAWC Research Paper, S UArmy War College, Carlisle Barracks, Pa, 6 February 1971.
11. 'Psyop Intelligence Notes: target analysis', 7th Psychological Operations Group, Army Post Office, San Francisco 96348, 21 March 1968.

Chapter 22 Special Psywar Techniques

1. James Price and Paul Jureidini, 'Witchcraft, sorcery, magic and other psychological phenomena and their implications on military and paramilitary operations in the Congo', SORO/CINFAC/6–64, 8 August 1964.
2. See reference 1 above, p. 1.
3. The title page of my copy of this document is missing.
4. 'Propitious and non-propitious dates in the Vietnam calendar (1964–67)', CRESS.
See also: 'A Vietnamese looks at Tet', JUSPAO Planning Office, 29 November 1966: ways of exploiting the ceremonies are reported in preparation 'for the coming 1967 Tet psychological warfare campaign'.
'Heroes of Vietnam', dates relating to historic events in Vietnam, supplied by Central Intelligence Organization from a captured VC document, dated 8 April 1963.
The more exotic abuses of customs included special packs of playing cards – consisting entirely of the ace of spades which was hated and feared by many Vietnamese (the card brought bad luck if they 'accidentally' picked it from a pack when offered them in a 'magician's trick').
5. Heda Jason, 'The narrative structure of swindler tales', Rand P–3788, February 1968.
Rolla Edward Park, 'Effects of graft on economic development: an examination of propositions from the literature', Rand P–4113, June 1969.
6. Daniel Ellsberg, 'The theory and practice of blackmail', Rand P–3883, *circa* 1968.
7. 'Low, medium and high altitude leaflet dissemination guide', Department of Army HQ, 7th Psychological Operations Group, Army Post Office, San Francisco 96248 (undated). 'The data contained in this book represents years of study conducted under government contract by Johns Hopkins University. It also reflects the years of field study and experimentation by Major David G. Underhill who has done extensive research into the subject of leaflet dissemination techniques while with the 7th Psychological Operations Group'.
8. A campaign of this type reached its height in Vietnam in the winter of 1972–3.
9. E. W. Schnitzer, 'East bloc forgeries: a weapon in the cold war' (translated from the German), Rand T–103, 24 October 1958.
Related: Natalie Grant, 'Communist psychological offensive: distortions in the translation of official documents', Research Institute on the Sino-Soviet Bloc, pamphlet series no. 1, 1961.
10. 'Debriefing of a CORDS officer – Asia agent training in Vietnam', Region 3, 1967–8.
11. 'Summary of VC propaganda leaflets, March through May 1964', 4th

summary of a series, 7th Psychological Operations Group (this one summarized 418 such leaflets).

12. Lorand Szalay, 'Persuasion overseas (Republic of Vietnam emphasis)', SORO, 1965, one of three similar documents produced on an 'urgent' basis for the army. The others were:
'Republic of Vietnam: selected vulnerabilities and propaganda opportunities' (U). 'A short guide to psychological operations in South Vietnam – 1965' (U). Includes detailed province-specific data.

13. See chapter 20, reference 11 above, Annex G1, H3 and H4.

14. See chapter 20, reference 11 above, section III, chapter 18, Annexes G2, G3 and G4.

15. See chapter 20, reference 12 above, 1–3.

16. *Psyop-Polwar Newsletter*, no. 11, 30 November 1969, 7th Psychological Operations Group.

17. *Guardian*, 7 February 1966.

18. See also: E. J. McGuigan, 'Feasibility of the use of acoustic generators as weapons', Frankford Arsenal, US Army Materiel Command, Missile Comand, Army Test and Evaluation Command, Aberdeen Proving Ground. D. A. Dobbins, 'Jungle acoustics', Project IL013001 AG1A.
For the more peaceable uses of psywar tapes, see: 'JUSPAO catalogue of tapes', HQ US Military Assistance Command Office of Psychological Operations Directorate, 3 December 1967. Includes: 'You are surrounded'; 'Soon you will be bombed by airplane', 'Surrender guidance'; 'Live to enjoy another Tet'.
Related: Martin Herz, 'Psywar against surrounded troops', *Military Review* August 1950.

19. See chapter 20, reference 12 above, pp. 1–3, 4. I have not gone into propaganda broadcasting in much detail because there is not a great deal that is new. The organization of broadcasting stations is covered briefly in chapter 20 in the section on US psyops organization. But for those interested in this aspect another fairly recent complete account is:
Jerome Clauser, 'VUNC: a description of a strategic radio broadcasting psychological operation', August 1971.

20. See chapter 20, reference 12 above, pp. 1–8, 1–9, and 1–10–11 respectively.

21. Maj. Gen. Lausdale (USAF Commandos), 'Psyaction by air', Memo, June 1963. The leaflet and loudspeaker campaign at San Isidro, Dominican Republic, 28 December 1962–13 January 1963 'stopped incipient revolt without further shedding of blood'.
For an account of an Egyptian attempt to fire artillery shells loaded with propaganda leaflets into Israeli positions on the Suez canal front, see: Guy Rais, 'Egypt fires poll propaganda at Israeli troops', *Daily Telegraph*, 1974.

22. Donald Tepas, 'Some relationships between behavioral and physiological

measures during a 48 hr period of harassment; a laboratory approach to psywar hardware development problems', Military Products Group Research Laboratory, Honeywell, Minneapolis, Minnesota (1st symposium, 1964).
23. 'US offers $100,000 to red who surrenders first MIG', *New York Times*, 28 April 1953.
24. See chapter 20, reference 12 above, pp. 1–4, 5.
25. (a) See chapter 20, reference 12 above, p. 1–6.
(b) 'A handbook for mobile motivation teams (armed) in Psyops', US Army Special Warfare School, Fort Bragg, North Carolina, ST33–154, March 1968.
For a view from the 'other side', see: M. V. Frunze *et al.*, 'Handbook for propagandists and agitators of the Army and Navy', Military Publishing House, Ministry of Defence, Moscow, 1968.
(c) Lt Col Neil Leva, 'A comparison of the relative combat efficiency of insurgent forces and counter-insurgent forces', USAWC Research Element (Thesis), US Army War College, Carlisle Barracks, Pa, 9 March 1970.
26. Reference 24(b) above.
27. See chapter 20, reference 12 above, p. 1–8.
28. *Psyop-polwar Newsletter*. 7th Psychological Operations Group, Army Post Office, San Francisco 96284, no. 11, 30 November 1969 – section headed, 'Flexagons, puppets and PRBs' (paper boats).
29. James Davies, 'Smith's gentle mind benders', *Daily Express*, 8 March 1977.
30. 'Utagoe: a psyop intelligence analysis of group singing as a highly effective propaganda technique exploited by Communists and other radical organizations in Japan and Okinawa', Topical Reports 179–69, 7th Psychological Operations Group, Army Post Office, San Francisco 96248, 24 November 1969.
31. Stuart Howard and William D. Hitt, 'Intercultural differences in olfaction', Battelle Memorial Institute, ARPA/Project AGILE Report, 2 May 1966.
Also: R. Huey Wright and Kenneth Michells, 'Human processing of olfactory information', Purdue University, Paper presented at Bionics Symposium, Aerospace Medical Division, US Air Force, Wright-Patterson Air Force Base, Ohio, 19–21 March 1963.
(Armies of the west, of course, have a long history of trying to capitalize scientifically on biological differences between races. The introduction of lactose into the diets of certain African and Asian tribes unable to digest it is one example that has been considered.)

Chapter 23 Psychological Aspects of Population Control

1. Paul Wehr, 'Non-violent resistance to occupation: Norway and Czechoslovakia', *Journal of Social Issues* (in press).
2. Rex Applegate, *Riot Control – Materiel and Techniques*, Stackpole Books, 1969.
3. Horst Menderhausen, 'Troop stationing in Germany: German public opinion', Rand RM–6172–PR, November 1969.
4. Adrian Jones *et al.*, '*A* selected bibliography of crowd and riot behaviour in civil disturbance', CRESS, 2nd edn., July 1968.
5. 'Report of the Internal Defense/Development Psychological Operations Instructors' Conference', Fort Bragg, North Carolina, 31 October–4 November 1966. With a list of seventy-eight participants from full colonel to 2nd lieutenant and from all branches of the services plus some surprising specialist agencies such as the US Army's Amphibious School, the Signal Center, the Finance School, Defense Language Institute, the Chemical Center, the School of the Americas.
6. Joseph F. Coates, 'Wit and humour: a neglected aid in crowd and mob control', *Crime and Delinquency*, vol. 18, no. 2, April 1972, pp. 184–91.
7. Robert Shellow, 'Reinforcing police neutrality in civil rights confrontations', *Journal of Applied Behavioral Science*, vol. 1, 1965, p. 243.
See also: S. F. Nadler, 'Social therapy of a civil rights organization', *Journal of Applied Behavioral Science*, vol. 4, 1968, p. 281.
8. *The Times* and the *Guardian*, 3 October 1973. See also reference 2 above, pp. 50ff.
Finally: C. A. Christner *et al.*, 'State-of-the-art-study on the pulsed light phenomenon', Battelle Memorial Institute, Report no. Bat–171–6, December 1964 (considers use of pulsed light to induce 'photogenic epileptiform' seizures in combat).

Chapter 24 Psychological Warfare: Some Final Considerations

1. Personal interview with Col John D. Howard, director of the psyop school at Fort Bragg, January 1974.
2. Albert Jenny II, 'Interpersonal influence processes in navy port calls', Human Sciences Research Incorporated, HSR–RR–66/4–As, 1966.
3. James Dodson *et al.*, 'The role of psychological operations in naval missions: an appraisal and recommendations', Human Sciences Research Incorporated, HSR–RR–68/3–Cs, June 1968. See especially Appendix 1: 'Potential naval psychological operations in diverse contexts', (this distinguishes eighteen specific 'scenarios' in which port calls may be appropriate).
4. Barry Bleckman and Stephen Kaplan, 'The uses of the armed forces as a political instrument', – an account appeared in the *Evening Standard* (London), 4 January 1977.

5. 'Personality assessment of communication effectiveness: conference proceedings', Technical Note 62–16, Contract SD–50, Task 8, IDA, April 1962 (Conference held October 1961).
6. I am grateful to Dr Smith for making available much of his material to me prior to publication.
7. Frank Kitson, *Low Intensity Operations*, Faber & Faber, 1971.
8. In all, the JUSPAO report examined twenty-six secret or confidential captured Viet Cong documents.
9. 'Psychological operations in the US mission', Memo for The Hon Ellsworth Bunker, 12 October 1967. Paragraph 6 of this account describes Westmoreland's high opinion of psyops effectiveness.
10. Col Emmett J. O'Brien, 'Defection: a military strategy for wars of liberation' (an individual research report), Military Intelligence USAWC Research Paper, US Army War College, Carlisle Barracks, Pa, 6 February 1971.
11. F. H. Lakin, 'Psychological warfare research in Malaya 1952–55', Army Operational Research Establishment, UK, Ministry of Defence, Paper to the 11th Annual US Army Human Factors Research and Development Conference, October 1965.
12. ibid.
13. Quoted in: James Dodson, 'Some theoretical observations on the assessment of psychological operations effectiveness', CRESS/CINFAC R–0796.
14. Lessing Kahn and Julius Segal, 'Psychological warfare and other factors affecting the surrender of North Korean and Chinese forces', Operations Research Office, 1953.
15. Personal interview, J. F. K. Center for Military Assistance, Fort Bragg, 14 January 1974.
16. See mainly reference 13 above, section 3 especially, but also:
E. F. Sullivan, 'Methods for estimating present and future insurgent strengths', US Naval Postgraduate School Thesis, October 1969.

Chapter 25 Conclusions

1. Marc Pilisuk, 'Military application of psychology: the ethical dilemma', Paper presented at the Annual Convention of the American Psychological Association, Montreal, 28 August 1973. An address to the panel: 'Is the military application of psychology consistent with the promotion of human welfare?'
2. *The Pentagon Papers (as published by the New York Times, based on investigative reporting by Neil Sheeham)*, Bantam Books Inc., 1971.
3. Carol Ackroyd *et al.*, *The Technology of Political Control*, Penguin Books, 1977.

4. Jerome D. Frank, *Sanity and Survival: Psychological aspects of war and peace*, Barrie & Rockliff, 1968.
5. Philip Agee, *Inside the Company: CIA Diary*, Penguin Books, 1975.
6. See reference 4 above, chapter 7.
7. See, for instance: Cdr William Robinson, Jr, 'An element of international affairs – the military mind', *Naval War College Review*, vol. 22, no. 3, November 1970, pp. 4–15.
In addition: E. G. Franch and Raymond Ernest, 'The relationship between authoritarianism and acceptance of military ideology', Personnel Research Laboratory Aerospace Medical Division, Lackland Air Force Base (no date).
J. McNeil, 'Military Mind vs Civilian Mind', *Journal of Conflict Resolution*, vol. 6, 1962, p. 277.
A. Mitchell, 'Needs of military commanders and staff', *Journal of Applied Psychology*, vol. 54, 1970, p. 282.
8. Adam Roberts, 'South Africa resorts to psywar', *New Society*, 21 April 1977.
For a more optimistic view, see: Robert Delaney, 'The psychological dimension in national security planning', *Naval War College Review*, vol. 25, no. 3, Jan-Feb 1973, pp. 53–9. Looks forward to a time when press coverage of wars and leaflet propaganda will be 'integrated'.
9. Michael T. Klare, *War without end: American planning for the next Vietnams*, Vintage Books, 1972.
10. 'Man in the 1990 environment. Volume 1', summary report, US Army Combat Development Command, Directorate of Organization, 6 July 1970. (The combined annual budget for ARPA, Army, Navy and Air Force Human Resources Research is now in the region of $40·5 million.)
Other 'scenarios': Donald Eberly, 'National service: alternative strategies', *Armed Forces and Society*, vol. 3, no. 3, 1977, pp. 445–55.
Lt Cdr Beth Coye *et al.*, 'Is there room for women in Navy management: an attitudinal survey', *Naval War College Review*, vol. 25, no. 3, 1973, pp. 69–87 (very equivocal).
'The new Army game', *Behavior Today*, 15 January 1973. On his appointment as chief of social processes at the US Army's RIBSS, David Segal foresaw the joint enlistment of married couples but did not think that the army would ever entirely eliminate the drudgery attached to much military work – 'drudgery', he said, 'is part of the heroic image of the military'; USARIBSS is also working on the idea of 'duty modules' each encompassing one crucial military duty. A number of these would provide a basic training – rather like 'credits' in a university degree course. The British navy already sections its ships when building them and foresees in the years to come a pattern when specialists will be flown to a ship to do a job and then flown home again – an end, in other words, to the long months at sea (interview with Edward Elliott, Ministry of Defence, 24 July 1974).

11. Lt Col James L. Scovel, 'Motivating the US soldier to fight in future limited wars', USAWC Research Element (Thesis), US Army War College, Carlisle Barracks, Pa, 25 March 1970.
USARIBBS is looking at different *methods* to research the soldier of the future – especially in the field of race relations, the quality of life, relations with and effects of the mass media, and so on.
Other related references: W. W. Haythorn *et al.*, 'Research on the psychiatric effectiveness of future weapons systems crews', Delaware University, Contract NONR 228504, August 1965.
Lt Gen William P. Yarborough, 'If we want a volunteer army', 9 November 1970 (a general paper arguing for the re-creation of elite units to attract more of the right sort of men).
Nancy Goldman and David Segal (eds.), *The Social Psychology of Military Service*, Sage Publications, 1976, especially pp. 215–80.
For thoughts about discipline problems in the future, see: 'Research on social-psychological factors underlying the idea of discipline', Stanford Research Institute, quoted in the *Capitol Times*, 25 July 1973.
In one study the 'potential' for dissidence was calculated to exist in 127 out of a group of 530 soldiers or roughly one in four. See: Howard Olsen and R. William Rae, 'Determination of the potential for dissidence in the US Army', Research Analysis Corporation Technical Paper 410, May 1971.

Index

Compiled by Gordon Robinson